A Maintenance Management Framework for Municipal Buildings in Developing Economies

The central aim of this book is to investigate and develop frameworks to aid effective maintenance management of municipal buildings in the education sector of developing economies. Using the South African education sector as a case study, this book provides readers with two major practical insights. Firstly, it focuses on the theoretical underpinnings of maintenance management research and introduces a maintenance management model through the development of a conceptual framework. This framework aids in explaining the factors underpinning the maintenance of municipal buildings but can also be used in the assessment and management of other public buildings. Secondly, the book highlights and addresses theoretical gaps in existing studies essential for the maintenance management of buildings in developing economies, providing a stimulus for future research.

The book will be of interest to researchers in construction management, building technology, estate management, civil engineering, architecture, and urban and regional planning. It is an essential manual for policymakers in the education sector, built environment, construction industry, facility maintenance, facility management, and consultants at government ministries, departments, and agencies (MDAs) charged with maintenance management of public infrastructures and assets.

Babatunde Fatai Ogunbayo received his Ph.D. from the University of Johannesburg in Construction Management, focusing on maintenance management and services. He is presently doing his post-doctoral research fellowship in the Department of Construction Management and Quantity Surveying, Faculty of Engineering and the Built Environment, University of Johannesburg, South Africa. He worked as a lecturer, teaching construction management modules at the honours level and supervising honours and postgraduate students on their dissertations in the same department. As a researcher, Ogunbayo has authored, co-authored, and presented at various international conferences. He has conducted research in several areas, including maintenance management and services, material testing, public–private partnership, housing, building component analysis, material testing and analysis, construction health, safety, welfare, and productivity. He is an associate member of the Association of South Africa

Quantity Surveyors (ASAQS) and a member of the South African Council for the Project and Construction Management Professions (SACPCMP). He has a candidate QS membership of the South African Council for the Quantity Surveyors Profession (SACQSP). He is also a corporate member of the Nigerian Institute of Building (MNIOB) and a full member of the Council of Registered Builders of Nigeria (CORBON).

Clinton Ohis Aigbavboa is a professor at the Department of Construction Management and Quantity Surveying, University of Johannesburg, South Africa. He worked as a quantity surveyor on several infrastructural projects in Nigeria and South Africa. Prof. Aigbavboa is currently the Chair of the Sustainable Human Settlement and Construction Research Centre at the University of Johannesburg. He is currently the editor of the *Journal of Construction Projects Management and Innovation* (accredited by the DoHET) and has received national and international recognition in his field of research.

Wellington Didibhuku Thwala is a professor at the Department of Civil Engineering, College of Engineering, Science and Technology, University of South Africa (UNISA) Pretoria, South Africa. Prof. Thwala has extensive experience in providing consultancy for project leadership and management of construction projects and teaching project management subjects at the postgraduate level. He has an extensive industry experience with a research focus on sustainable construction, leadership, and project management. He is the editor-in-chief of *Journal of Construction Projects Management and Innovation* and serves as an editorial board member of various reputable international journals.

Routledge Research Collections for Construction in Developing Countries

Series Editors: Clinton Aigbavboa, Wellington Thwala, Chimay Anumba, and David Edwards

A Maintenance Management Framework for Municipal Buildings in Developing Economies

Babatunde Fatai Ogunbayo,
Clinton Ohis Aigbavboa, and
Wellington Didibhuku Thwala

Routledge
Taylor & Francis Group

LONDON AND NEW YORK

First published 2023
by Routledge
4 Park Square, Milton Park, Abingdon, Oxon OX14 4RN

and by Routledge
605 Third Avenue, New York, NY 10158

Routledge is an imprint of the Taylor & Francis Group, an informa business

© 2023 Babatunde Fatai Ogunbayo, Clinton Ohis Aigbavboa, and Wellington Didibhuku Thwala.

British Library Cataloguing-in-Publication Data
A catalogue record for this book is available from the British Library

Library of Congress Cataloging-in-Publication Data
Names: Ogunbayo, Babatunde Fatai, author. | Aigbavboa, Clinton, author. | Thwala, Wellington, author.
Title: A maintenance management framework for municipal buildings in developing economies / Babatunde Fatai Ogunbayo, Clinton Ohis Aigbavboa, and Wellington Didibhuku Thwala.
Description: Abingdon, Oxon ; New York, NY : Routledge, [2023] | Series: Routledge research collections for construction in developing countries | Includes bibliographical references and index. |
Identifiers: LCCN 2022038859 (print) | LCCN 2022038860 (ebook)
Subjects: LCSH: Public buildings--Maintenance and repair--Standards--Developing countries. | School buildings--South Africa--Maintenance and repair--Case studies.
Classification: LCC TH3351 .O39 2023 (print) | LCC TH3351 (ebook) | DDC 690/.502880968--dc23/eng/20221018
LC record available at https://lccn.loc.gov/2022038859
LC ebook record available at https://lccn.loc.gov/2022038860

ISBN: 978-1-032-37184-9 (hbk)
ISBN: 978-1-032-38370-5 (pbk)
ISBN: 978-1-003-34468-1 (ebk)

DOI: 10.1201/9781003344681

Typeset in Times New Roman
by MPS Limited, Dehradun

Contents

Preface

This book explores maintenance management and conceptualises a maintenance management framework for the maintenance of municipal buildings in developing economies. While there has been much documentation on maintenance management in the context of the developed countries as evidence in the body of knowledge, relatively little has been written about maintenance management for municipal buildings in developing economies. There is also no reference book that conceptualises a maintenance management framework for municipal buildings of developing economies within the education sector.

The central aim of this book is to study maintenance management and conceptualise a maintenance management framework for municipal buildings of developing economies within the education sector with a Delphi technique case study of the South African education sector. This research book provides readers with two major practical insights. Firstly, it focuses on the theoretical underpinnings of maintenance management research and thereafter conceptualises a maintenance management framework for municipal buildings within the developing economies' education sector through the development of a conceptual framework. The maintenance management conceptual framework, among other things, aids in explaining the main and sub-attributes underpinning maintenance management of municipal buildings within the education sector, as well as providing the assessing maintenance management of other public buildings. Secondly, the reference book highlights and addresses theoretical gaps in existing maintenance management studies essential for the maintenance management of developing economies' municipal buildings within the education sector.

This book's conceptual framework provides insight into the multi-facet factors influencing maintenance management of municipal buildings in developing economies based on South Africa's example. The book is therefore of interest to researchers in building construction, construction management, quantity surveying, project management, building design, facility management, quality management, civil engineering, estate management and valuation, housing, urban and regional planning studies. It is an essential manual for owners (government/private) and maintenance managers of municipal buildings within the education sector. It is also useful for government ministries, departments and agencies (MDA), policymakers, corporate bodies, entrepreneurs, industries, and non-governmental agencies as they plan for and implement maintenance processes and operations designed to enhance the effective maintenance management of their proposed and in-use buildings and other public infrastructures towards meeting their useful life and satisfying the immediate needs of its users. It offers researchers further insight into maintenance management as it has been considered from a multi-faceted dimensional concept. The authors confirm that the text utilised in this work reflects original work and, where necessary, the material has benefited from relevant context-setting/referencing.

Acknowledgements

We are very grateful to God Almighty, who has given us life, strength, and grace to author this book. We also appreciate the support we received from our families and the following people and institutions; the Construction Industry Development Board, Cidb Center of Excellence at the University of Johannesburg, and the Delphi study experts for their time and for bringing their knowledge and experience to bear in the Delphi Study. We are also grateful to the Head of Department, Construction Management and Quantity Surveying, University of Johannesburg, South Africa, and the Postgraduate School (PGS) for the diverse support and assistance.

1 Introduction and General Background

Introduction

Like any other structure, municipal buildings deteriorate with age at various life cycle stages. The growth in the educational system in most developing economy countries, including South Africa, due to the liberalisation of the education sector has taken place against the mounting pressure on existing municipal buildings in the educational sector. This has led to an explosion in the yearly recruitment of academic and administrative staff and student intake, with little or no expansion of the available municipal buildings or their maintenance in the education sector. Thus, the full potential of many municipal buildings in the education sector is never wholly realised. Nonetheless, there is a dearth of empirical studies that investigate and develop frameworks to aid the effective maintenance management (MM) of municipal buildings in the education sector of developing economies.

Thus, this book investigates and develops a cohesive framework that will aid the effective MM of municipal buildings in developing economies using the South African education sector as a study. Primarily, the research modelled the extent that organisational maintenance policy, maintenance budget factors, human resources management, training factors, monitoring and supervision, maintenance information systems (MISs), communication among stakeholders, and the maintenance culture factors (which were classified as the exogenous variables) predict the MM of municipal buildings. A conceptual integrated, holistic MM framework was developed based on the theory developed from the literature review and the findings of the Delphi study. MM is the set of attributes (factors) that will aid maintenance managers of municipal buildings in the education sector to effectively maintain the municipal buildings (academic and residential) to the satisfactory level of their users. Results from the investigation pertained to three broad areas. Firstly, the results related to the theory of MM studies. The findings were that the study addressed the lack of theoretical information about which factors are most significant in predicting the MM of municipal buildings in the education sector. Also, the findings revealed the theory that the MM of municipal buildings in the education sector is multi-faceted and that the latent variables

DOI: 10.1201/9781003344681-1

lead to effective MM of municipal buildings outcome variables which could be used to measure the MM of buildings. The second set of findings is related to the Delphi study.

The findings from the Delphi study indicated that the MM of municipal buildings could be an eight-factor model defined by the influence of organisational maintenance policy, maintenance budget factors, human resources management, training factors, monitoring and supervision, MISs, communication among stakeholders, and the maintenance culture factors. The study's contribution to the body of knowledge is significant because it addresses the lack of theoretical information (historical literature data) about which factors are most significant in predicting the MM of municipal buildings in the education sector of developing economies. The current framework advances that the MM of municipal buildings is an eight-factor construct, with the inclusion of two new variables, namely communication among stakeholders and the maintenance culture factors. Previous studies have tried to model MM using other variables without the inclusion of the present two additional variables. Hence, this book has shown that more than one factor influences the MM of municipal buildings.

The book recommends that governmental, non-governmental organisations, education institutions (public and private), and education sector policymakers consider the empirically tested constructs as they plan for and implement MM processes and operations designed to enhance the effective MM of municipal buildings in the developing economies' education sector. Similarly, the validated conceptual framework of the MM of municipal buildings will provide a reference for researchers who may study MM in the future.

General Background

Municipal buildings are an important component of life daily. Without municipal buildings such as educational buildings, hospitals, and other public buildings, the economy and society could not function (Chinowsky, Schweikert, & Hayles, 2014). Educational buildings, as one of the important municipal buildings, are exclusively used for a college or school established by the relevant government institutions, boards, or any other competent authority involving assembly for instruction or education. Therefore, their environment must be accorded the highest premium for effective functioning and productivity. Moreover, educational institutions are responsible for providing educational opportunities through well-developed curricula that aid students in obtaining academic and professional competencies. This depends on students and staff being supported and not frustrated by inadequate building facilities or a dysfunctional built asset environment. However, the educational buildings, like any other similar structures, at various life cycle stages deteriorate with age. According to Okolie (2011), the deterioration rate of all these educational buildings mainly depends on the construction method, environmental conditions, type of materials used, and available maintenance arrangement.

The proliferation of courses and rapid expansion in universities without adequate regard for resource constraints have placed a large burden on the educational institutions, especially their buildings (Ojogwu & Alutu, 2009). Okolie (2011) states that this has led to an unhealthy learning environment with decayed and dilapidated building infrastructure. Ojogwu and Alutu (2009) lament that students and teachers have become apathetic towards and disinterested in learning goals. Ojogwu and Alutu (2009) noted that the reasons for this state of educational buildings are mainly caused by a lack of an effective maintenance system. This has seriously undermined the goals and objectives of national policy on education.

Additionally, this has exposed the functional inadequacies of the educational buildings and therefore poses a tremendous challenge to the educational institutions regarding academic capacities, building infrastructure, funding, environmental concerns, and productivity. Confirming this, Okebukola (2002) reports that the educational buildings in most developing economies are overstretched, thus presenting a recipe for rapid decay in the face of dwindling funds for maintenance. Thus, the full potential of many educational facilities is not wholly realised except through the arrangement of a preventive MM approach that will help in the effective maintenance of the buildings.

Since buildings form a significant part of infrastructural facilities in the municipalities and the educational environment, this challenge calls for an effective MM system in order to improve the value of constructed buildings. This is because modern trends in teaching and learning demand a paradigm shift from staff teaching to student learning. However, this cannot be achieved in an environment with dysfunctional municipal buildings in the education sector (Ojogwu & Alutu, 2009). Given the fact that the education sector is in urgent need of improved infrastructural development, especially building facilities. There is a need to address this problem by providing a clear theoretical understanding of the basic constructs and related concepts of MM of buildings and its application in construction and municipal building maintenance within the education sector of developing economies.

Maintenance Management

The process of deterioration and decay of the fabric and the services begins the moment a building is completed and occupied; hence there is a requirement to undertake maintenance activities to ensure that the building performs to an acceptable level (Williamson, Williams & Gameson, 2010). Maintenance, as defined by British Standard (EN 13306: 2001), is the combination of all technical, management, and administrative engagements during the life cycle of a building to restore it to or retain it in a state in which it can perform the required function. Lateef (2009) defined maintenance as the services provided or undertaken for a structure (building) after completion in order to replace it to a standard that will make its components stable without upsetting its functional ability throughout its entire life span. Olatubara (2007) postulated that maintenance comprises all works relating to redecoration, repairs, and replacement carried out on a structure (building) and its auxiliary facilities to prevent damage and injuries. It helps increase its economic life usefulness and improve its functionality and beauty (Olatubara, 2007). These definitions indicate that the essence of maintenance in a building structure is to meet its required functional performance and the expectation of the users. Thus, maintenance is not merely to rectify defects. It is also needed to prevent them.

In other words, maintenance ensures a building meets its required and desirable performance over its life cycle. Puķīte and Geipele (2017) state that building maintenance aims to retain an investment's value by protecting it right from the preliminary stage. However, Wordsworth and Lee (2001) opined that decisions on maintenance execution, despite its importance, are mostly compromised between finance availability and the physical needs of users. However, Emetarom (2004) indicates that with the growth of more socially responsible factors within building structures, it is important to provide more sustainable maintenance systems throughout their life cycle to maintain a maintainable building. Salisu (2001) posited that the maintenance of an institutional building structure should be primarily vested with its maintenance unit to boost users' productivity, needs, and expectations. This could be achieved by satisfactorily enhancing the outdoor and indoor environments with quality maintenance service delivery on the building.

Nonetheless, the MM unit of an institution is a managerial team put in place to administer and coordinate maintenance activities within the institution so that the available facility is in good shape to meet the purpose for which it was constructed. Campbell and Jardine (2001) postulated that the set goals of MM of these institutional buildings are achieved through directing, organising, and planning of organisations' available resources. The management team may subsume and precede purposive management by directing and leading the organisation in achieving its set goals. Thus, Duffuaa, Raouf, and Campbell (2000) assert that the maintenance function in a managerial organisation is achievable through the combination of good administrative tendencies, better managerial skills, and technicality. This is carried out in the life cycle of an institutional building, which is intended to retain it to perform its required functions.

Campbell and Jardine (2001) posit that MM can be summarised as managing all facilities owned by an organisation to maximise their return on investment in a facility. On the other hand, Duffuaa et al. (2000) indicate that MM is a simple input and output system where equipment, tools, management, and manpower represent the input while working reliably to reach the planned operation and equipment configured well represents the output. Duffuaa et al. (2000) state that good maintenance planning, stable maintenance control, and a viable maintenance organisation are required activities that will make MM systems functional.

In developed nations, the use of MM for the municipal maintenance of facilities is more of a profession (Telfer, 2005). In managing their municipal buildings, MM has been adopted as a strategy to improve and stabilise the performance of their facilities (Telfer, 2005). Accordingly, Edum-Fotwe, Egbu, and Gibb (2003) assert that MM provides a safe environment for work which is vital to the success of any municipality through an effective combination of available resources for day-to-day activities. This is critical to the performance of available facilities within a municipality, irrespective of scope and size. Today, the scope of MM, with increasing competitiveness within the municipality, business, and learning environments, is taking on a completely new meaning and dimension. With this new demand, maintenance managers are now required to adopt more systematic approaches in carrying out their maintenance functions through the total asset planning process involved in their MM planning (Pukīte & Geipele, 2017). In reality, many organisations, businesses, and educational sectors still do not value the essence of MM in their organisation. However, maintenance must be started immediately after a building has been completed and is ready for use by establishing a MM unit that oversees the sustainability of the components of the building, its auxiliary facilities, and services (Arditi & Nawakorawit, 1999). Pinjala, Pintelon, and Vereecke (2006) and Alsyouf (2009) state that not many municipalities view MM's effectiveness as strategic importance despite their municipality building deterioration, failures, and damages or their performance below expectations.

Karia, Asaari, and Saleh (2014) state that MM should be viewed as a preventive management philosophy that will provide a platform to retain assets' value, improve life quality, and improve risk cost and productivity in organisations. In achieving a good preventive maintenance system, it is the responsibility of the MM unit, such as the physical planning unit, building department, asset management department, logistics, administration department, facility department, maintenance department, and property division to manage the maintenance and operations of the available physical facilities of their respective municipalities. Conversely, when fully implemented, MM offers favourable effects on the municipality through increased productivity, stable operational cost, and stable facility. Additionally, Karia, Asaari, and Saleh (2014) postulated that the concept of MM in building maintenance systems for an organisation includes inspections, cleaning, equipment maintenance, repairs, and replacement of a building component or fixing the malfunctioning component in a building. Jones (1993) states that in carrying out this function, the MM unit must be involved in administrative and technical action through viable supervision, aiming to restore components of the structure to perform their expected functions.

As a result of the importance of MM in the preservation of municipal buildings, some nations and government institutions around the world set up maintenance departments and units to oversee the effective MM of their facilities, including buildings. Some even go to the extent of establishing maintenance agencies such as the Federal Road Maintenance Agency (FERMA) in Nigeria, the Ghana Infrastructural Investment Fund (GIIF), the National Infrastructure Maintenance Strategy (NIMS) in South Africa, the British Maintenance Guide (British Standard Institution, 1986, 1993), the Hong Kong building maintenance scheme (Hon, Chan & Yam, 2012) among others. These maintenance agencies were founded and funded to eliminate health and safety hazards and avoid unnecessary expenses on major corrective maintenance on municipal infrastructure (Construction Industry Development Board, 2013).

Additionally, some countries and states in the world have public building maintenance policies and laws that are aimed, among others, to encourage continuous improvement of municipal buildings to make them more efficient and effective. A typical example of some of these policies is the Government Immovable Asset Management Act No. 19 of 2007 (Parliament of the Republic of South Africa, 2007), the NIMS Act No. 108 of 1996 (Republic of South Africa, 2006), the Norms and Standards for Education Facilities (Republic of South Africa, Department of Basic Education, 2018), the Kwazulu-Natal Maintenance Strategy (KZN Department of Education, 2016) and the Ontario Building Code Act 1992: Property Maintenance and Repair (Canada, Ontario Municipality, 2013), among others. All these indicate the importance some nations, states, and government institutions ascribe to their municipal buildings to be effectively maintained for efficiency.

However, there has been conjecture around different views in literature on MM; nonetheless, little is known of studies that holistically organise the various MM attributes into frameworks and examine their relative contribution to effective MM of municipal buildings, especially in the developing economies' education sectors. In affirmation, Lateef (2009) and Xaba (2012) assert that mainly all the studies on effective MM have firms and/or industries in the developed countries as their focus, while little is known of that of developing countries. As a result, there is inadequate empirical evidence on how effective MM of buildings could be achieved (Olanrewaju, Idrus & Khamidi, 2011). There is also a lack of comprehensive theory that explains the effective MM of municipal buildings in the education sector of developing economies (Okolie, 2011; Cloete, 2014).

Therefore, it is against this backdrop that this study seeks to organise the attributes that determine the effective MM of municipal buildings into a framework and investigate their relative contribution to the educational sectors of the developing economies, using the South African education sector as a case study. South Africa was chosen for this study partly because previous studies that sought to address issues relating to the MM of municipal buildings in the education sector largely focused on performance evaluation of the municipal buildings and primarily concentrated largely on user needs/ requirements within the organisational context. A typical example of such studies is that of Okolie (2011). In a single study, other equally influential factors have not been studied together with users' needs for the MM of municipal buildings. Thus, there is a growing need for all-inclusive frameworks and models as well as data for decision-making for municipality maintenance managers and policymakers in the educational sector of most developing economies (Quayson & Akomah, 2016).

According to Karia et al. (2014), several studies on MM in various disciplines, especially within the domain of the manufacturing industries, have modelled the attributes of effective MM using statistical, econometric, and/or correlation analysis; for instance, Cholasuke, Bhardwa, and Antony (2004), Pinjala et al. (2006), Alsyouf (2009) and Lind and Muyingo (2012). Nevertheless, little is known about the use of the Delphi survey in conceptualising the determinants of an effective MM, especially in the education sector of the developing economies' municipalities. Thus, this study conceptualises a MM framework for municipal buildings in the education sector of developing economies using the Delphi survey (DS).

The Research Problem Statement

The liberalisation of the educational sector in most developing economies has led to an explosion in the yearly recruitment of academic and administrative staff and the students' intake, with little or no expansion of the available municipal building in the education sector. As a result, available municipal physical facilities hardly meet the educational institutions' facility requirements for staff and students within the educational system in developing economies, including South Africa. The available municipal buildings in the education sector are overstretched, with the rapid decay of facilities and the swift rise in both student and staff intakes with dwindling funds for the maintenance unit to develop a better MM framework in handling the overstretched facilities. Even with the private involvement in educational provisions presently, the maintenance activity on available municipal buildings takes place in a context that does not create a fully integrated maintenance approach towards managing its structures and auxiliary facilities and services. However, considering the credence given to education in South Africa, developing a MM framework for municipal buildings within the education sector will help the sector develop a functional maintenance system for the educational buildings.

Additionally, previous studies have suggested that the full potential of many municipal buildings is never wholly realised in developing economies because there is a lack of an effective MM system that will help in viable maintenance planning and scheduling implementation. Nevertheless, there is a dearth of empirical studies that investigate and develop a framework to aid the effective MM of municipal buildings in the education sector of developing economies. Moreover, in common with previous studies on MM, there is still a misperception of the attributes that determine effective MM. For this reason, very few researchers have organised these variables into a framework to be able to examine and analyse the relationships produced amongst them.

Supporting this view, previous studies have not adopted a more holistic approach in addressing the MM of municipal buildings in the education sector. Equally, the method used in these studies may not always have been completely successful. An observable sign of this inadequacy is the existence of conflicting and sometimes even inconsistent research results about the factors that determine MM. Nonetheless, this may be due to the variances in samples used in the previous studies, as the sample for most studies might not be representative of the population under study and the way the key variables were defined. It may be because of the subjective nature of the MM studies or how the data was analysed in the previous studies. Hence, the current study is determined to overcome these problems to understand better the constructs that determine effective MM.

Consequently, in this study, the problem that has been addressed could be stated as follows:

> Given that the previous models on MM established in the developed economies cannot be relied upon in developing economies, the findings of what determines effective MM of municipal buildings in developing economies' education sector are rarely known from previously conducted research. However, with the lack of research into the overall impact and influence of the direct and holistic active involvement of MM constructs and the absence of a MM model for municipal buildings. It is pertinent to note that effective maintenance of the municipal buildings in the educational sector is unlikely.

Aim of the Research

This research aims to develop a cohesive MM framework for municipal buildings in developing economies using the educational building in the South African education sector as a case study. The framework will benefit the municipalities of developing economies to achieve effective MM of their buildings. Additionally, the framework aids in understanding and explaining the factors that determine effective maintenance and guide maintenance managers towards the effective MM of educational buildings and their auxiliary facilities and services.

Primarily, the framework organises and examines the relationship between organisation maintenance policy, maintenance budget, human resources management, monitory and supervision, training factors, maintenance information systems, communication among stakeholders, and maintenance culture factors on the effective MM of the municipal buildings in the developing economies, using South African education sector. However, it is worthwhile to mention that communication among stakeholders and maintenance culture factors are the new construct peculiar to this framework as they have not been considered in the previous models on effective MM of buildings generally.

Research Questions

Based on the research problem statement, the following are the research questions:

RQ1. To what extent is the MM of municipal buildings in the South African education sector influenced by organisational maintenance policy and maintenance budget?

RQ2. To what extent is the MM of municipal buildings in the South African education sector influenced by human resources management and training?

RQ3. To what extent is the MM of municipal buildings in the South African education sector influenced by monitoring and supervision?

RQ4. To what extent is MM of municipal buildings in the South African educational sector influenced by maintenance information systems?

RQ5. To what extent is the MM of municipal buildings in the South African educational sector influenced by communication among stakeholders?

RQ6. To what extent is MM of municipal buildings in the South African educational sector influenced by maintenance culture?

Research Objectives

In providing answers to the study research questions and achieving its aims, its objectives were set as follows:

RO1. To establish the factors that influence municipalities in the attainment of MM of the municipal buildings in the education sector;

RO2. To investigate and establish the current theories and literature that have been published on MM of buildings and to identify the gaps that need consideration;

RO3. To determine the factors/attributes (main and sub-variables) that are perceived to be of importance and of major impact in obtaining effective MM of municipal buildings, and to determine whether the factors that have aided in obtaining effective MM in other geographical contexts are the same in the South African education sector; and

RO4. To develop a holistic maintenance management framework (MMF) for municipal buildings in the South African education sector.

The Objectives RO1 and RO2 were achieved through a literature review on MM this study conducted. The focus of the literature review was to achieve a theoretical understanding of the subject matter. The **Objectives RO3 and RO4** were achieved through the Delphi survey.

Research Methodology Employed

Research methodology is the basic guidelines and techniques used to collect and analyse data effectively. It also provides the starting point for choosing an approach made up of ideas, theories, concepts, and definitions of the topic (Aigbavboa, 2014, Ogunbayo et al., 2022). One of the most important outcomes of research is the identification of methodological traditions. This, in turn, aids in identifying data collection techniques that could be considered for use in research. In this book, a qualitative approach to research was employed.

Qualitative Research

Qualitative research is primarily exploratory research, with an approach used to uncover trends in opinions and thoughts that delve deeper into the problem, thereby understanding underlying opinions, reasons, and motivations. Aigbavboa (2014) opined that qualitative analysis is a theory developed through the exegesis of available evidence against a theoretical background. Qualitative measurement is often binary as it is interested in the presence or absence of phenomena or works implicitly with simple scaled (Ogunbayo, Ohis Aigbavboa, Thwala, & Akinradewo, 2022). It seeks to understand and interpret the meaning of a situation or event from the perspectives of the people involved and as the people involved understood it. Generally, qualitative research is inductive – that is, the process of inferring a generalised conclusion from particular instances – rather than deductive in its approach (Saunders et al., 2007). Thus, it generates theory from an interpretation of the evidence, though against a theoretical background. The qualitative method used in this current research is the structural and semi-structured interview (using an interview guide). This was made possible by the Delphi techniques. The findings from this section of the study helped in validating the findings. Additionally, the Delphi findings were used to resolve issues surrounding the MM of municipal buildings and other MM issues in the study setting (South Africa) through the consensus (agreement and disagreement) that was reached in the views of the experts the Delphi study engaged. The Delphi technique is discussed in detail in this book.

Method of Obtaining Data (Empirical Measure)

A literature review on maintenance, MM, facility management, maintenance systems, parties to MM, and other maintenance materials related to the study was reviewed to provide a background to the study. Other various sources that were reviewed include books, articles in accredited journals, newspaper cuttings, theses, published and unpublished works such as dissertations, and web-based publications on previous relevant works on the research theme. The Delphi survey method was employed in collecting qualitative data. Concerning the Delphi survey method, the data needed to be collected was the prediction of the likelihood of MM factors and other MM issues surrounding MM of municipal buildings in the South African education sector. The method was used for the second stage of the study to identify the main attributes that bring about MM and to determine whether the attributes that determine the MM of municipal (including educational) buildings and facilities in other geographical contexts as identified from the literature are the same within South Africa (developing economies).

Similarly, the Delphi technique was used to explore the extent to which these main attributes and sub-factors impact or influence the effective MM of municipal buildings in the South African educational sector. This data was obtained using a structured questionnaire interview. The panellists that formed the experts in the Delphi study were asked to complete the questionnaires, and the consensus was reached on the rated likelihoods and influences of various factors. The Delphi process involved a three-round iterative process, with the main aim of getting experts to reach a consensus on the questions raised in the structured questionnaires. The panel of experts was also encouraged to give reasons for their dissenting views.

Data Sources

The Delphi study required data regarding the rating of the influence and likelihood of the factors that determine effective MM and other issues regarding effective MM of municipal buildings in the South African education sector. This was obtained from the expert panels. Equally, data from the questionnaire survey was obtained from the built environment professionals, faculty, and top-level maintenance managers of municipal buildings in the education sector within the study area.

Data Analysis

Data obtained under the Delphi study was analysed in a spreadsheet software program, Microsoft Excel. Data analysis expected output for the study was from a set of descriptive statistics such as mean, mean, standard deviations, and their respective derivatives.

Result

The result of this study related to the relationship between the endogenous variables (organisational maintenance policies, maintenance budget factors, human resources management, monitoring and super-vision, training factors, maintenance information systems, communication among stakeholders, and maintenance culture) and the exogenous variables (MM) are presented in charts and tables of values delineating the extent of the attributes and sub-attributes on the overall effective MM of municipal buildings in the South African education sector. By utilising data from Delphi, a holistic, integrated MM framework for municipal buildings in South African (developing economies) education sector was developed.

Delphi Specific Objectives

Specifically, the Delphi survey sought to achieve the following objectives:

DSO1. To identify the attributes (main and sub-) that determine effective MM of municipal buildings and to examine whether the attributes that influence the MM of municipal buildings in other geographical contexts are the same in the South African education sector;

DOS2. To identify the key factors to effective MM and determine the relative influence of each of the factors in the maintenance of municipal buildings in the South African education sector; and

DOS3. To identify the measuring indicators (outcomes) of effective MM and to establish the relative influence of the indicators in the MM of municipal buildings in the South African education sector.

The main output expected from the Delphi studies is the identification of the factors with the significant influence that will aid educational institutions in achieving effective MM of the municipal buildings within the South African education sector, as well as a conceptual framework defining attributes/factors that determine the effective MM of municipal buildings in the South African education sector.

Delimitation and Limitation of the Study

Geographically, this research was limited to the municipal building of the South African (a developing economy) education sector. It is worth em-phasising that municipal buildings in educational institutions are not homogeneous; there are several key units apart from their basic academic function, while there are structures used for administrative, residential, and other purposes. Thus, this research is limited to South African education and the built environment sector. This is because South Africa has a rela-tively stable educational system and well-organised built environment in-dustry, and a plethora of existing relevant literature suggests that municipal buildings have been faced with a dearth of abandonment, deterioration, and poor management, among others, just as their other counterparts in some developing economies (Xaba, 2011; Ferim, 2013; Frey,2018; Phathela, & Cloete, 2018; Ogunbayo, Ohis Aigbavboa, Thwala, & Akinradewo, 2022).

Moreover, municipalities were of concern because earlier studies have established those municipal buildings in the education sector of developing economies lack an appropriate MM system compared to the municipal buildings in developed economies (Olanrewaju et al., 2011; Okolie, 2011; Cloete,2014; Ogunbayo et al., 2018). There is a dearth of studies that address the problems, namely assisting municipalities in understanding the attributes of effective MM of municipal buildings in the education sector and also aiding them in the maintenance process and subsequently sustaining it (Ogunbayo and Aigbavboa, 2019; Olanrewaju et al. 2011).

Also, the study focuses on municipal buildings since they are buildings accessible to the public and funded by public sources. They represent significant taxpayers' money, and thus maintaining these buildings is important (Ampofo et al., 2020, Ogunbayo & Mhlanga, 2022). Also, built environment professionals, faculty, and top-level maintenance managers in the education sector were considered in the study, and the buildings considered were municipal buildings in the education sector. This study focused on developing an integrating MM framework that will aid educational institutions in the education sector in South Africa to obtain effective MM.

The scope of this book comprised the MM of municipal buildings in the South African education sector. The study emphasises the influence of the endogenous variables (organisational maintenance policies, maintenance budget factors, human resources management, monitoring and supervision, training factors, MISs, communication among stakeholders, and maintenance culture factors) on the exogenous variable (MM). The result of the study was presented in the form of tables and charts.

This book understudied and analysed the MM apparatus of the South African education sector towards avoiding facility decay, deterioration, and abandonment of their educational buildings (academic and residential buildings). As a result, this study was limited to the sampled views of South African built environment professionals. It is worth noting that most of the study's sampled views were also involved in teaching maintenance modules and members of the MM committee or units of the institutions. Also, they are members of professional bodies in the engineering and built environments, such as the Association of South African Surveyors (ASAQS), the South African Council for the Quantity Surveying profession (SACQSP), South African Council for planners (SACPLAN), South African Council for the Property Valuers Profession (SACPVP), the South African Council for the Project and Construction Management Professions (SACPCMP), Association of Construction Project Managers (ACPM), Institute for Landscape Architecture in South Africa (ILASA), Consulting Engineers South Africa (CESA), and South African Institute of Architects (SAIA). Again, the education sector in South Africa has many similarities with other developing economies. As such, the findings would be a true archetype of what may be prevailing in other education sectors of developing economies in effectively maintaining their municipal buildings within the education sector.

In this study, MM is related to the attributes and conditions (hereinafter called attributes, determinants, or factors) that influence educational institutions to obtain effective MM of the municipal buildings in the education sector; and an integrated MM framework is a model or framework that holistically explains the factors that determine effective MM, together with the relative contribution of each of the attributes/factors towards the MM of municipal buildings in the education sector. Nevertheless, municipal buildings are synonymous with educational buildings and may be used interchangeably in this research to mean the same unless otherwise stated.

Equally, the MM framework that this study develops is municipalities centred. Hence, it could be termed a *municipality-centred integrated framework for MM of municipal buildings in the South African education sector.* Finally, the study discussion focused on the implication of its findings on the MM of municipal buildings. Necessary recommendations were made to improve the existing maintenance strategy of MM of municipal buildings in the South African education sector and other developing economy economies.

Significance of the Book

Over the past three decades, significant theoretical and empirical frameworks have been developed in the field of MM as a whole; however, only a few studies have researched the state of MM in municipal buildings in developing economies, especially in South Africa. The desperate need to investigate MM is linked to evidence revealing an improved level of building users' well-being, productivity, performance, and health benefits associated with an adequate and effective preventive MM system. The book intends to add to the existing body of knowledge on MM systems by developing an integrated MM framework for municipal buildings in developing economies. The research addresses this shortfall by making use of the Delphi study.

When applied correctly, the conceptual integrated MM framework will prolong the usage of the municipal buildings. Also, it provides a benchmark for municipal buildings' maintenance process and usage. The conceptual framework will help develop an early warning system for identifying damaged components of buildings and using quality replacement material for maintained buildings. Additionally, the conceptual framework will provide a maintenance template that will bridge the communication gap between stakeholders (maintenance personnel and users) involved in the maintenance process and operations. It will also help to improve stakeholders' attitudes to the maintenance process and the maintained buildings.

Furthermore, the study anticipates that the research findings would benefit stakeholders of educational institutions and the government in incorporating an integrated MM system before developing educational and other municipal buildings. Economically, the conceptual framework may influence the budgetary decisions in the education sector of the South African economy owing to the benefit associated with a pre-design provision of a preventive-based integrated maintenance system versus a post-occupant corrective maintenance design. Finally, the research will help reduce maintenance costs and resource wastage in the maintenance process of municipalities and educational institutions.

Thus, this book contributes to the existing knowledge on MM by establishing the factors that determine the effective MM of municipal buildings in the education sector of developing economies using South Africa as a case study and by establishing the influence of the factors (exogenous variables) on the MM of municipal buildings (endogenous variable). Moreover, the framework was centred on municipal buildings; therefore, the respondents were top-level maintenance managers and experts in both academic and built environment professions in South Africa.

Additionally, instead of using existing MM models that previous studies have conceptualised in developed countries, this study rather evaluated communication among stakeholders and maintenance culture, which are the constructs peculiar to this book's framework in addition to organisational maintenance policy and maintenance budget factors among others, which previously models have already considered. Nonetheless, the study currently employed the Delphi method in conceptualising the MM of municipal buildings in the South African education sector. Thus, the innovative method and the outcome variable measures used in this current study contribute to the existing body of knowledge on the determinants of effective MM.

Hence, this book will add new knowledge to the attributes determining the effective MM of the municipal building. The innovative methodology and the outcome variable measure also contribute to the existing body of knowledge on the effective MM of municipal buildings. Additionally, the generic literature review on MM and theories on MM expand the frontier of existing knowledge by providing synthesised literature that will guide the effective MM of municipal buildings in the South African education sector and other developing economy economies.

Motivation for the Book

Barde (1994) and Collier et al. (2010) described developing economies as countries with low or medium human development index (HDI), low living standards, low per capita income, widespread poverty, underdeveloped industry, outdated infrastructure, and are facing long periods of recession. The motivation for this book is to contribute to the body of knowledge on the MM of municipal buildings in developing economies, using the South African education sector as a case study. This was done to achieve more refined sustainable economic growth and development for the educational system in developing economy countries. Thus, the quaternary sector of the developing economy must undertake studies on ensuring sustainability in educational institutions and their buildings as they house the innovation and economic development of tomorrow. There are effects of lack of maintenance of municipal buildings in the education sector, such as dilapidation, deterioration, and abandonment, which should be taken into consideration by stakeholders and the government, which could affect the users of educational buildings.

This is partly attributable to the dearth of effective MM frameworks that will aid the maintenance of educational buildings and other public assets in developing countries to obtain an effective MM system. As a result, this book hopes to fill this gap by developing an integrated MMF for municipal buildings of the developing economies' educational institutions using the South African education sector as a case study.

Alignment to International Imperative

The book is aligned with the Paris Agreement and the 2030 Agenda for Sustainable Development, which supports the Sustainable Development Goals (SDGs) developed by the UN member states. As stated in the SDGs, one of its core elements is to ensure the healthy lives of all people and a better state of development for improvements and amelioration of living standards (UNDP, 2016). In all the listed SDGs, infrastructure appears both as an explicit goal and an implicit means to implement and achieve other SDGs.

Mainly this book is aligned with SDGs 9: industry, innovation, and infrastructure. The key point of SDG 9 deals with building a resilient infrastructure based on objective design, maintenance, restoration of the building, and infrastructure. Also, the book aligned with SDGs 4: quality education. The essence of the goal is to ensure inclusive and equitable quality education, promote lifelong learning opportunities, and target and demand the construction and upgrading of learning facilities. Equally, the study aligned with SDGs 11: sustainable cities and communities. The core relevance of this goal is to make cities and human settlements inclusive, resilient, safe, resilient, and sustainable. The targets of SDGs 11 are related to infrastructure planning sustainability, i.e., buildings and other public utilities which require sustainable infrastructure development to reach this goal.

Ethical Consideration

Ethical issues were a key consideration in embarking on this study. The principle of voluntary participation was strictly upheld. This required that the population was not persuaded to participate in the research. Additionally, participants in the study were only involved in the research where informed consent had been established. An effort was made to help protect the privacy of research participants by ensuring confidentiality in not making available identifying information to anyone who was not directly involved in the study. Confidentiality was further enhanced by keeping participants anonymous throughout the study.

Structure of the Book

This book is organised as follows.

Chapter 1: Introduction and General Background

This chapter presents information on the background of the study, the main research problem, and a general description of the study, stating the study's aim and objectives. Furthermore, the chapter presents a description of the methods that were employed to conduct the research and ethical considerations.

Chapter 2: Generic Overview of Maintenance Management

This chapter discusses generic literature on maintenance definitions, the concept, aim, purpose, types, operations, and generators, amongst others, related to the effective MM of buildings. The chapter presents a survey of related literature from books, theses, conference papers, journal articles, and Internet searches from reputable databases such as Scopus, Google Scholar, ResearchGate, and institutional repositories. The chapter is presented relative to the guiding background for the study's theoretical and conceptual perspective of MM research.

Chapter 3: Policy, Planning, and Performance Measurement for Maintenance Management

The chapter reviews the literature concerning maintenance organisation, MM strategies models, maintenance policy, forms of maintenance policy, maintenance policy evolution, maintenance policy objectives, maintenance policy instruments, and independent performance variables in MM studies. The review primarily focuses on policy, planning, and performance measurement for MM.

Chapter 4: Theories, Models, and Concepts in Maintenance Studies

This chapter reviews the literature in relation to the theoretical and conceptual perspective of MM research. It includes discussions on previous MM theories and models, subjective definitions of MM, the relative nature of a MM study, the complexity of a MM study, and methodological issues in the study of MM. Also, various models of MM of buildings are highlighted and conceptualised.

Chapter 5: Maintenance Management Research Theories

This chapter presents a framing overview of MM research theories. Highlighting, among others, some theories with complementary perspectives on the maintenance and management of buildings, such as total quality management (TQM), the balanced scorecard (BSC), requirements management theory (RMT), and change management theory (CMT). Also, maintenance policy, the evolution of maintenance, policy legal framework, and forms of maintenance policy are presented. Similarly, the objectives of the maintenance policy and maintenance policy instruments that enable the intentions of the maintenance policy to be realised are also included.

Chapter 6: Gaps in Maintenance Management Research

This chapter addresses the gaps observed in MM research, which were not evaluated as an all-inclusive construct in the previous models. These gaps form the additional new constructs in the conceptual framework of the current study. The identified gaps and how to achieve them in the effective MM of municipal buildings in the education sector are discussed.

Chapter 7: Maintenance Management of Municipal Buildings in Developing Economies – An African Experience

This chapter presents maintenance policies and other maintenance issues, including the maintenance of educational buildings in developing economies, with Ghana and Nigeria as examples. Additionally, the role played by various bodies such as the government together with other maintenance organisations and departments involved in the maintenance of buildings in the two developing countries is also presented. A comprehensive background review on maintenance activities within these two developing countries and a summary of lessons learned are also inculcated.

Chapter 8: Maintenance Management of Municipal Buildings in the South African Education Sector

This chapter comprises the background of South Africa, an overview of MM studies in the South African education sector, the evolution of maintenance policies and laws in South Africa, maintenance policy and legal framework in the maintenance of educational buildings in the South African education sector, and challenges associated with the implementation of maintenance policy and laws in South Africa. A summary of lessons learned is also inculcated.

Chapter 9: Methodological Framework for Developing a
Maintenance Management Conceptual Model

This chapter presents a detailed description of the methodology, including methods and the tools used to collect data in this current study. This chapter also describes the research design and how the collected data was treated. Finally, a description of the population, the sampling design, and the interpretation of results are included in this chapter.

Chapter 10: The Outcome of the Delphi Study

This chapter presents the outcomes of the Delphi study. The objectives that governed the Delphi study are also presented.

Chapter 11: An Integrated Maintenance Management Conceptual
Model for Municipal Buildings in the Developing Economies

This chapter theorises a MM conceptual model for maintaining municipal buildings in developing economies. Similarly, the chapter highlights the theoretical framework of the variables selected for constructing the integrated MM conceptual model for the municipal buildings in the South African education sector.

Chapter 12: Conclusion and Recommendations

This chapter concludes the study and details recommendations relative to the conclusions drawn from the study. Also, recommendations for future research are offered in this chapter. Similarly, the book's contribution, limitations, and future studies have been highlighted in this section.

Summary

This chapter introduced the subjects of the book. Among others, it gives an insight into an effective MM, the aim and objectives, and research significance. The next chapter investigates a generic overview of MM of buildings and other related literature in line with the objectives guiding this study.

References

Aigbavboa, C. O. (2014). *An integrated beneficiary-centered satisfaction model for publicly funded housing schemes in South Africa* (Published doctoral dissertation, University of Johannesburg, Johannesburg).

Alsyouf, I. (2009). Maintenance practices in Swedish industries: Survey results. *International Journal of Production Economics, 121*(1), 212–223.

Ampofo, J. A., Amoah, S. T., & Peprah, K. (2020). Examination of the current state of government buildings in senior high schools in Wa Municipal. *International Journal of Management & Entrepreneurship Research, 2*(3), 161–193.

Arditi, D., & Nawakorawit, M. (1999). Issues in building maintenance: Property managers' perspective. *Journal of Architectural Engineering, 5*(4), 117–132.

Barde, J. P. (1994). Economic instruments in environmental policy: Lessons from the OECD experience and their relevance to developing economies. OECD Development Centre Working Papers, No. 92, OECD Publishing, Paris, 10.1787/754416133402

British Standard [BS] (2001). EN 13306: 2001. *Maintenance terminology.*

British Standard Institute (1993) *Glossary of Terms used in Terotechnology*, BS 3811, Milton Keynes. BSI.

British Standards Institution [BSI] (1986). *Guide to building maintenance management.* BS 8210: 1986. BSI.

Campbell, J. D., & Jardine, A. K. (2001). *Maintenance excellence: Optimizing equipment life-cycle decisions.* New York: Marcel Dekker Inc.

Chinowsky, P., Schweikert, A., & Hayles, C. (2014). Potential impact of climate change on municipal buildings in South Africa. *Procedia Economics and Finance, 18*, 456–464.

Cholasuke, C., Bhardwa, R., & Antony, J. (2004). The status of maintenance management in UK manufacturing organisations: Results from a pilot survey. *Journal of Quality in Maintenance Engineering, 10*(1), 5–15.

CIDB. (2013). *Standard for developing skills on infrastructure contracts.* Board Notice 180 of 2013. Pretoria: Government Gazette 36760, 23.

Cloete, N. (2014). The South African higher education system: Performance and policy. *Studies in Higher Education, 39*(8), 1355–1368.

Collier, P., Van Der Ploeg, R., Spence, M., & Venables, A. J. (2010). Managing resource revenues in developing economies. *IMF Staff Papers, 57*(1), 84–118.

Duffuaa, S. O., Raouf, A., & Campbell, J. D. (2000). *Planning and control of maintenance systems.* New York: Wiley and Sons, pp. 31–32.

Edum-Fotwe, F. T., Egbu, C., & Gibb, A. G. F. (2003). Designing facilities management needs into infrastructure projects: Case from a major hospital. *Journal of Performance of Constructed facilities, 17*(1), 43–50.

Emetarom, U. G. (2004). Provision and management of facilities in primary schools in Nigeria: Implications for policy formulation. In *Paper presented at the annual national congress of Nigerian Educational Administration and Planning (NEAP).* University of Ibadan. 28th 31st, 2004.

Ferim, V. (2013). *African solutions to African problems. The African union ten years after: Solving African problems with Pan-Africanism and the African Renaissance* (pp. 143–155). Pretoria: Africa Institute of South Africa.

Frey, B. B. (Ed.). (2018). *The SAGE encyclopedia of educational research, measurement, and evaluation.* Sage Publications.

Hon, C. K., Chan, A. P., & Yam, M. C. (2012). Empirical study to investigate the difficulties of implementing safety practices in the repair and maintenance sector in Hong Kong. *Journal of Construction Engineering and Management, 138*(7), 877–884.

Jones, B. P. (1993). *The British standards institution.* In Proceeding of International Conference on Power System Technology: IEE Colloquium on Metering Standards and Directives, London, 5–5 April 1993, pp. 111–115.

Karia, N., Asaari, M. H. A. H., & Saleh, H. (2014). Exploring maintenance management in service sector: A case study. In: Proceedings of *International Conference on Industrial Engineering and Operation Management*, Bali, 7–9 January 2014, pp. 3119–3128.

Kwazulu-Natal (2016). KZN*maintenance strategy*. Republic of South Africa: Department of Education.

Lateef, O. A. (2009). Building maintenance management in Malaysia. *Journal of Building Appraisal*, *4*(3), 207–214.

Lateef Olanrewaju, A., Idrus, A., & Faris Khamidi, M. (2011). Investigating building maintenance practices in Malaysia: A case study. *Structural Survey*, *29*(5), 397–410.

Lind, H., & Muyingo, H. (2012). Building maintenance strategies: Planning under uncertainty. *Property Management*, *30*(1), 14–28.

National Universities Commission (NUC) (2020). *Manual of Accreditation Procedures for Academic Programmes in Nigerian Universities* (MAP). Abuja: National Universities Commission.

Ogunbayo, B. F., & Aigbavboa, O. C. (2019). Maintenance requirements of students' residential facility in higher educational institution (HEI) in Nigeria. In *IOP Conference Series: Materials Science and Engineering* (Vol. 640, No. 1, p. 012014). IOP Publishing.

Ogunbayo, B. F., Aigbavboa, C. O., Thwala, W., Akinradewo, O., Ikuabe, M., & Adekunle, S. A. (2022). Review of culture in maintenance management of public buildings in developing countries. *Buildings*, *12*(5), 677.

Ogunbayo, B. F., Ajao, A. M., Alagbe, O. T., Ogundipe, K. E., Tunji-Olayeni, P. F., & Ogunde, A. (2018). Residents' facilities satisfaction in housing project delivered by public-private partnership (Ppp) in Ogun State, Nigeria. *International Journal of Civil Engineering and Technology (Ijciet)*, *9*(1), 562–577.

Ogunbayo, B. F., Ohis Aigbavboa, C., Thwala, W. D., & Akinradewo, O. I. (2022). Assessing maintenance budget elements for building maintenance management in Nigerian built environment: A Delphi study. *Built Environment Project and Asset Management*, 12(4), 649–666.

Ogunbayo, S. B., & Mhlanga, N. (2022). Effects of training on teachers' job performance in Nigeria's public secondary schools. *Asian Journal of Assessment in Teaching and Learning*, *12*(1), 44–51.

Ojogwu, C. N., & Alutu, A. N. G. (2009). Analysis of the learning environment of university students in Nigeria: A case study of University of Benin. *Journal of Social Sciences*, *19*(1), 69–73.

Okolie, K. C. (2011). *Performance evaluation of buildings in Educational Institutions: A case of Universities in South-East Nigeria* (Published doctoral dissertation, Nelson Mandela Metropolitan University, Port Elizabeth, South Africa).

Olatubara, C. O. (2007). *Fundamentals of housing. Housing development and management, A book of readings*. Ibadan: Artsmostfare Prints, pp. 70–106.

Okebukola, P. (2002). The state of university education in Nigeria. Abuja: National University Commission Report.

Ontario Municipality (2013). *Property maintenance and Repair*, Ontario Building Code Act 1992. Canada: Government printer.

Phathela, A. V., & Cloete, C. E. (2018). The impact of the Government Immovable Asset Management Act (GIAMA) on the Department of Public Works, South Africa. RELAND*: International Journal of Real Estate & Land Planning*, *1*, 285–291

Pinjala, S. K., Pintelon, L., & Vereecke, A. (2006). An empirical investigation of the relationship between business and maintenance strategies. *International Journal of Production Economics*, *104*(1), 214–229.

Puķīte, I., & Geipele, I. (2017). Different approaches to building management and maintenance meaning explanation. *Procedia Engineering*, *172*, 905–912.

Quayson, J. H., & Akomah, B. B. (2016). Maintenance of residential buildings of selected public institutions in Ghana. *African Journal of Applied Research (AJAR)*, 2(1).

Republic of South Africa (2006). *The National Infrastructure Maintenance Strategy Act No. 108 of* South Africa.

Republic of South Africa (2007). *Government Immovable Asset Management Act No 19 of 2007*, Parliament.

Republic of South Africa (2018). *Norms and Standards for Education Facilities.* Department of Basic Education.

Salisu, R. A. (2001). *The Influence of School Physical Resources on Students Academic Performance* (Unpublished M.Ed thesis, University of Lagos, Department of Educational Administration).

Saunders, M., Lewis, P., & Thornhill, A. (2007). *Research methods for business students.*(4th ed.). England: Pearson Education Limited.

Telfer, A. (2005). *Hotel supply chain*: A strategic approach. Rome: European Hotel Manager Association.

United Nations Development Programme (UNDP) (2016). "Human Development Data (1990-2015): New York: United Nation Report. http://hdr.undp.org/en/data.

Williamson, A., Williams, C., & Gameson, R. (2010). The consideration of maintenance issues during the design process in the UK public sector. *In Proceedings of 26th Annual Association of Researchers in Construction Management Conference*, Leeds, UK: 6–8, pp. 1091–1100.

Wordsworth, P., & Lee, R. (2001). *Lee's building maintenance management*. London: Blackwell Science.

Xaba, M. I. (2011). The possible cause of school governance challenges in South Africa. *South African Journal of Education, 31*(2), 201–211.

Xaba, M. I. (2012). A qualitative analysis of facilities maintenance—A school governance function in South Africa. *South African Journal of Education, 32*(2), 215–226.

2 Generic Overview of Maintenance Management

Introduction

The contribution of maintenance management (MM) cannot be underestimated in operation and preservation of buildings and their components. This chapter gives a generic overview of the MM of buildings. It seeks to highlight the background on maintenance, the approach to maintenance, maintenance operations in buildings, maintenance organisation, the concept of building maintenance, and building defects, among others.

Maintenance

Definition

Maintenance refers to all works relating to repairs performed on any building to enhance its value, functionality, beauty and to prevent damage and injury toward increasing the useful economic life of the building. As defined by Cobbinah (2010), maintenance refers to the measures taken to preserve buildings' structural, functional, and aesthetic features to retain investment value in them over a long period. Maintenance purposes ensure continued peak performance of residential facilities throughout the design life, optimise productivity, and increase occupants' satisfaction with minimum resources (Abd-Wahab et al., 2015). However, Uma, Obidike, and Ihezukwu (2014) maintain that the inadequate application of maintenance strategy towards maintenance requests has led to very low output in the use of assets provided. The level of maintenance concerning the building and its auxiliary services govern the health of a building throughout its life cycle (Amaratunga & Baldry, 2002). Buildings cannot remain new forever; there is a need to draft a good maintenance programme for every newly constructed public or private building. Green and Turrell (2005) contend that maintenance ensures a building meets its required desirable performance over its life cycle. Both definitions show that the importance of maintenance in a structure (building) is to meet its required functional performance and the users' expectations.

DOI: 10.1201/9781003344681-2

In a building, maintenance is a condition that keeps a building fulfilling its purpose and presenting an attractive external appearance. Fianchini (2007) observed that maintenance involves an array of activities in keeping a building in the state to enable it to continue fulfilling its functions effectively and safely. Maintenance needs proper planning and co-ordination of activities geared towards preserving and restoring the building and its auxiliary facilities and services for top performance. Olatubara and Adegoke (2007) contend that maintenance activities are the logical follow-ups to ensure the continuing efficient functioning of a building and all its auxiliary facilities and services. Hoffman (2007) defined maintenance as the work undertaken to keep, restore or improve every facility, i.e., every part of a building, its services, and surroundings to a currently acceptable standard, and to sustain the utility and value of the building. British Standards (BS 3811:1974) define maintenance as "a combination of any action carried out to retain an item in or restore it to an acceptable condition." A more functional definition is that "maintenance is synonymous with controlling the condition of a building" (Lee & Seo, 1981; Brennan & Norris, 2000). Adeniji (2003) and Ab Wahab, Basari, and Samad (2013) maintained that maintenance is undertaken to restore, keep or improve every facility, i.e., every part of the structure, surroundings, and services to an accepted standard. Similarly, Chew, Tan, and Kang (2004) define it as the required processes and services undertaken to preserve the building's fabric and services after completion. Maintenance can render a building and its auxiliary facilities and services a healthy environment for a living (Leung & Fung, 2005).

It can be deduced from the above definitions that maintenance within a building serves as a series of activities taken to ensure the intended functions; the most desirable elements of a building structure are taken care of, as well as its auxiliary facilities and services. With the growth of more socially responsible factors within building structures, it is important to provide a more sustainable maintenance system for the building throughout its lifecycle to provide a maintainable building. As noted by Chanter and Swallow (2007), maintenance is inevitable in a building because materials used for its construction deteriorate over time owing to usage and exposure to the elements of climate. Nevertheless, Chew et al. (2004) and Too (2012) submitted that maintenance of buildings is often regarded as a cost burden, with most building owners reluctant to spend money to preserve their condition. Leung and Fung (2005). advance that maintenance is a way for potential gain toward facility stability for the users.

Concept of Building Maintenance

The maintenance of buildings has been characterised in different works of literature as work undertaken to keep, restore, or improve every building component and its auxiliary facilities and services. Lateef (2009) and Cobbinah (2010) maintain that the vital role building maintenance plays in the services provided or undertaken for a structure (building) after completion will make its components stable. It makes the building serves its required functions without upsetting its functionality (Ogunbayo et al., 2022). Brennan et al. (2000) state that building maintenance is about ensuring that the building meets its functional performance towards improving user productivity. Adeniji (2003) and Wahab, Basari, and Samad (2013) noted that the maintenance of a building is often neglected in the aspect of general building management. Ab Wahab et al. (2013) described building maintenance as a process of keeping the building and its other attached components to a continuous, usable standard. Ahuja and Khamba (2008) state that building maintenance is a service carried out to enhance building function and other attached components to preserve and protect the building. Additionally, Sherwin (2000), Chew et al. (2004), and Too (2012) maintained that maintaining a building helps to preserve it and other attached components to its original status. Also, Leaman, Stevenson, and Bordass (2010) postulated that building maintenance would increase economic value, especially in rentable buildings.

Likewise, Brennan and Norris (2000) postulated that building maintenance is a combination of actions to restore or retain a building and its components to an acceptable standard. Ogunbayo, Aigbavboa, Akinradewo, Oguntona, Thwala (2022) opined that over the years, the requirements for good practice in the maintenance of building stock had been established. Also, building maintenance in the built environment has been recognised as one of the key areas in which significant improvement must be achieved (Márquez, 2007). Additionally, Garg and Deshmukh (2006) state that the maintenance of buildings could either be based on a plan or a response toward assessing the needs and priorities of maintenance works that would be carried out within the building. The need for building maintenance arises where a condition that could endanger users and the existence of the building is revealed after a thorough assessment.

Sherwin (2000), Alsyouf (2007), and Too (2012) hypothesised that one major problem in the maintenance of buildings is the fact that building owners are usually unwilling to spend money to preserve the condition of their buildings. Conversely, Taillandier, Sauce, and Bonetto (2011) recognised effective building maintenance as a potential profit generator for owners and should not be viewed as a waste of financial resources. Globally, building maintenance remains an issue, especially in developing countries (Son, 1993).

Rational for Building Maintenance

The need for the maintenance of buildings arises when many defects that could endanger the building and its users are revealed after assessment (Adeniji, 2003). Nonetheless, the rationale for building maintenance is to increase the safety, reliability, quality, and availability of a building (Adeniji, 2003). Maintenance efforts can increase higher service/product quality and labour costs reduction, among others. Mills (1994) observed that maintenance has a great impact on services; it increases user satisfaction (effectiveness) and resource utilisation (efficiency) in service.

Odudu (1994) posits that the main aim of maintenance is to secure the economy in money, material, and time, achieve efficient building functioning, and secure user satisfaction with better performance. Also, Adeniji (2003) and Ab Wahab et al. (2013) contend that the rationale for building maintenance is to keep it up to an acceptable standard. Additionally, Ahuja and Khamba (2008) opined that maintenance is carried out in a building to enhance its function and facility. However, Chapman and Beck (1998) postulated that the ability of the building to support the activities efficiently and effectively within and around it through planned maintenance is critical to the users. Equally, Sherwin (2000) and Too (2012) contend that maintaining a building helps preserve all its components to their original status. This is affirmed by Olanrewaju, Fang and Tan (2018), namely that maintenance is a continuous activity. However, Alner and Fellows (1990) and Pukīte and Geipele (2017) list the rationale for undertaking maintenance work to include the following:

- Retaining aesthetic value;
- Preserving building structure;
- Preserving the value of an investment;
- Maintaining an acceptable quality standard;
- Prolonging the life span of a building structure;
- Upgrading the quality and standard of a building; and
- Attracting higher rental value in case of commercial and residential use.

Similarly, Ikpo (1998) adds that the rationale for building maintenance includes the perpetuation of the building to avoid danger to life and property. Also, it ensures the satisfaction of legal requirements for economic reasons, such as preventing heavier future expenses or justifying an increase in rental value (Ikpo, 1998). Jones (2002) states that one of the overriding reasons for building maintenance is to enhance the profitability of the use to which the building is put. On the other hand, these reasons may centre on the reservation of building elements as far as practicable to serve their purposes efficiently. In most developing countries, based on the present state of a depressed economy, the alternative to an existing dilapidated building is renovation and refurbishment. Martin (1993) sees maintenance as embracing all actions which inject a new or better life into a system. He further opined that maintenance helps to bring a building back to its original standard, near to it or well above such standard. It has not yet been possible to build a structure that could be classed as "maintenance-free" (Odudu, 1994). This is because many of the materials, elements, and composites used in building exhibit different reactions and behaviours when exposed to environmental features and man-made conditions of use (Odudu, 1994). This, in essence, means that the building, like our body, requires constant checkups to revitalise and strengthen it. Many

Table 2.1 Average life span of some building components

Building Components	Average Life Span
Reinforced concrete frame	100 years
Reinforced concrete floors	100 years
Timber floors	60 years
Timber roof construction	60–80 years
Asbestos roofing sheets	25–30 years
Roof tiles	50–60 years
Corrugated roofing sheets	20–30 years
Bituminous roofing felt	20 years
External rendering	50 years
Internal rendering	60 years
Clay floor tiles	60 years
Cement and sand paving	40 years
Sanitary appliances	25 years
Electrical installation	35 years
External painting	5 years
Internal painting	7 years
Electrical installation	35 years

Source: Olatubara and Adegoke (2007).

building components need to be replaced or refurbished from time to time as the need arises so that they can function satisfactorily (Thomas, 2005). Table 2.1 shows the average life span of some of the maintained building components and their expected life span.

What to Maintain in a Building?

Building maintenance focuses on keeping, restoring, and upgrading all components that make housing safe and healthy for human habitation to the original functional proficiency and a currently acceptable standard (Ogunbayo & Aigbavboa, 2019). Olatubara and Adegoke (2007) opined that building maintenance transcends the focus on the building structure alone. It includes the maintenance of all other auxiliary facilities, utilities, equipment, and services, both internal and external of the building, to cover its entire immediate neighbourhood. This, in essence, covers the building structure, the internal facilities, and equipment like a plumbing system, electrical fittings, and sewers. Also, external facilities, utilities, and services such as sewage systems, drainage systems, access roads, street lighting, refuse and waste management, water mains, etc., will have to be maintained.

Approach to Maintenance

In maintaining a building, selecting the type of approach to retain its economic life is important. Anderson and Edwards (2000) observed that approach to maintenance is an essential aspect of managing the building and its components that are often neglected. Anderson and Edwards (2000) state that proper selection and understanding of the type of approach to maintain a building will lead to a drop in maintenance cost with higher profitability. Whereas, poor selection and lack of knowledge of the type of approach to maintenance will lead to higher expenditures on maintainable facilities (Anderson & Edwards, 2000).

However, Swanson (2001) asserts that proactive and aggressive maintenance strategies are the best approach to the maintenance of buildings because of their positive influence on building performance. The findings of the study by Cholasuke et al. (2004) showed that the approach to maintenance is a process that leads to continuous improvement of the maintenance process. Cholasuke et al. (2004) observed further that effective selection of a suitable approach to maintenance contributed to the efficient maintenance and management of the building. Pinjala et al. (2006) postulated that in maintenance organisations' the approach to maintenance is guided by maintenance policies, planning, and control system. Accordingly, the approach to maintenance, as stated in BS 3811 (1993), can be categorised into two main types: planned and unplanned (see Figure 2.1).

Furthermore, Lind and Muyingo (2012) postulated that the building maintenance approach includes planned and unplanned maintenance, as described next.

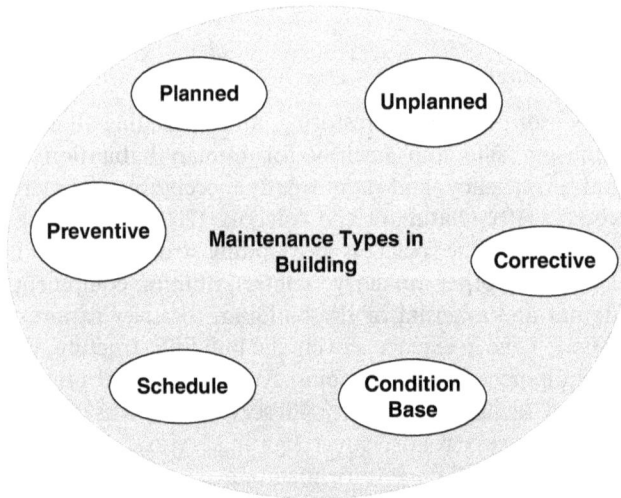

Figure 2.1 Maintenance types in building.

Source: Adapted from British Standard 3811 (1993).

Planned Maintenance

According to Fielden (1997) and Onaro (1997), planned maintenance (PM) is a type of maintenance approach that is executed in expectancy of failure in a building. EN 13306 (2001) and Seeley (1987) described planned maintenance as the maintenance work organised and carried out with forethought, control, and use of records to a predetermined plan. The PM, as opined by Olatubara and Adegoke (2007), is classified into two major types and includes preventive and corrective maintenance. However, the main objective of PM remains the same regardless of the type chosen for any maintenance operations (Olatubara & Adegoke, 2007).

Preventive Maintenance

Ben-Daya, Duffuaa, Raouf, Knezevic, and Ait-Kadi (2009) postulated that preventive maintenance is a maintenance approach directed at preventing failure in buildings to ensure their continued operational efficiency. It is a maintenance approach that is executed at predetermined intervals to pre-scribed criteria to reduce the probability of failure on the performance de-gradation of a building (Ben-Daya et al., 2009). Hallberg (2009) noted that an organisation's maintenance strategy is important in achieving an effective preventive approach toward maintenance. Additionally, Arditi and Nawakorawit (1999) claimed that when the serviceability condition of the building is recognised, preventive maintenance is an approach used to improve the performance of the facilities of the building. Arditi and Nawakorawit (1999) further postulated that preventive maintenance pre-vents failures, repairs, and replacement without the incident of any specific fault, and it could be planned in-house within an organisation or through outsourcing service, or by contract. Equally, Dubbs (1992) states that the significant factor in the success of the preventive approach in maintenance is the availability of knowledge, skills, and providers with special equipment. Moreover, Hui (2005) also asserts that other key factors to the preventive maintenance approach depend on the type, age, and size of the building and the design and condition of the components. Hauer, Bombach, Mohr, and Masse (2000) highlight the importance of a preventive approach to main-tenance, including the following:

- To access the condition and inventory of buildings;
- To evaluate maintenance project and the capacity to execute;
- To strategically plan for maintenance in both the long and short term;
- To outline the operating structure for preventing maintenance operation;
- To improve the capability of maintenance managers and workers; and
- To encompass suitable maintenance personnel in decision-making and communicating users' maintenance needs in the building.

Nevertheless, Hui (2005) states that the preventive approach to maintenance could be further subdivided into scheduled maintenance and condition-based maintenance.

Scheduled Maintenance

According to Agbola and Adegoke (2007), scheduled maintenance is a preventive maintenance approach carried out at a predetermined time interval, the number of operations, and so on. The scheduled maintenance activities include regular inspection, cleaning, testing, and routine checks to pre-empt components from breaking down or total collapse before carrying out repairs.

Condition-Based Maintenance

Condition-based maintenance (CBM) is a preventive maintenance approach that starts as a result of awareness of the condition of a building component through continuous monitoring or routine inspection (Agbola & Adegoke, 2007; Arditi & Nawakorawit, 1999). The CBM is a form of maintenance executed in response to a considerable deterioration of the building components. Nonetheless, Olatubara and Adegoke (2007) posit that the CBM is another form of preventive maintenance initiated as a result of knowledge of the condition of a building from routine inspection or through continuous monitoring.

Corrective Maintenance

Corrective maintenance (CM), as noted by Karia, Asaari, and Saleh (2014), is a form of planned maintenance that originated as a result of knowledge of the state of a building through constant monitory and routine inspection. Karia et al. (2014) posit that CM is a preventive maintenance approach that concentrates more on repairing any damaged part of a building. This repair could be carried out immediately or delayed to a later date, and it can even be planned in a long-term maintenance plan (Karia et al., 2014).

Unplanned Maintenance

According to Agbola and Adegoke (2007), an unplanned maintenance (UM) maintenance approach is a type of maintenance that results from an unanticipated breakdown or damage to a component, which may be due to external forces. Agbola and Adegoke (2007) state further that sudden damages cause this type of maintenance approach in a building due to some natural forces such as the ripping off part of a building during heavy storms. Also, British Standard (EN 13306: 2001) affirms that UM is maintenance work carried out to no predetermined or initial plan. Furthermore, this type of maintenance is usually unplanned, unexpected, and unavoidable (Seeley, 1987).

Other Maintenance Categories

Mills (1994) opined that other categorisations of maintenance are sub-divided into predictable and avoidable. Predictable maintenance is regular periodic maintenance work that may be necessary to retain the performance characteristics of a building and its auxiliaries' services and other facilities that require replacement or repair after it has achieved its useful life span (Olatubara & Adegoke, 2007). Predictable maintenance could be related to planned maintenance (Olatubara & Adegoke, 2007). On the other hand, avoidable maintenance is regarded as work carried out to rectify failures caused by poor design, workmanship, wrong installation, and the use of defective material. This is similar to unplanned maintenance (Olatubara & Adegoke, 2007).

Maintenance Generator in Buildings

Olatunji, Sher, and Gu (2010) opine that the estimated life of a building is between 50 and 90 years. However, Ogunbayo et al. (2018) state that maintenance work in a building is generated by different factors, including weathering corrosion, structural and thermal movements, old age, poor workmanship, natural disasters, and fire. According to Puķīte and Geipele (2017), the intensity and frequency of maintenance of work to be carried out on a building are a function of some factors which are more prominent in generating maintenance work in buildings, amongst which are environmental, human, materials, age, and use of the building. Some of these factors are briefly discussed in the following sections.

Environmental/Climate Factors

The performance of a building in any climatic environment is related to the dynamic equilibrium established between the external actions and the internal actions to which the building is subjected, together with the building's internal resistance (durability) (Olatubara & Adegoke, 2007). Additionally, Musa (2019) posits that the combination of sunshine, wind pressure, driving rain pressure, and humidity changes accelerate building structure deterioration. For example, the effects of wind on a building depend on its intensity, direction, and exposed surface area of the structure. In terms of materials used for building construction, sunshine affects materials such as timber, plastics, bitumen, paints, and concrete. Driving rains may keep walls moist and encourage moisture penetration. Moreover, pipes and other metals may corrode, wood may decay and rot, and damage may be caused by animals and insects taking advantage of the moist conditions. Other environmental factors include flooding, tornados, earthquakes, pollution, frost, and salt crystallisation (Musa, 2019).

Design and Construction Factors

Designers may deliberately use a low-cost or high-maintenance building component, and then the level of maintenance is already anticipated and can be planned for. However, Musa (2019) observed that when innovative and untested designs are produced at the expense of sound construction technology when proper site investigation is not carried out. This can also occur when contractors or maintenance officer uses poor quality materials or bad workmanship (Musa 2019). The resultant maintenance problems may manifest at any time in the life of the building. If these defects occur during the defects and liability period, the contractor is called upon to effect the repairs (Ogunbayo & Aigbavboa, 2019).

Use Activities Factors

Olatubara and Adegoke (2007) state that usage activities are another maintenance generator in a building. The usage activities include both human and mechanical wear and tear resulting from the usage of the building and its components. For example, paint fades, and door and window hinges wear out. Additionally, accompanying the normal wear and tear is the vandalism of components (Musa, 2019).

Social Value/Modernisation Factors

Olatubara and Adegoke (2007) postulated that social value or modernisation involves changing social tastes to prevent the structure's physical, functional, or economic obsolescence. This maintenance generator demands that work be carried out more frequently, which is functionally necessary. Musa (2019) states further that it involves conversion, alteration, renovation, and refurbishment to meet the tenant and user's demands, new use of building structure and its component, or to satisfy the current aesthetic needs of the owner or occupants. It may be due to new standards that must be met in line with modernisation and changing trends.

Accident Factors

Maintenance can also be generated through accidents in the building. This is because, in any building (Olatubara & Adegoke, 2007), structural maintenance could also be generated through damages that occur from accidents brought about by whatever reason, both natural and artificial or man-made, and it will have to be restored to an acceptable condition (Musa, 2019). Accidents' maintenance includes fire, flood, landslides, and earthquakes.

Maintenance Operations in Building

As Karia et al. (2014) noted, maintenance operations are carried out in a building to combat the progressive deterioration of building elements, components, and site and neighbourhood-related facilities and services such as access roads, water supply systems, electricity, and waste management systems. Consequently, Olatunbara and Adegoke (2007) affirm that maintenance operations include servicing, rectifying, replacing, and rehabilitating all or any part of the building components. This was further discussed below:

Servicing

Servicing is a cleaning operation undertaken at regular intervals of varying frequency and is sometimes termed day-to-day maintenance (Seeley, 1987). The required level and frequency of service will depend on the usage, age, and sophistication of facilities and equipment (Karia et al., 2014). The frequency of cleansing will particularly vary with different elements. For instance, the floor may have to be swept daily and polished weekly, while windows may have to be washed monthly and painted for redecoration and protection every four years. Similarly, while the grass may have to be cut more frequently during the rainy season, such frequency may not be necessary during the dry season.

Rectification

This essentially entails correcting inherent defects in building components, site/neighbourhood facilities, and services (Pinjala et al., 2006). The need for rectification may be due to shortcomings in the design, inherent faults in or unsuitability of components, damaged goods in transit or during installation, and incorrect assembling of parts (Pinjala et al., 2006). According to Seeley (1987), rectification represents a fruitful point at which to reduce the cost of maintenance because it is avoidable. Rectification work could be substantially reduced using performance specifications and the development of installation codes.

Replacement

This is the process of exchanging bad or damaged components with new ones. This is inevitable because service conditions cause materials to decay at different rates owing to usage and age (Cholasuke et al., 2004). Thus, most replacements are as a result of normal wear and tear rather than as a result of damage or physical breakdown (Cholasuke et al., 2004). This is why each element, component, or material has an acceptable useful economic life which often involves subjective judgement of the aesthetic of change. The measurement of a material's durability or length of life is highly technical owing to the complex and varying nature of the environment and the difficulty of determining how much material may change before it is discarded. However, the replacement frequency can be substantially reduced using high-quality materials and components and good workmanship (Ogunbayo, Ohis Aigbavboa, Thwala, & Akinradewo, 2022).

Renovation and Rehabilitation

Maintenance operations involve the renovation, rehabilitation, or refurbishment. This essentially consists of work done to restore a structure to improve it to the original design in conformity with the current dictates (Cotts & Lee, 1992). Therefore, renovation, rehabilitation, or refurbishment may involve a substantial modification or addition to the building structure and the upgrading of the existing facilities and services as well as the addition of new ones (Langston, Feng, Yu & Zhao, 2008). Building maintenance may involve not only the individual dwelling and its component and facilities but also rehabilitation and upgrading of neighbourhood facilities and services (Cotts & Lee, 1992).

Building Defects

Ratay (2005) defines a defect as a lack of something necessary for completeness or shortcoming. Accordingly, Street (2005) described a defect as the non-conformity of a component with a standard specified by standard characteristics. Ahzahar, Karim, Hassan, and Eman (2011) posit that any problem that decreases the value of a building is a defect. Ratay (2005) and Ahzahar et al. (2011) hypothesised that building defects could be through any of the following:

- Defects in structure through cracks or collapse;
- Faulty plumbing;
- Faulty electrical wiring and/or lighting;
- Faulty drainage systems;
- Inadequate insulation or soundproofing;
- Faulty ventilation, cooling, or heating systems; and
- Inadequate fire protection/suppression systems.

Additionally, apart from the causes of defects in buildings listed above, Ratay (2005) asserted that fungus, wood rot, dry rot, termites, land movement, and earth settlement might trigger building defects.

Building Failure

As Ratay (2005) states, building failure occurs when a structure loses its ability to perform its projected (design) function during its use. Ahzahar et al. (2011) and Brown (2001) postulated that building failure is the termination of the ability of an item or system to perform an intended or required function. Yacob et al. (2019) states that overstressing a building could lead to foundation failure and structural instability due to over-imposing loads above the bearing capacity of the structural component. Building failure could also be noted in a building through overstressing at earlier stages of usage through the development of deformation and fractures. Brown (2001) postulated that the following are factors that influence failure in a building:

Climate Conditions

Building and its other service facilities react to weather swiftly, predominantly external building components wide-open to external causes such as rain, wind, and solar radiation, including atmospheric pollution and ultraviolet light. (Ahzahar et al., 2011). This implies that buildings tend to weather rapidly, particularly the external part of the building that is exposed to external causes such as atmospheric pollution, rain, wind, and solar radiation, including ultraviolet light (Brown, 2001). Other symptoms associated with the climatic effect on a building include defective plastered rendering, erosion of mortar joints, fungal stains, harmful growth, and peeling paint (Ahzahar et al., 2011).

Location of Buildings

The location of the building is one of the leading factors that influence the failure of a building. A building within a waterlogging (near sea or rivers) tends to have more common building defects than a building located in an area of solid ground (Yacob et al., 2019). This is because the water from the ground will cause dampness penetration and structural instability for a building located within a wet area (Brown, 2001; Yacob et al., 2019).

Construction Material

The choice of material could also cause defects and failure in a building (Brown, 2001). Therefore, understanding the nature of the building materials and the perfect diagnosis of defects in the MM of the building is essential (Brown, 2001). Buildings, like older people, are vulnerable to all sorts of diseases. Thus, there is a need for maintenance officers to familiarise themself with construction materials and to have a deep understanding of the preservatives of materials and other building components (Ogunbayo & Aigbavboa, 2019).

Building Type and Change in Usage

Most defects and failures are caused in buildings by the change in usage and space, where the effect of the new changes on the existing structure is not considered (Brown, 2001). Additionally, Ahzahar et al. (2011) posit that most buildings were built based on a design and to hold certain loads, whereas, in the case of any additional load or attachment, they might not withstand it. Olatubara and Adegoke (2007) observed further that these changes affect the appearance of the buildings and apply to the existing building fabric.

Maintenance of Building

Continuous building maintenance activities play a major role in preventing building defects. Olanrewaju (2012) noted that ignoring the constant maintenance could cause several defects, leading to structural failures within the building. During the maintenance inspection of a building, adequate checking for signs of abnormal deterioration and defects is necessary to secure the structural stability of the building (Ogunbayo Aigbavboa, Amusan, Ogundipe, & Akinradewo, 2021). Also, Ahzahar et al. (2011) state that regular maintenance through constant inspection should be carried out on a building to avoid structural failure. Olatubara and Adegoke (2007) opined that it is significant that buildings continue to be properly maintained to ensure that they can function as efficiently and effectively as possible in supporting the delivery of a wide range of services. Olanrewaju (2012) posited that lack of adequate maintenance of buildings, apart from causing failure, also leads to poor building performance.

Faulty Design

Common design errors such as reducing the size of columns, demolition works, and foundation adjustment due to faulty design could lead to the instability of the building, which can cause defects and failure (Ahzahar et al., 2011). Brown (2001) states that defect and failure can result from a misjudgement leading to the assumption that is not consistent with the building structure and its components.

Faulty Construction

Faulty construction is one of the leading factors that lead to failure in the building (Olatubara & Adegoke, 2007). This, according to Brown (2001) and Ahzahar et al. (2011), is caused by shoddy workmanship, poor supervision, poor management, use of substandard material, and the use of wrong construction methods. All this will lead to defects and failure of the building structure with little or known notice at the construction stage (Olatubara & Adegoke, 2007). According to Olanrewaju (2012), faulty construction happens during construction due to construction contractor performance or material used. It is one of the common causes of early deterioration in building structures. The impact of faulty construction will reduce the service life of a building structure (Brown, 2001).

Corruption

corruption is one of the greatest problems in the maintenance process and operation (Olatubara & Adegoke, 2007). Ahzahar et al. (2011) observed that although corruption is generally assumed within the construction system, its scale and form are difficult to establish. Nonetheless, Brown (2001) maintains that in the maintenance process, corruption occurs mostly through the project identification stage, designing stage, tendering process, financing, or execution. Brown (2001) state further that act of corruption in each of the maintenance processes may involve consultants, contractors, sub-contractors, and project owners, among others. The Chartered Institute of Building (2008) asserts that corruption in the maintenance process through inflation of project costs, irregularities in the selection process, and bribery always leads to maintenance output being authorised as questionable and the maintained structure being defective.

Supervision Issues

The quality of supervision mainly influences the overall performance and efficiency of maintained buildings during maintenance operations (Owolabi, Tunji-Olayeni, Peter, & Omuh, 2014). Nevertheless, Chew, Tan, and Kang (2004) observed that lack of quality monitoring and control is one of the key causes of rework of maintenance work due to the failure of components of buildings maintained. Additionally, Amusan et al. (2014) posited that the overall maintenance of the efficiency and performance of the quality of buildings are impacted through effective monitoring and control. Chew et al. (2004) noted that to avoid failure of any part of the building maintenance, the performance of the maintenance supervisors will depend on an effective communication system among individual workers working together and effective maintenance operation planning.

Causes of Deterioration in Buildings

According to Micheal, Omole, Aderomu, Babalola, and Mosaku (2018), a maintenance-free building is highly necessary but unrealistic. The Chartered Institute of Buildings (2008) postulated that in respect of the quality of material used in the construction of buildings, its components deteriorate with time. Chew et al. (2004) noted that buildings and their attached components deteriorate with time owing to different factors, including wear and tear, environmental effects, and usage. Thus, Amusan et al. (2014) suggested that, left to themselves, buildings and their attached components will, in the long run, become unreliable, inefficient, and fail. Moreover, Micheal et al. (2018) estimated that within ten years of construction, the physical condition of buildings depreciates at a rate varying between 6% and 10%. However, Chew et al. (2004) posit that the rate of buildings' deterioration from year to year is aggravated when maintenance is ignored. Micheal et al. (2018) identified agents of building deterioration that determine maintenance within a building as follows:

Wear and Tear process

Wear and tear are natural phenomena that occur in buildings that sometimes defy solutions. Nevertheless, Micheal et al. (2018) observed that the rate of wear and tear could only be controlled through effective MM procedures and the provision of a standard maintenance manual for the personnel of the maintenance organisation.

Lack of Technical Ability

According to Amusan et al. (2014) and Micheal et al. (2018), maintenance processes and operations recently involved using different information systems in dealing with complex maintenance designs, specifications, and maintenance techniques, among others. Micheal et al. (2018) observed that there is a lack of technical training in the use of different techniques, other new areas, and advanced methods of operations toward effective maintenance that will reduce deterioration in a building. Nevertheless, Olatubara and Adegoke (2007) note that maintenance organisations require the input of well-trained and experienced personnel to reduce deterioration in building management.

Problem of Replacement Material and or Equipment

Micheal et al. (2018) posit that lack of maintainability of building components due to the non-availability of replacing material or equipment is another leading factor in the rapid deterioration of building components. Brown (2001) states that the non-consideration of sustainable or replacement materials or required equipment at the design stage to simplify maintenance work is a major factor that aggravates the rate of deterioration. Brown (2001) posits further that the unavailability of replacement materials or required equipment for maintenance operations at the time of need will increase the deterioration rate in the building.

Vandalism

According to Okoye (2014), users' actions or behaviour often constitute a great source of deterioration in buildings. Micheal et al. (2018) state that building components could become vandalised or deteriorate due to users' actions or constant usage. Ogunbayo and Aigbavboa (2019) opine that vandalism has its roots in the social fabric of society. Chew et al. (2004) and Okoye (2014) asserted that the aesthetic value of a building is impaired by vandalism. Moreover, Iyagba (2005) posits that vandalism, besides causing deterioration in a building, also reduces the building's life span. Similarly, Amusan et al. (2014) noted that apart from vandalism caused by human usage due to overcrowding and vandalism, the deterioration could also be caused by destructive animals such as rodents and insects.

Environmental Stress Effects (Sick Building Syndrome)

According to Nduka, Ogunbayo, Ajao, Ogundipe, and Babalola (2018), the impact of environmental agencies on the deterioration cannot be over-emphasised. Iyagba (2005) posits that environmental agencies such as in-dustrial pollution, climatic conditions, and chemical agents such as sulphates and chlorides all influence stress on the building, which results in its dete-rioration. Moreover, Nduka et al. (2018) posit that the identified stressors act based on the orientation of the building and its external elements, which lead to gradual deterioration of the building. Okoye (2014) asserts that the sub-sequent effect of these stressors on the building is known as sick building syndrome.

Construction Design Deficiency and Building Components Interdependence

The maintainability of buildings is often affected by how some components in the building are designed and located (Brown, 2001; Ahzahar et al., 2011). Assaf, Al-Khalil, and Al-Hazmi (1995) maintain that deficiency in design construction has led to the deterioration in most maintenance organisations. Improvising components of buildings due to a lack of replacement parts should be avoided because of the high risk of deterioration of the building and its components (Iyagba, 2005). However, Ogunbayo and Aigbavboa (2019) opine that the building should be maintainable to sustain a concept or design of its maintenance.

Selection and Quality of Material

Different materials abound for new construction and replacement purposes for maintenance purposes. However, Micheal et al. (2018) advised that the responsibility of selecting quality material for building maintenance to avoid deterioration lies on the designers, constructors, owners, merchants, and suppliers with assured warranties, maintenance manuals, and documents before the application. Moreover, Tsang (2002) states that collecting the materials, manuals, and documents is not just a matter of collecting them. It is important to adhere to the instructions in these manuals strictly. Assaf et al. (1995) postulate further that this will not only prolong the life of the building but reduce the impact of deterioration with minimal maintenance rates, making the building function optimally. Recently, in the design phase of the maintenance process, there has been a clamour for selecting recyclable, eco-friendly materials for enhanced sustainable maintenance practice (Tsang, 2002).

Introduction of New Construction and Maintenance Concepts

Micheal et al. (2018) opine that introducing new construction and maintenance concepts has created a new direction for the maintenance process of buildings. Divakar and Subramanian (2009) posited that the penetration of new technology, construction methods, design concepts, and systems of maintenance of infrastructure has led to an economical and environmentally valid construction product. Elmualim, Valle, and Kwawu (2012) observed that a synergy approach among built environment stakeholders is needed in the maintenance process to avoid deterioration in a building. Assaf et al. (1995) state that the application of the new construction and maintenance concepts should be valid, flexible, environmentally compatible, and sustainable.

Management Concept

The management concepts evolved in the industrial era, and through today's technological age, they have continued to evolve. Santos (1999) indicates that it is a known fact that all great business builders from time immemorial till today understand the significance of management in their business, which guides their actions and decisions. The word 'management' has been described with different meanings, while some people see it as a process of demanding the performance of a specific function or task (Santos, 1999). However, the term' management' posited by French, Kast, and Rosenzweig (1985) is mental work performed by people in organisational content. French, Kast, and Rosenzweig (1985) maintain that the mental work in management includes harmonising financial resources with human and material in realising organisational goals. Additionally, Egunjobi and Alabi (2007) described management as a process of getting things done through others for proper accountability.

According to Amaratunga (2000), the word "management" does not have a universally accepted standard definition. Nonetheless, Pearce, Kramer, and Robbins (1997) observed that management is an activity that has been used in practice for many business concepts as far back as memory stretches. Martin and Bartol (1998) define management as the process of engaging in the four major functions of planning, organising, leading, and controlling in achieving organisational goals. Naylor, Baker, and Szuba (2004) maintain that management aims to achieve organisational objectives by working with others to obtain the most from limited resources. Mintzberg (2009) opined that management means getting things done through other people. According to Montana and Charnov (2008), management is to accomplish the organisation's objectives by working with other people. Bateman, Snell, and Konopaske (2017) contend that management deals with assessing and harmonising people

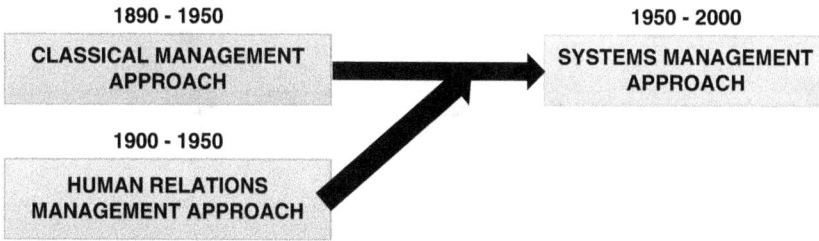

Figure 2.2 Evolution of management approaches.
Source: Adapted from Amaratunga (2001).

and resources to achieve organisational goals. Furthermore, Kinicki et al. (2011) postulated that management is the pursuit of organisational goals by integrating the work through planning, organising, leading, and controlling humans and resources.

However, Pearce et al. (1997: 152) observed that the first early definition of "management" was provided by Fayol (1916), namely that "to manage is to forecast and plan, to organise, to command, to co-ordinate and to control." Cole and Kelly (2015) asserted that management is a procedure that assists an organisation in achieving their projected objectives through planning, organising, and controlling their resources, including gaining the commitment of their employees. Cole et al. (2015) further observed that many organisations often pay no attention to the importance of understanding the management history as they pay more attention to occurring present actions. To help identify the main trends in the development of management theory (see Figure 2.2). It is important to categorise management into various approaches, namely the classical approach, human relation approach, and system approach, based on organisational views, structure, and structure management (Amaratunga, 2001).

Management Process

According to Bateman et al. (2017) management process (see Figure 2.3) is the attainment of organisational goals through planning, organising, directing, and controlling organisational resources. Daft (2004) indicates that the management concept follows seven processes: forecasting, planning, organising, motivation, controlling, co-ordinating, and communicating.

The first three processes are referred to as the thinking processes, while the other three are referred to as the doing processes (Kaehler & Grundei, 2019). This is because forecasting, planning, and organising take place before any physical work is carried out; as a result, they are referred to as the thinking processes while motivating, co-ordinating, and controlling involve carrying out the work and are known as the doing processes. The seventh process,

BASIC RESOURCES (Input)	FUNDAMENTAL MANAGEMENT FUNCTION (Process)	MANAGEMENT PROCESS OBJECTIVES
• Men and women • Machinery • Materials • Money • Resources	• Forecasting • Planning • Organising • Motivating • Coordinating • Controlling • Communicating	• End results • Goals • Output
PROCESS OF MANAGEMENT		

Figure 2.3 Process of management.

Source: Adapted from Bateman et al. (2017).

called communicating, is at the heart of the thinking and doing processes (Kaehler & Grundei, 2019). The processes can be briefly elucidated as follows.

Forecasting

This means looking into future to try and assess the possible trends of events that are likely to influence the conditions of the working situation in such a way that steps may be taken to overcome any constraints before or as soon as they arise (Bateman et al., 2017). Good forecasts should guide maintenance operation, activities, and resource management (Daft, 2004). Forecasting is an essential part of maintenance operations because there is a need to prepare future maintenance tasks for good estimates of future maintenance workload (Bateman et al., 2017). With a good forecast, the maintenance process will be easy for maintenance managers (Bateman et al., 2017).

Planning

Planning is the thinking which determines what course of action should be taken to achieve a particular objective. Planning assists organisations on how to develop a schedule for their activities. Planning is also concerned with making the best use of resources. The resources are often referred to as the five Ms of management, including manpower of labour, materials, machines, money, and management expertise of methods (Bateman et al., 2017).

Organising

This is the state between planning and putting the plan into action. It can also be described as an arrangement for all resources to be available exactly as and when required (Kaehler et al., 2019).

Motivating

In the management process, motivating is a way of inducing or encouraging people to work well and give their best. This is because, without the willing co-operation of other members of the construction or maintenance team, the aims of management are unlikely to be achieved (Kaehler et al. 2019).

Co-ordinating

Co-ordinating within the management process ensures the unity of active individuals, workgroups, and departments. It brings harmony in carrying out the different activities and tasks to achieve organisational goals efficiently. According to Daft (2004), integrating all resources will enable the smooth running of every task involving each member of an organisation.

Controlling

In the management process, controlling involves the methods by which it is possible to establish and determine whether work is being carried out as scheduled so that a comparison can be made against the plan and variations are noted and analysed (Kaehler et al., 2019).

Communicating

In the management process, communicating involves sharing, transmitting, or imparting information in such a manner that it is clear to all concerned, free of ambiguity and fully understood (Egunjobi & Alabi, 2007). Communicating is an organisation's most important management process at both vertical and horizontal levels (Kaehler et al., 2019). Communication guides management to run organisations effectively by following the mission, vision, rules, process, regulations, KPIs, and implementation to attain the desired result.

The seven processes are often considered as the cornerstone of effective and efficient management within an organisation.

Significant of Maintenance Management of Buildings

The Centre for Facilities Management (1992) described MM as a procedure through which maintenance organisations sustain structured quality support services to meet the organisation's objectives at the best cost. Furthermore, Puķīte and Geipele (2017) maintained that MM is a legal and technical set of operations required to maintain the buildings and preserve their usable condition and attached component. The significance of MM is for the maintained buildings to perform optimally (Olanrewaju & Abdul-Aziz, 2015; Uma, Obidike, & Ihezukwu, 2014). Nevertheless, Chanter et al. (2007) observe that maintenance work is inevitable in a building. Several studies on maintenance have indicated the significant role that maintenance play in keeping buildings in a safe condition (Ikpo, 1998; Sherwin, 2000). One of the overriding reasons for MM of buildings is to minimise the cost of enhancing the profitability of the use to which the building is put and, at the same time, minimise the degree of care on the building (Puķīte & Geipele, 2017).

The Royal Institution of Chartered Surveyors (2011) considered MM to consist of three distinctive but interconnected areas, namely the management of support services, the management of the building, and the management of MISs. Thompson and Kathawala (1991) posit that a standard MM department should play the role of the following four functions: real estate and development, building operations and maintenance, facility planning, and general office services. Adeniji (2003) opined that the need for MM of building arises when defects that could endanger the building and the users' well-being are discovered through effective managerial assessment. Miller (2007) opined that the prevalence of building facilities input had created a need for a more operational MM system of buildings. Pintelon and Gelders (1992) posited that the MM of buildings is not an isolated process; it is a system based on set organisational maintenance goals, as well as the maintenance policy of the organisation. Also, Vanneste and Van Wassenhove (1995) opined that the most desirable situation for effective building maintenance is the complete integration of MM into the organisational system. According to Mat Noor and Eves (2011), to boost the users of buildings' productivity and satisfy their immediate environmental and physical needs, the MM team of buildings is responsible for improving the indoor environment quality through viable service delivery. Waggaman (1992) observed that the maintenance of buildings through a better MM system is a matter of urgency. Price and Pitt (2012) opine that the MM response to maintenance requests is vital to the viable maintenance of a building and its auxiliary services. Spedding and Holmes (1994) postulated that built asset management, organisation (people and processes), strategic property management, valuations, and contract procedures are the five main groups of activities within the practice of MM of buildings and its process (see Figure 2.4).

Nevertheless, in the MM of buildings, the building functions, users' perception of the building's condition together with its relevance to their needs is the area that has to be of great consideration in a maintenance organisation (Laccase, Vanier & Kyle, 1997). Lateef (2009) opined that one key function of the MM of the building is to meet the users' requirements and the building itself with effective usage of the organisation's resources. Barrett (1992) categorises the organisational content of MM into three: resources, policy, and application (see Figure 2.5).

However, Chapman and Beck (1998) opine that users' needs are one of the essential prerequisites for the MM of the building. Prahalad and Hamel (1990) argued that for maintenance organisations to sustain their competitiveness, they need to create value-based MM systems that are better than those of their rivals in the maintenance industry. Fernandez, Labib, Walmisley, and Petty

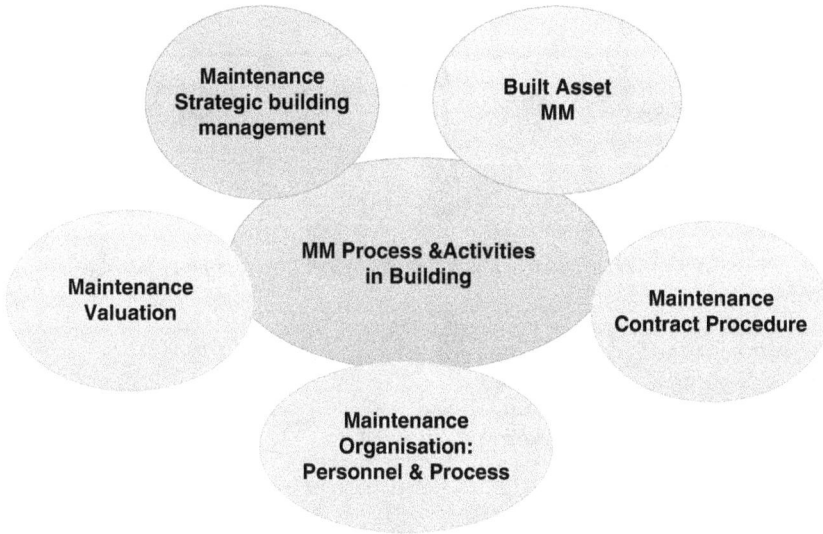

Figure 2.4 Maintenance management process and activities in building.

Source: Adapted from Spedding and Holmes (1994).

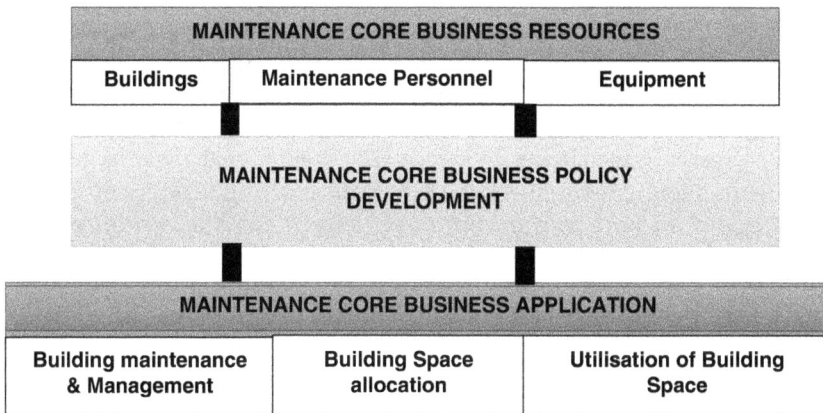

Figure 2.5 Maintenance Core business development.

Source: Adapted from Barrett (1992).

(2003) argued that maintenance organisations could only have an operational edge among their competitors when they deploy a more effective MM style that enables them to offer superior service to users. Márquez (2007) postulated that this shows that the purposes of formulating maintenance policies and plans within an organisation are to develop a good management strategy for

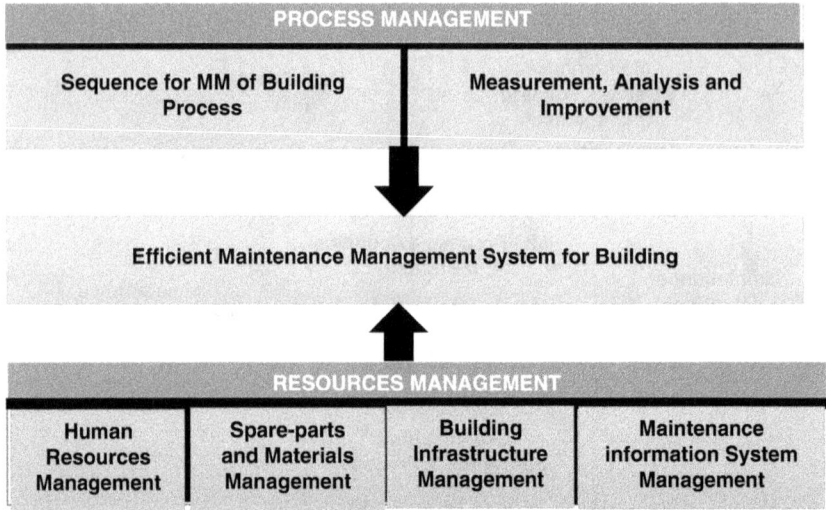

Figure 2.6 Resources and processes in the maintenance management system.
Source: Adapted from Barberá et al. (2012).

maintenance operations. However, Barberá, Crespo, Viveros, and Stegmaier (2012) postulated that in creating a viable MM framework, all aspects related to managing resources, including humans, components, and available and necessary materials, need to be given careful consideration (see Figure 2.6).

Amaratunga (2001) states that MM under viable human resources will bring together a range of stakeholders' functions that will benefit employees and the organisation. Hence, Márquez (2007) submitted that the MM of buildings is more proficient when there are suitable and efficient human resources that support personnel through the best combination of cost, efficiency, and quality of the organisation. Atkin and Brooks (2000) maintain that MM is not only about maintaining the buildings; it involves the management of the organisation's resources and the processes it supports. Spedding and Holmes (1994) opine that the significance of the MM within the maintenance process is to support the core business process of the organisation through workable human resources management. Amaratunga and Baldry (2002) contended that the significance of MM in a building is to make sure that buildings and their attached components are performing well. However, Atkin and Brooks (2000) assert that the significance of MM of buildings includes the following:

- High cost-effectiveness of the physical assets of an organisation;
- An easy change in the future use of space;
- To increase the organisational culture;

- To enhance the individuals' well-being; and
- Delivery of effective and responsive services.

Maintenance Management System in Buildings

The Value-Based Maintenance Management System

According to Lateef (2009), in the MM of building, maintenance organisations critically consider organisational resources toward maintenance processes and operations. In attaining better management of these resources, Lateef (2009) sustains that there is a need for an alternative value MM system based on a philosophy and principle that is entirely consistent with the theory and practice of value (management). As noted by Bateman and Snell (2009), value encompasses capital associated with how effectively and efficiently an organisation's operations meet organisational set goals. This shows that value is the relationship between the life-cycle cost of a building and the primary function of a building (Bateman and Snell, 2009). Also, Ab Wahab et al. (2013) viewed value as a process in which organisations are positioned by facilitating a network of relationships. However, the value-based maintenance management system (VMMS) extends beyond funds' availability and the current condition-based maintenance but encompasses maintenance needs as a factor of production (Lateef, 2009).

Additionally, Lavy (2008) states that VMMS focuses on a more effective and efficient strategy of creating value for the maintenance organisation, building, and users. Also, Sherwin (2000) stated that VBMM systems enhance productivity, profitmaking, and saving through factors of production. The study of Ab Wahab et al. (2013) revealed that a VBMM system provides an organised approach to enhancing planning, organising, directing, and controlling maintenance programmes. The merit of a VBMM system is that, in principle and practice, the building life cycle maintenance can be initiated at all levels (Ab Wahab et al., 2013).

Lateef (2009) submitted that in VBMM systems, the cost of implementing the maintenance programme is very low, but it is essentially high through production intervention (see Figure 2.7 below). As shown in Figure 2.8, as the maintenance programme passes through VMMS practices to the

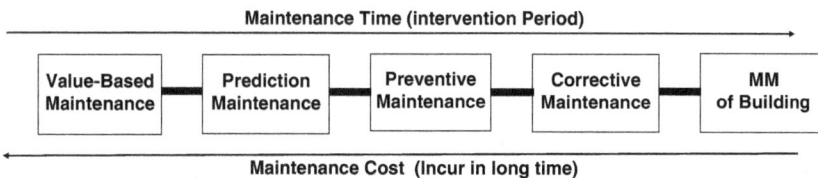

Maintenance Time (intervention Period)

Value-Based Maintenance	Prediction Maintenance	Preventive Maintenance	Corrective Maintenance	MM of Building

Maintenance Cost (Incur in long time)

Figure 2.7 Time-cost intervention concept in maintenance management of building.
Source: Adapted from Lateef (2009).

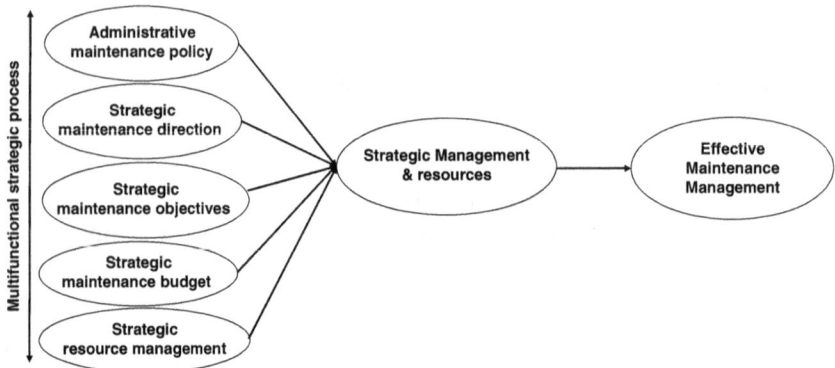

Figure 2.8 Strategic maintenance management system.

Source: Author's review (2022).

corrective approach, it shows clearly the declines in the possibility of positive savings; in contrast, there are increases in the cost of implementing the maintenance programme (Lateef, 2009). The VBMM system submitted by Lateef (2009) is user-focused through an effective and efficient organisational strategy.

Strategic Maintenance Management System

In maintenance activities, operation and management are two distinct and important functions in the MM of buildings. Adamu and Shakantu (2016) state that maintenance activities' required skills vary from those needed for managerial input. Pintelon and Parodi-Herz (2008) pointed out that technical skill is required purely in the maintenance operational aspect, while the decision-making of maintenance activities is dealt with through the management aspect of the maintenance process. Lee and Scott (2009) observed that in maintenance activities, maintenance personnel were more concerned with technical issues, with less attention being given to the strategic plans and goals of the maintenance department of an organisation. Buys (2004) and Olagunju (2011) posited that MM had moved away from simple technical functions to a multifunctional process that includes important management units (strategic and operations management) of an organisation.

A strategic maintenance management system (SMMS), as postulated by Tsang (2002), is the process that needs the input of top management's analysis of the environment in which the organisation operates before formulating a strategy and plans for the implementation and control of maintenance operations. Tapsir and Usman (2005) assert that an SMMS is more centred on the executive management domain of an organisation for maintenance decisions that guide maintenance operations and strategies. Lee and Scott

(2009) opine that concerning MM, the main function of the SMMS is formulating organisational maintenance policies that will guide maintenance managers in preparing maintenance programmes and strategy choices. Nevertheless, Garg and Deshmukh (2006) affirm that without a strategic plan, maintenance operations are inadequate and have less value for deploying and implementing resources in an organisation. Additionally, Lee and Scott (2009) posit that formulation of a maintenance policy should be guided in determining the strategic direction for a maintenance budget and resources within the maintenance organisation.

Yahya and Ibrahim (2010) opine that in SMMSs, the main maintenance operational objectives must align with the organisation's objectives. Also, Chanter and Swallow (2007) proclaim that in an SMMS, the position of maintenance unit objectives always depends on the organisation's strategic objectives. Smith and Hinchcliffe (2003) postulated that an SMMS is a maintenance system that is cost-efficient and that focuses on maintenance resource management. Smith and Hinchcliffe (2003) concluded that for effectiveness, SMMS adapts performance evaluation techniques through organisational results based on strategic planning.

Maintenance Information System for Maintenance Management of Buildings

Maintenance Information System

The arrangement in which information is kept in maintenance activities is important. Where information is kept only in record books, it might be difficult to quickly identify when to do maintenance or what to do in certain circumstances. This type of documentation in which information is kept in record books is the most common in developing economies.

However, a maintenance organisation needs a good and reliable information system to enable it to function effectively (Lateef Olanrewaju, Idrus, & Faris Khamidi, 2011). Olatunbara and Adegoke (2007) state that information is required for prediction, comparison, and knowledge or instruction. In the case of effective predictions, data are usually collected over time to establish possible causes of problems (Olatubara & Adegoke, 2007). In some cases, data are needed to make a comparison in terms of job performance and effectiveness. In maintenance operations, data might be required to acquire knowledge to explain what has to be done, why, and how it is to be done. Marquez and Gupta (2006) postulated that it guides maintenance managers and personnel to access maintenance operations and equipment data. Yu, Lung, and Panetto (2003) state that the MIS requires large quantities of data sets and is available in appropriate recordings in terms of location, timeline, and other requisite calibrations. Marquez and Gupta (2006) posit that data collected within maintenance organisations will be transformed into information that will guide maintenance decisions within the maintenance process and procedures used to prioritise maintenance actions.

Maintenance information systems allow for proper monitoring and controlling of buildings and other attached components (Marquez & Gupta, 2006). Olatubara and Adegoke (2007) assert that MISs within the maintenance process is more significant, especially where the components of building to maintain are complex. Additionally, Marquez and Gupta (2006) argue that MIS within a maintenance organisation provides critical tools that are useful to three maintenance activities: effective information processing capability, collaboration among expert systems, and communication tools and maintenance decision support. Yu et al. (2003) emphasised that MISs are not only about operation concepts alone but also help support negotiation and co-operation among stakeholders based on sharing complementary knowledge. Yu et al. (2003) observed that the MIS is a monitoring technology focusing more continuously on actions, operational, and potential tactical decisions that will significantly expand MM efficiency.

Nevertheless, Olatubara and Adegoke (2007) posit that MISs are largely missing in developing countries because of the poor culture of record keeping. Nevertheless, Fatokun (1997) states that two types of documentation are required for MIS management within a maintenance organisation. The selection of these documents, as noted by Fatokun (1997), is based on building maintenance records and maintenance manuals. Additionally, Yu et al. (2003) maintain that in addition to providing building construction records for developing MIS for effective maintenance processing. Also, the availability of maintenance manuals through their original manufacturers on all building components will greatly support MIS development to facilitate routine checks.

Maintenance Information Feedback in Maintenance Management of Buildings

In ensuring a consistent approach, the basic information collected during maintenance should be processed to meet the particular requirements of decision-makers at different stages in the maintenance operation of the building (Marquez & Gupta, 2006). The important maintenance information feedback in buildings is made to the owners, the maintenance managers, the designers of the building maintained, and the users of the maintained facilities (Olatubara & Adegoke, 2007).

To the owners of maintained buildings, feedback is more on expenditure and results, especially meeting their needs and expectations (Lee, Ni, Djurdjanovic, Qiu & Liao, 2006). Furthermore, owners of maintained buildings are more concerned with the total amount to be spent over a period of time rather than with the exact costs of individual items unless these happen to be large and non-recurring (Lee et al., 2006). To the managers, however, the feedback is basically a technicality. The feedback expected within the maintenance process by the maintenance managers has to do mainly with the planning and controlling of maintenance resources (Marquez & Gupta, 2006). Additionally, Yu et al. (2003) opined that in achieving effective MM of buildings, maintenance organisations are more concerned with feedback on maintenance strategy, maintenance cost implications, personnel performance, the effectiveness of organisation policy, and the quality of material and work executed, among others.

However, Olatubara and Adegoke (2007) postulated that the design of maintenance information feedback should be based on knowledge of the client's needs and expectations, organisation maintenance policy, information on the behaviour of the users of the building, maintenance characteristics of alternative materials, design solutions, knowledge of how maintenance and cleaning will be carried out and information on the maintenance cost.

Maintenance Information System Software

The main purpose of MM software is to keep records that will guide and report on the MM of building activities. However, in the maintenance process, there is a need for a software application that can take into account information in real-time and possible anomalies in behaviour coming from different sensors and other information sources that will allow for early diagnosis and the possibility to plan effective maintenance actions (Lee et al., 2006). Moreover, Garcia Sanz-Bobi and Del Pico (2006) developed an E-maintenance system software to address the predictive intelligence tools that can monitor degradation and ultimately optimise asset utilisation in the facility. Moore and Starr (2006) developed a system that automatically prioritises maintenance jobs that arise from corrective base maintenance using cost-based criticality (CBC). The CBC is software that draws together various types of information to support the MIS for operational effectiveness (Neelamkavil, 2009). Additionally, Neelamkavil (2009) postulated that with cost information and risk factors, CBC weights each incident flagged by condition monitoring alarms, allowing for optimised prioritisation of maintenance activities.

Moving forward, Trappey, Sun, Trappey, and Ma (2011) developed software that tackled engineering asset management functions and processes with diagnosis and prognosis expertise in a collaborative environment integrated with a service centre. The maintenance software was designed in such a way that the collaborative maintenance chain jointly combines asset operation sites (maintenance demanders), a first-tier collaborator (maintenance provider), a service centre (maintenance co-ordinator and the system provider), and the maintenance supplier. However, Trappey et al. (2011) posit that to achieve automation of negotiation and communication among maintenance organisations, including other core stakeholders, the multi-agent system (MAS) technique should be applied. Ko (2009) presents another maintenance information software tool called radio frequency identification (RFID) technology for building maintenance systems. The main aim of the RFID is to provide a data management module to collect building usage and maintenance data and to display the collected data graphically, where a statical module is then established. Neelamkavil (2009) maintained that the RFID is set in three modules: data management, statistics, and scheduling. This enables users to implement maintenance work using tablet PCs attached with RFID devices. In his study, Neelamkavil (2009) identified other available MIS software tools, such as asset management for power generation, asset management software, and preventive maintenance software.

Maintenance Information System Benefit in Maintenance Management of Buildings

According to Trappey et al. (2011) and Ko (2009), maintenance managers and maintenance officers of buildings and their auxiliary facilities and services can derive the benefit of using the MIS to achieve the following:

- To check specifications during supervision of buildings;
- To identify and correct faulty construction and methods;
- To eliminate less durable materials and retain the durable ones;
- To guide in the evaluation of the running cost of the building and its auxiliary facilities and services; and
- To predict a likely maintenance guide for the building.

Nevertheless, Cato and Mobley (2002) and Marquez (2007) opined that the MIS provides functionality that is normally grouped into modules or subsystems for specific activity sets. Cato and Mobley (2002) assert that the following activities are performed through an MIS in the maintenance process and operation toward effective MM, and they include but are not limited to:

- Invoices matching and accounts payable;
- Human resources;
- Purchasing;
- PM plan;
- Development and scheduling and receiving;
- Work order creation, scheduling, execution, and completion;
- Equipment/asset bill of materials creation and maintenance;
- Equipment/asset and work order history;
- Inventory control, and
- Tables and reports

Summary

This chapter reviewed the literature concerning the MM of buildings. The review also includes background on maintenance, the approach to maintenance, and maintenance operations in buildings. The chapter revealed that the term MM encompasses a broad spectrum of definitions. The chapter affirmed that maintenance in a building aims to retain the value of an investment in the building by protecting it from the preliminary stage. The chapter reviewed the MM systems applicable in building maintenance. It revealed that the SMMS is the common maintenance system used for maintenance processes and operations. It also shows that the SMMS is based on understanding the relationship between executive management at strategic and operational levels of maintenance personnel. The chapter pointed out that in MM operation in a building, data are needed to compare job performance and acquire knowledge to assist in explaining what must be done, why, and how it is to be done.

References

Ab Wahab, Y., Basari, A. S. H., & Samad, A. (2013). Building maintenance management preliminary finding of a case study in ICYM. *Middle East Journal of Science Research*, *17*(9), 1260–1268.

Abdul Lateef Olanrewaju, A. (2012). Quantitative analysis of defects in university buildings: User perspective. *Built Environment Project and Asset Management*, *2*(2), 167–181.

Abd-Wahab, S. R. H., Sairi, A., Che-Ani, A. I., Tawil, N. M., & Johar, S. (2015). Building maintenance issues: A Malaysian scenario for high-rise residential buildings. *International Journal of Applied Engineering Research*, *10*(6), 15759–15776.

Adamu, A. D., & Shakantu, W. M. (2016). Strategic maintenance management of built facilities in an organisation. *International Journal of Economics and Management Engineering*, *10*(4), 1104–1107.

Adeniji, W. A. (2003). Maintenance of Housing Stock Relevance to Housing Development. *A paper Presented at NIOB National Conference Housing Development*, Ikeja, 1–2 April 2003.

Agbola, T., & Adegoke, S. A. (2007). Economics in housing. In *Housing development and management. A book of readings*. Department of Urban and Regional Planning, Faculty of Social sciences (pp. 107–147). Nigeria: University of Ibadan.

Ahuja, I. P. S., & Khamba, J. S. (2008). Total productive maintenance: Literature review and directions. *International Journal of Quality & Reliability Management*, *25*(7), 709–756.

Ahzahar, N., Karim, N. A., Hassan, S. H., & Eman, J. (2011). A study of contribution factors to building failures and defects in construction industry. *Procedia Engineering*, *20*, 249–255.

Alner, G. R., & Fellows, R. F. (1990). Maintenance of local authority school building in UK: A case study. In *Proceedings of the International Symposium on Property Maintenance Management and Modernisation, Singapore*, 7–9 March 1990, pp. 90–99.

Amaratunga, D. (2000). Assessment of facilities management performance. *Property Management*, *18*(4), 258–266.

Amaratunga, D., & Baldry, D. (2002). Performance measurement in facilities management and its relationships with management theory and motivation. *Facilities*, *20*(10), 327–336.

Amaratunga, R. G. (2001). *Theory building in facilities management performance measurement: Application of some core performance measurement and management principles* (Published doctoral dissertation, University of Salford).

Amusan, L. M., Owolabi, J. D., Tunji-Olayeni, P. F., Peter, N. J., & Omuh, I. O. (2014). Assessing the effectiveness of maintenance practices in public schools. *European International Journal of Science and Technology, 3*(3), 103–109.

Anderson, J., & Edwards, S. (2000). Addendum to BRE methodology for environmental profiles of construction materials, components, and buildings. *DETR Framework Project: support for government policies on sustainable development.* Watford, UK: Centre for Sustainable Construction, BRE Building Research Establishment.

Arditi, D., & Nawakorawit, M. (1999). Issues in building maintenance: Property managers' perspective. *Journal of Architectural Engineering, 5*(4), 117–132.

Assaf, S. A., Al-Khalil, M., & Al-Hazmi, M. (1995). Causes of delay in large building construction projects. *Journal of Management in Engineering, 11*(2), 45–50.

Atkin, B., & Brooks, A. (2000). *Total facilities management.* London: Blackwell Science.

Alsyouf, I. (2007). The role of maintenance in improving companies' productivity and profitability. *International Journal of Production Economics, 105,* 70–78.10.1016/j.ijpe.2004.06.057

Barberá, L., Crespo, A., Viveros, P., & Stegmaier, R. (2012). Advanced model for maintenance management in a continuous improvement cycle: Integration into the business strategy. *International Journal of System Assurance Engineering and Management, 3*(1), 47–63.

Barrett, P. (1992). Development of a post-occupancy building appraisal model. In Barrett, P. (Ed.), *Facilities management: Research directions.* London: RICS Books.

Bateman, T. S. & Snell, S. (2009). *Administración: Liderazgo y colaboración en un mundo competitivo* No. Sirsi) a 458252.

Bateman, T. S., Snell, S. A., & Konopaske, R. (2017). *Management: Leading & collaborating in a competitive* (pp. 299–303). New York, NY, USA: McGraw Hill Companies Inc.

Ben-Daya, M., Duffuaa, S. O., Raouf, A., Knezevic, J., & Ait-Kadi, D. (Eds.). (2009). *Handbook of maintenance management and engineering* (Vol. 7). London: Springer.

Brennan, B. & Norris, M. (2000). *Repair and Maintenance of Dwellings.* Housing Unit. Ireland: An-Taonad Tithlochta Press.

British Standard [B.S.] (2001). *Maintenance terminology.* EN 13306: 2001.

British Standard Institute (1993). *Glossary of terms used in terotechnology.* BS 3811: 1993. Milton Keynes, BSI.

British Standard Institute (1974). *Glossary of general terms used in maintenance organization.* BS 3811: 1974. London, BSI.

Brown, R. W. (2001). *Practical foundation engineering handbook.* New York: McGraw

Buys, N. S. (2004). *Building maintenance management systems in South African tertiary institutions* (Unpublished doctoral dissertation, University of Port Elizabeth).

Cato, W. W., & Mobley, R. K. (2002). *Computer-managed maintenance systems in process plants: A step-by-step guide to effective management of maintenance, labor, and inventory in your operation.* Houston: Gulf Publishing Company.

Centre for Facilities Management (1992). *An overview of the facilities management industry* (Part 1). Strathclyde: Strathclyde Graduate Business School.

Chanter, B., & Swallow, P. (2007). *Building maintenance management* (p. 134). Oxford: Blackwell.

Chapman, K., & Beck, M. (1998). Recent experiences of housing associations and other registered social landlords in commissioning stock condition surveys. In *Proceedings of RICS Conference COBRA*, London, 4–6 September 1998, pp. 32–37.

Chew, M. Y. L., Tan, S. S., & Kang, K. H. (2004). Building maintainability—Review of state of the art. *Journal of Architectural Engineering, 10*(3), 80–87.

Cholasuke, C., Bhardwa, R., & Antony, J. (2004). The status of maintenance management in UK manufacturing organisations: Results from a pilot survey. *Journal of Quality in Maintenance Engineering, 10*(1), 5–15.

Cobbinah, P. J. (2010). *Maintenance of buildings of public institutions in Ghana. Case study of selected institutions in the Ashanti Region of Ghana* (Published master thesis, Kwame Nkrumah University of Science and Technology, Kumasi).

Cole, G. A., & Kelly, P. (2015). *Management theory and practice.* Boston, Massachusetts: Cengage Learning.

Cotts, D. F., & Lee, M. (1992). *The facility maintenance handbook.* Washington, D.C.: American Management Association.

Daft, R. L. (2004). Theory Z: Opening the corporate door for participative management. *Academy of Management Perspectives, 18*(4), 117–121.

Daft, L. R. (1989). *Organization theory and design,* 3rd edn. New York: West Publishing Company.

Divakar, K., & Subramanian, K. (2009). Critical success factors in the real-time monitoring of construction projects. *Research Journal of Applied Sciences, Engineering, and Technology, 1*(2), 35–39.

Dubbs, D. (1992). Balancing benefits of outsourcing vs. in-house. *Facilities Design & Management, 11*(8), 42–44.

Egunjobi, L., & Alabi, M. (2007). Housing facilities supply and management. In *Housing development and management. A book of readings.* Department of Urban and Regional Planning, Faculty of Social sciences (pp. 350–390). Nigeria: University of Ibadan.

Elmualim, A., Valle, R., & Kwawu, W. (2012). Discerning policy and drivers for sustainable facilities management practice. *International Journal of Sustainable Built Environment, 1*(1), 16–25.

Fatokun, T. O. (1997). Prospect control of schemes of maintenance, repairs and improvement works. Paper presented at a 2-Day Workshop on *Residential Property Maintenance Planning and Management*, organized by Estate Surveyors Registration Board of Nigeria with First State Communication.

Fayol, H. (1916). General principles of management. *Classics of Organization Theory, 2*(15), 57–69.

Fernandez, O., Labib, A. W., Walmisley, R., & Petty, D. J. (2003). A decision support maintenance management system: Development and implementation. *International Journal of Quality & Reliability Management, 20*(8), 965–979.

Fianchini, M. (2007). Fitness for purpose: A performance evaluation methodology for the management of university buildings. *Facilities, 25*(3), 137–146.

Fielden, B. M. (1997). *Conservation of historic buildings.* London: St. Edmundsbury Press Ltd.

French, W. L., Kast, F. E., & Rosenzweig, J. E. (1985). *Understanding human behavior in organizations.* New York City: Harper & Row.

Garcia, M. C., Sanz-Bobi, M. A., & Del Pico, J. (2006). SIMAP: Intelligent system for predictive maintenance: Application to the health condition monitoring of a wind turbine gearbox. *Computers in Industry*, *57*(6), 552–568.

Garg, A., & Deshmukh, S. G. (2006). Maintenance management: Literature review and direction. *Journal of Quality in Maintenance Engineering*, *12*(3), 205–238.

Green, D., & Turrell, P. (2005). School building investment and impact on pupil performance. *Facilities*, *23*(5/6), 253–261.

Hallberg, D. (2009). *System for predictive life cycle management of buildings and infrastructures* (Published doctoral thesis, KTH Research School—HIG, University of Gävle, Gävle, Sweden).

Hauer, J., Bombach, V., Mohr, C., & Masse, A. (2000). *Preventive maintenance for local government buildings: A best practices review*. USA: ERIC Clearinghouse.

Hoffman, D. (2007). Prognostics and health management (PHM)/condition based maintenance (CBM). In *IEEE reliability society 2007 annual technology report*, Phoenix, 15–19 April.

Hui, E. Y. Y. (2005). Key success factors of building management in large and dense residential estates. *Facilities*, *23*(1/2), 47–62.

Ikpo, I. J. (1998). Application of the Weibull distribution technique in the prediction of the times between failures (MTBF) of building components. *Nigerian Journal of Construction Technology and Management*, *1*(1), 79–87.

Iyagba, R. O. A. (2005). The menace of sick buildings – A challenge to all for its prevention and treatment. In *An inaugural lecture delivered at University of Lagos*, Lagos, 2005, pp. 1–10.

Jones, B. P. (1993). *The British standards institution*. In Proceeding of International Conference on Power System Technology: IEE Colloquium on Metering Standards and Directives, London, 5–5 April 1993, pp. 111–115.

Jones, K. (2002). *Best value in construction*. Oxford: Blackwell, 280–300.

Kaehler, B., & Grundei, J. (2019). The concept of management: In search of a new definition. *HR Governance* (pp. 3–26). Cham: Springer.

Karia, N., Asaari, M. H. A. H., & Saleh, H. (2014). Exploring maintenance management in service sector: A case study. In *Proceedings of International Conference on Industrial Engineering and Operation Management*, Bali, 7–9 January 2014, pp. 3119–3128.

Kinicki, A., Williams, B. K., Scott-Ladd, B. D., & Perry, M. (2011). *Management: A practical introduction*. New York City: McGraw-Hill Irwin.

Ko, C. H. (2009). RFID-based building maintenance system. *Automation in Construction*, *18*(3), 275–284.

Lacasse, M. A., Vanier, D. J., & Kyle, B. R. (1997). Towards integration of service life and asset management tools for building envelope systems. In *Proceedings of 7th Conference on Building Science and Technology: Durability of Buildings Design, Maintenance, Codes and Practice*, pp. 153–167. March 20- 21, Toronto, Ontario, Canada.

Langston, C., Feng, C. C., Yu, M. X., & Zhao, Z. Y. (2008). The sustainabilityimplications of building adaptive reuse. In Proceedings of Criocm 2008 International Research Symposium on Advances of Construction Management and Real Estate, Beijing, China, November 20-22, pp. 1–11.

Langston, C., & Lauge-Kristensen, R. (2013). *Strategic management of built facilities*. London: Routledge.

Lateef, O. A. (2009). Building maintenance management in Malaysia. *Journal of Building Appraisal, 4*(3), 207–214.

Lateef Olanrewaju, A., Idrus, A., & Faris Khamidi, M. (2011). Investigating building maintenance practices in Malaysia: A case study. *Structural Survey, 29*(5), 397–410.

Lavy, S. (2008). Facility management practices in higher education buildings. *Journal of Facilities Management,* 6(4, 303–315.

Lee, H. H. Y., & Scott, D. (2009). Overview of maintenance strategy, acceptable maintenance standard and resources from a building maintenance operation perspective. *Journal of Building Appraisal, 4*(4), 269–278.

Leaman, A., Stevenson, F., & Bordass, B. (2010). Building evaluation: practice and principles. *Building Research & Information,* 38, 564–57710.1080/09613218.2010. 495217

Lee, J., Ni, J., Djurdjanovic, D., Qiu, H., & Liao, H. (2006). Intelligent prognostics tools and e-maintenance. *Computers in Industry, 57*(6), 476–489.

Lee, P. S., & Seo, K. S. (1981). A study for the estimation of the maintenance-repair cost of the buildings. *Journal of the Architectural Institute of Korea, 98*(1), 63–68.

Leung, M., & Fung, I. (2005). Enhancement of classroom facilities of primary schools and its impact on learning behaviors of students, *Facilities, 23*(13/14), 585–594.

Lind, H., & Muyingo, H. (2012). Building maintenance strategies: Planning under uncertainty. *Property Management, 30*(1), 14–28.

Márquez, A. C. (2007). *The maintenance management framework: Models and methods for complex systems maintenance.* London: Springer-Verlag.

Marquez, A. C., & Gupta, J. N. (2006). Contemporary maintenance management: Process, framework, and supporting pillars. *Omega, 34*(3), 313–326.

Martin, D. C., & Bartol, K. M. (1998). Performance appraisal: Maintaining system effectiveness. *Public Personnel Management, 27*(2), 223–230.

Martin, L. L. (1993). *Total quality management in human service organizations* (p. 69). Newbury Park: Sage.

Mat Noor, N., & Eves, C. (2011). Malaysia high-rise residential property management: 2004–2010 trends and scenario. In *Proceedings of the 17th Annual Pacific-Rim Real Estate Society Conference*: Pacific Rim Real Estate Society, Gold Coast, Australia, 4–6 April 2011, pp. 1–7.

Michael, A., David, O., Peter, A., Olufemi, B., & Olusoji, M. (2018). Appraisal of the state of Health of Residential Building Facilities in a Private University in Nigeria. *International Journal of Engineering Technologies and Management Research, 5*(4), 153–167.

Miller, B. A. (2007). *Assessing organizational performance in higher education.* USA: John Wiley and Sons, Inc.

Mills, E. D. (1994). Design and building maintenance. In E. D. Mills (Ed.), *Building maintenance and preservation: Guide for design and management* (2nd ed., pp. 1–15). London: Butterworth Heinemann.

Mintzberg, H. (2009). *Managing.* San Francisco, CA: Berrett-Koehler Publishers.

Montana, P. J., & Charnov, B. H. (2008). *Management,* Vol. 333, New York: Barron's Educational Series, Inc..

Moore, W. J., & Starr, A. G. (2006). An intelligent maintenance system for continuous cost-based prioritisation of maintenance activities. *Computers in Industry, 57*(6), 595–606.

Musa, M. A. (2019). Assessing the effects of floor levels on daylight distribution in mid-rise office buildings in composite climate of Nigeria. In *Simulation for Sustainable Built Environment Conference*, New Cairo, Egypt, 28–30 November 2019, pp. 1–10. IOP Conference Series: Earth and Environmental Science, Vol. 397, No. 1, 012023. IOP Publishing.

Naylor, B. J., Baker, J. A., & Szuba, K. J. (2004). Effects of forest management practices on red-shouldered hawks in Ontario. *The Forestry Chronicle*, *80*(1), 54–60.

Nduka, D., Ogunbayo, B. F., Ajao, A. M., Ogundipe, K., & Babalola, B. (2018). Survey datasets on sick building syndrome: Causes and effects on selected public buildings in Lagos, Nigeria. *Data, in Brief*, *20*, 1340–1346.

Neelamkavil, J. (2009). *A review of existing tools and their applicability to facility maintenance management*. Canada: Institute for Research in Construction Report # RR-285.

Odudu, W. O. (1994). Maintenance Management Culture. In *A Paper Presented at the Seminar by European Economic Community, National Planning Commission, and Federal Ministry of Works and Housing in Collaboration with Yaba College of Technology on High Rise Building in Nigeria: Problem and Prosperity*. Lagos, Yaba.

Ogunbayo, B. F., Aigbavboa, C. O., Thwala, W., Akinradewo, O., Ikuabe, M., & Adekunle, S. A. (2022). Review of culture in maintenance management of public buildings in developing countries. *Buildings*, *12*(5), 677.

Ogunbayo, B. F., & Aigbavboa, O. C. (2019). Maintenance requirements of students' residential facility in higher educational institution (HEI) in Nigeria. In *Proceedings of 1st International Conference on Sustainable Infrastructural Development*, Ota, 24–28 June 2019, IOP Conference Series: Materials Science and Engineering, 640, (1) 012014. IOP Publishing.

Ogunbayo, B. F., Aigbavboa, C. O., Amusan, L. M., Ogundipe, K. E., & Akinradewo, O. I. (2021). Appraisal of facility provisions in public-private partnership housing delivery in southwest Nigeria. *African Journal of Reproductive Health*, *25*.

Ogunbayo, B., Aigbavboa, C., Akinradewo, O., Oguntona, O., & Thwala, D. (2022). Validation of Maintenance Management Co re Elementsfor Higher Educational Institution Buildings in developing Countries Education Sector. In Maciejko, A. (eds) *Human Factorsin Architecture, Sustainable Urban Planning and Infrastructure*. AHFE(2022) International Conference. AHFE Open Access, vol 58. AHFE International, USA. http://doi.org/10.54941/ahfe1002367

Ogunbayo, B. F., Ajao, A. M., Alagbe, O. T., Ogundipe, K. E., Tunji-Olayeni, P. F., & Ogunde, A. (2018). Residents' facilities satisfaction in housing project delivered by public-private partnership (PPP) in Ogun State, Nigeria. *International Journal of Civil Engineering and Technology (IJCIET)*, *9*(1), 562–577.

Ogunbayo, B. F., Ohis Aigbavboa, C., Thwala, W. D., & Akinradewo, O. I. (2022). Assessing maintenance budget elements for building maintenance management in Nigerian built environment: A Delphi study. *Built Environment Project and Asset Management*, *12*(4), 649–666.

Okoye, C. O. (2014). Residents 'partnering in public housing basic infrastructure provision and maintenance: A strategy for satisfactory public housing provision. *IOSR Journal of Environmental Science, Toxicology and Food Technology*, *8*(1), 73, 79.

Olagunju, R. E. (2011). Development of Mathematical Models for The Maintenance of Residential Buildings In Niger State, Nigeria (Unpublished doctoral dissertation,

Ph.D (Architecture) Thesis, Department of Architecture, Federal University of Technology, Minna, Nigeria.

Olanrewaju, A. L., & Abdul-Aziz, A. R. (2015). Building maintenance processes, principles, procedures, practices, and strategies. In *Building maintenance processes and practices* (pp. 79–129). Singapore: Springer.

Olanrewaju, A., Fang, W. W., & Tan, Y. S. (2018). Hospital building maintenance management model. *International Journal of Engineering and Technology, 2*(29), 747–753.

Olatubara, C. O., & Adegoke, S. A. O. (2007). Housing maintenance. In *Housing development and management. A book of readings.* Department of Urban and Regional Planning, Faculty of Social sciences (pp. 391–318). Nigeria: University of Ibadan.

Olatunji, O. A., Sher, W., & Gu, N. (2010). Building information modeling and quantity surveying practice. *Emirates Journal for Engineering Research, 15*(1), 67–70.

Onaro, A. N. (1997). Procurement Arrangement for Rehabilitation Projects. In *A paper presented at the Seminar of Nigerian Institute of Quantity Surveyors*, Lagos, March 2.

Pearce II, J. A., Kramer, T. R., & Robbins, D. K. (1997). Effects of managers' entrepreneurial behavior on subordinates. *Journal of Business Venturing, 12*(2), 147–160.

Pinjala, S. K., Pintelon, L., & Vereecke, A. (2006). An empirical investigation of the relationship between business and maintenance strategies. *International Journal of Production Economics, 104*(1), 214–229.

Pintelon, L. M., & Gelders, L. F. (1992). Maintenance management decision-making. *European Journal of Operational Research, 58*(3), 301–317.

Pintelon, L., & Parodi-Herz, A. (2008). Maintenance: An evolutionary perspective. In *Complex system maintenance handbook* (pp. 21–48). London: Springer.

Prahalad, C. H., & Hamel, G. (1990). The core competence of the corporation. *Harvard Business Review, 68*(3), 295–336.

Price, S., & Pitt, M. (2012). The influence of facilities and environmental values on recycling in an office environment. *Indoor and Built Environment, 21*(5), 622–632.

Puķīte, I., & Geipele, I. (2017). Different approaches to building management and maintenance meaning explanation. *Procedia Engineering, 172*, 905–912.

Ratay, R. T. (2005). *Structural condition assessment.* American Society of Civil Engineers, Reston, VA. Co-publisher: John Wiley & Son Inc. (Wiley ISBN: 0-471-64719-5) 2006, Soft Cover, 700 p.

Royal Institution of Chartered Surveyors (2011). *RICS Valuation Standards: Global and UK.* RICs. Sage publications.

Santos, A. D. (1999). *Application of flow principles in the production management of construction sites* (Published doctoral dissertation, University of Salford).

Seeley, I. H. (1987). *Building maintenance.* London: Macmillan Publishers Limited.

Sherwin, D. (2000). A review of overall models for maintenance management. *Journal of Quality in Maintenance Engineering, 6*(3), 138–164.

Smith, A. M., & Hinchcliffe, G. R. (2003). *RCM –Gateway to world-class maintenance.* Burlington: Elsevier Butterworth-Heinn.

Spedding, A., & Holmes, R. (1994). Facilities management. *CIOB Book of Facilities Management.* London: Longman Scientific & Technical.

Son, L.H. (1993). *Building maintenance technology.* London: Macmillan International Higher Education.

Street, R. A. (2005). *Hydrogenated amorphous silicon*. Cambridge, United Kingdom: Cambridge University Press.

Swanson, L. (2001). Linking maintenance strategies to performance. *International Journal of Production Economics*, 70(3), 237–244.

Tapsir, S. H., & Usman, F. (2005). Service life planning for affordable housing design: A challenge to Malaysia Construction Industry. In *Proceedings of the 2005 World Sustainable Buildings Conference*, Tokyo, 27–29 September 2005, pp. 1–9.

Taillandier, F., Sauce, G., & Bonetto, R. (2011). Method and tools for building maintenance plan arbitration, *Engineering, Construction and Architectural Management*, 18(4), 343–362.

The Chartered Institute of Building. *Corruption in Construction Industry*. Retrieved from http://www.ciob.org.uk on July 2008.

Thomas, S. J. (2005). *Improving maintenance and reliability through cultural change*. Connecticut: Industrial Press Inc.

Thompson, M., & Kathawala, Y. (1991). Management of the maintenance function revisited: Emphasis on the electric utility industry. *International Journal of Quality & Reliability Management*, 8(6).

Too, E. (2012). Infrastructure asset: Developing maintenance management capability. *Facilities*, 30(5/6), 234–253.

Trappey, A. J., Sun, Y., Trappey, C. V., & Ma, L. (2011). Re-engineering transformer maintenance processes to improve customized service delivery. *Journal of Systems Science and Systems Engineering*, 20(3), 323–345.

Tsang, A. H. (2002). Strategic dimensions of maintenance management. *Journal of Quality in Maintenance Engineering*, 8(1), 7–39.

Uma, K. E., Obidike, C. P., & Ihezukwu, V. A. (2014). Maintenance culture and sustainable economic development in Nigeria: Issues, problems, and prospects. *International Journal of Economics, Commerce and Management*, 2(11), 1–11.

Vanneste, S. G., & Van Wassenhove, L. N. (1995). An integrated and structured approach to improve maintenance. *European Journal of Operational Research*, 82(2), 241–257.

Waggaman, J. S. (1992). *Strategies and consequences: Managing the cost of Higher Education*, ERIC. Washington D.C.: George Washington University, Retrieved from http://www.ed.gov/databases/ERIC_Digest/ed347959.html on 14 October 2000.

Yacob, S., Ali, A. S., & Au-Yong, C. P. (2019). Establishing relationship between factors affecting building defects and building condition. *Journal of Surveying, Construction and Property*, 10(1), 31–41.

Yahya, M. R., & Ibrahim, M. N. (2010). Strategic and operational factors influence on building maintenance management operation process in office high rise buildings in Malaysia. *1st International Conference on Sustainable Building and Infrastructure*, Kuala Lumpur, Malaysia, 5–17 June 2010.

Yu, R., Iung, B., & Panetto, H. (2003). A multi-agents-based E-maintenance system with case-based reasoning decision support. *Engineering Applications of Artificial Intelligence*, 16(4), 321–333.

3 Policy, Planning, and Performance Measurement for Maintenance Management

Introduction

Maintenance of buildings requires proper planning and policies to achieve set maintenance objectives. This chapter provides insight into maintenance policy and planning in maintenance operations. It seeks to highlight the maintenance organisation's objectives and responsibilities, maintenance management (MM) strategies models, forms of maintenance policy, objectives of maintenance policy, and instrumentations. It also focuses on the performance measurement of MM, independent performance variables in MM, building performance, and users' productivity.

Maintenance Organisation

According to Fagbenle and Oluwunmi (2010), arranging resources together, including people, materials, and technology, to attain the organisation's goals and strategies is referred to as an organisation. In other words, the organisation structure is the formal arrangement in which the various parts of an organisation's structure are arranged (Fagbenle & Oluwunmi, 2010). Haroun and Duffuaa (2009) postulated that one common function in any organisational structure is allocating tasks through the division of labour and coordinating the performance results. However, Daft (1989) states that in any organisational structure, no one best structure meets the needs of all circumstances within the organisation's operations. Schermerhorn (1996) observed that ineffective organisational structure through poor management has led to poor enterprises' profitability. This has led to many organisations, including maintenance organisations, developing new methods that can efficiently foster management processes within their organisation. Cholasuke, Bhardwa, and Antony (2004) assert that in achieving organisational stability, the management of maintenance organisations needs to consider buildings and other maintainable assets as an integral function of their organisation. Pinjala, Pintelon, and Vereecke (2006) postulated that in achieving organisational effectiveness, especially in maintenance organisations, there is a need for satisfactory and reliable equipment as demanded by the operation needs

DOI: 10.1201/9781003344681-3

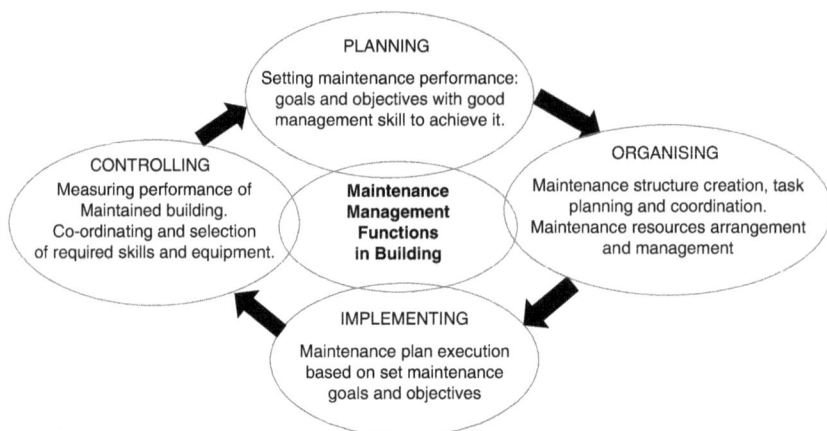

Figure 3.1 Maintenance management function in building.

Source: Adapted from Pinjala et al. (2006).

(see Figure 3.1). There is no universally accepted methodology for designing a maintenance system for meeting organisational set goals (Schermerhorn, 2007). However, Chelson, Payne, and Reavill (2005) claimed that a viable maintenance organisation should define its maintenance policies and procedures. Similarly, a maintenance approach and strategy might differ from organisation to organisation due largely to technological advancement, production size, and application of different maintenance systems (Halevi, 2001). Daft (1989) observed that designing the maintenance system toward realising organisational set objectives and goals is based on experience and judgement supported by effective decision tools and techniques.

Nonetheless, Chelson, Payne, and Reavill (2005) sustain that in designing an organisational maintenance policy, there is a need for effective planning and good management decisions on the maintenance task to be performed as well as identifying skills together with the provision of resources to undertake the maintenance task. Nevertheless, Karia, Asaari, and Saleh (2014) opined that the maintenance policy of an organisation is one of the basic and integral parts of the MM function (MMF). Pinjala et al. (2006) opined further that the MMF is a management function that comprises planning, organising, implementing, and controlling maintenance activities. The maintenance department, physical planning units, and facility managers are responsible for managing the operations and maintenance of all the physical facilities, including buildings, through an effective MMF (Chelson et al., 2005; Karia et al., 2014).

The management of these maintenance organisations provides maintenance policies for the effective management of resources such as hardware, material selection and usage, assets, capital, and personnel, among others (Hallberg, 2009). Lind and Muyingo (2012) maintain that once there is a maintenance policy in place in an organisation, the function of the management is to ensure that maintenance tasks are executed based on the laid-down maintenance policies to meet up with set maintenance goals and objectives of the organisation. For operation tasks to be effective and efficient, an organisational maintenance policy should have a clear mission, an implementation strategy for setting goals and objectives, a corporate culture, job clarifying, delegation of authority, and a chain of command, among others (Lind & Muyingo, 2012).

Maintenance Organisation Objective and Responsibility

According to Vosloo and Visser (2002), maintenance organisations' objectives and responsibilities are heavily influenced by the organisation's type, size, and structure. Fagbenle and Oluwunmi (2010) state that the organisational maintenance policy can also impact it. Lee and Scott (2009) emphasise that in attaining their responsibilities, maintenance organisations seek several objectives such as profit maximisation, the benchmark for the quality level of service, effective cost control, a clean and safe environment, and human resource development.

Nevertheless, Pinjala et al. (2006) noted that maintenance organisations' objectives are influenced heavily by maintenance activities. In other words, the maintenance objectives must be in line with the organisation's objectives, whereas the main responsibility of MM is to provide services that aid an organisation in realising its set objectives (Karia et al., 2014). Additionally, Haroun and Duffuaa (2009) state that the specific responsibilities vary from one maintenance organisation to another; however, on a general note, they include the following:

- Utility operation and energy converse;
- New plant commission;
- Spare parts and materials' conservation and control;
- Performance of efficient and effective maintenance activities; and
- Prioritisation of equipment and assets.

According to Visser (2002) and Bradley (2002), the maintenance organisation's structure is determined by the maintenance capacity. Haroun and Duffuaa (2009) posit that the maintenance capacity of a maintenance organisation is impacted by the level of capacity planning, centralisation, decentralisation, in-house or out-sourcing adopted in the maintenance organisation's operations as discussed next.

Capacity Planning

As noted by Visser (2002) and Bradley (2002), maintenance capacity planning of an organisation is critical in determining the essential resources for maintenance, including the required administration, crafts, equipment, tools, and space to execute the maintenance operations efficiently based on the objectives of the organisation. One critical aspect of maintenance capacity planning is effectively predicting the required numbers of skilled craftsmen or personnel to carry out maintenance operations (Bradley, 2002). However, Borkowski, Pawlowski, and Makowiecki (2011) observed that in meeting maintenance expectations, it is difficult to determine the exact number of required artisans or personnel because the maintenance load is uncertain. Visser (2002) posits that precise forecasts are necessary for future maintenance work requests in determining the maintenance capacity. Bradley (2002) observed that the numbers of available artisans and personnel in most maintenance organisations were reduced below their expected needs owing to a lack of effective forecasting. As noted by Visser (2002), this has accumulated uncompleted maintenance processes and operations. Nevertheless, Borkowski et al. (2011) submitted that this backlog could also be cleared if the maintenance load is less than the capacity planned.

Centralised Maintenance Organisation Model

HajShirmohammadi and Wedley (2004) state that the maintenance services and functions are directly supplied to the organisation from a centrally administered location in a centralised maintenance system. The central maintenance manager coordinates all craft and related maintenance functions in the central maintenance organisation system. Niebel (1994) opined that the benefit of a centralised maintenance organisation is that it is more flexible and uses better resources. Visser (2002) maintains that it provides efficient and effective line supervision and job training and permits purchasing modern equipment. Additionally, the centralised maintenance system enables departmental, functional goals, in-depth skill development, and economies of scale. HajShirmohammadi and Wedley (2004) stated that this maintenance structure is suitable for small to medium maintenance organisations. Lee and Scott (2009) posit that the disadvantage of the centralised maintenance process in maintenance organisations is difficulty in supervising crafts owing to time constraints. Niebel (1994) further asserted that the response time of the centralised system to environmental changes is slow, which may cause delays in decision-making; thus, this may affect coordination among departments toward achieving organisation goals.

Decentralised Maintenance Organisation Model

According to Haroun and Duffuaa (2009), decentralised maintenance assigns organisational departments to specific areas or units. Borkowski et al. (2011) state that each production area manager can manage and supervise its maintenance functions through this system. Campbell (1995) observed that decentralised maintenance reduces the flexibility of the maintenance system. Visser (2002) sustains that in decentralised maintenance, manpower utilisation is less efficient, and there is a reduced range of skills available. However, the strength of the decentralised maintenance organisation is that it facilitates effective coordination among the maintenance unit and other departments (HajShirmohammadi & Wedley, 2004). Additionally, it allows the organisation to accomplish adaptability and efficiency in a centralised overhaul group (HajShirmohammadi & Wedley, 2004).

Centralisation and decentralisation could be combined for a compromise solution in the maintenance process (Campbell, 1995). This type of combination or hybrid is generally referred to as a cascade system. Borkowski et al. (2011) emphasised that whatever exceeds the capacity of each organised maintenance area in the cascade system is channelled to a centralised unit. In this way, the benefits of both systems may be realised. HajShirmohammadi and Wedley (2004) postulated that in actual practice, maintenance operations in most organisations are based on the mixture of these two maintenance organisation systems.

In-House vs. Outsourcing for Building Maintenance Capacity

One important area in MM is the consideration of the sources for building the maintenance capacity. Therefore, Tsang (2002) identified in-house direct hiring, outsourcing, or a combination of both as the main sources for building the maintenance capacity. Accordingly, Bradley (2002) postulated that strategic considerations and economic and technological factors are the criteria for selecting sources for making the maintenance capacity.

In selecting sources for building maintenance capacity, Visser (2002) and Bradley (2002) identified the following criteria that can be employed:

- Dependability and availability of the source on a long-term basis;
- Ability of the source to accomplish the maintenance objectives set by the organisation;
- Cost availability and control for short- and long-term investment;
- Experience and expertise of personnel in maintenance work; and
- Understanding regulatory bodies' specifications for maintenance.

Input–Output Model for Maintenance Organisation Systems

According to Tsang (2002), resources input to a maintenance organisation system for effectiveness includes materials, labour, information, materials, spares, tools, money, and external services, whereas the expected output includes profits, maintainability, safety, and availability. This transformation process as a summary of a maintenance organisation was well described in the input–output model (see Figure 3.2).

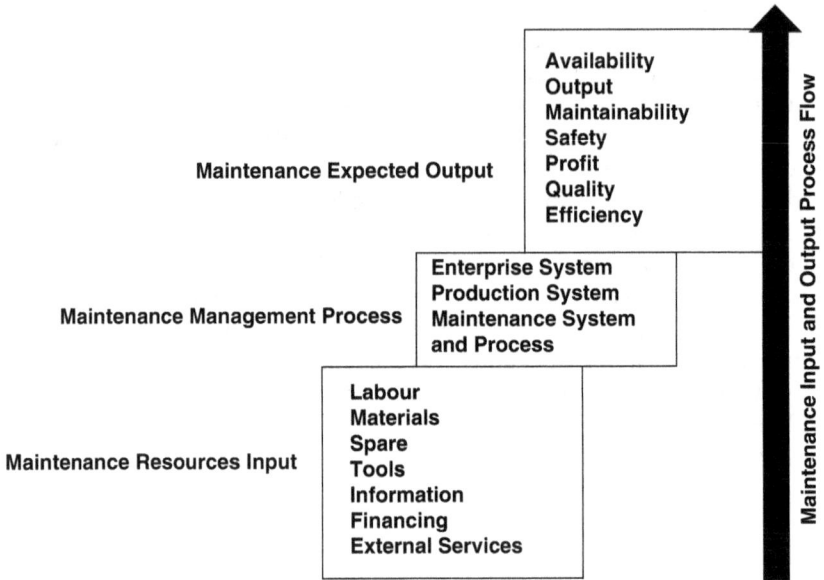

Figure 3.2 Maintenance input–output process.

Source: Adapted from Tsang (2002).

Tsang (2002) describes the model as a transformation process summarising a maintenance organisation system. The details of the model showed that the way maintenance is executed will influence the cost of production, quality of work, volume, and the availability of production facilities, together with the safety of the maintenance operation. Visser (2002) noted that the profitability of the maintenance organisation would be determined through this process because in making maintenance decisions, the use of external service providers has always been an option. Tsang (2002) maintained that in designing an organisational maintenance policy, there is a need to consider external resources, as shown in Figure 3.2. Tsang (2002) postulated that a maintenance organisation policy could be built around four strategic dimensions as informed by the model. The strategy includes the following:

- Maintenance operations option should be a choice between in-house capability and outsourced services;
- Maintenance functions and tasks should be structured to suit maintenance purposes;
- Equipment that supports maintenance activities should be identified; and
- There should be a better understanding of the procedure suitable for the maintenance operations and processes.

Criteria for Maintenance Organisational Effectiveness

According to HajShirmohammadi and Wedley (2004), maintenance organisation is subjected to frequent changes owing to the desire for organisational efficiency. This desire for excellence and uncertainty always makes maintenance managers switch from supporting centralised and decentralised maintenance systems (Campbell, 1995). However, Campbell (1995) states that in establishing a maintenance organisation, there is a need for an objective method that caters to factors that impact the organisation's effectiveness in the maintenance operation.

HajShirmohammadi and Wedley (2004) stated that most maintenance organisations design and redesign their policy to fix a perceived maintenance problem. Hence, Bradley (2002) and HajShirmohammadi and Wedley (2004) hypothesised that reasons for enacting changes to organisational maintenance policy toward effectiveness include the following:

- To improve the maintenance design of the organisation;
- To monitor the maintenance performance of the organisation
- To increase accountability;
- To ensure effective management of the maintenance resources;
- To manage operation time effectively;
- To understand the business aspect of the organisation;
- To set clear maintenance objectives and goals for the organisation; and
- To show detailed job descriptions for personnel.

However, Fagbenle (1998) advised that rather than developing the organisational maintenance policy to solve a specific maintenance problem, it is essential to establish criteria that will guide the maintenance organisation in meeting its set maintenance objectives. Visser (2002) identified the following criteria as significant in the development of an organisational maintenance policy toward effective MM:

- Roles and responsibilities should be clearly defined and assigned;
- Identification of suitable procedures, strategies, and processes is necessary;
- The positions of the organisation in meeting the set maintenance objectives should be stated;
- Line of operational information flow from both top-down and bottom-up should be defined;
- Motivation and organisation culture statement should be stated;
- Continuous improvement built-in structure should be established.
- Techniques for effective maintenance work coordination should be optimised;
- Supporting training should be provided for personnel;
- Specifications from regulatory agencies should be understood; and
- Maintenance finance methods should be identified.

Maintenance Management Strategies Models

Basic Terotechnology

The basic terotechnology model is a maintenance strategy that originated from UK government work. The model was developed when the maintenance costs were rising, and the service availability of the maintained systems was unacceptably low (Coetzee, 1997). According to Sherwin (2000), the model showed a need for feedback on information in the maintained system's life cycle at several points. Al-Najjar (2001) states that the model's main advantage is combining engineering, management, and financial elements to pursue economic life cycle maintenance costs.

In designing the model, its originator, led by Dennis Parkes, did not exactly mention optimisation as such, but because of their experience, did advise the revision of the schedule (Al-Najjar, 2001). Nonetheless, Sherwin (2000) observed that the optimum position is not clear in the basic terotechnology model and judging the sensitivity of cost rate to PM interval without calculation is very difficult. The demerits of the model, as noted by Sherwin (2010), show that it includes feedback loops necessary and desirable for reliability engineering management and the development of machinery for manufacturing; however, it negates costs for effectiveness.

Advanced Terotechnological Model – Need for Integrated IT System

Gabbar, Yamashita, Suzuki, and Shimada (2003) postulated that in developing the advanced terotechnological model for MM, it becomes necessary to acknowledge dependencies and connections that were always there but were not previously specifically brought into policy calculations and company planning. Nonetheless, this model was developed to accommodate the profit aspect of MM (Gabbar et al., 2003). Sherwin (2000) acknowledged that the development of the advanced model terotechnology was formed from life cycle cost (LCC)- based to life cycle profit (LCP)-based, which might seem minor. Nevertheless, Sherwin (2000) noted that the model allows the maintenance function to be better placed and is seen as contributing to profits rather than unnecessarily spending money.

However, Hipkin and De Cock (2000) state that LCP-based will remain just an unworthy objective unless the maintenance organisation's information technology system is sufficiently integrated to cope with the detailed and unambiguous information to feed the mathematical models and other decision-guiding procedures, calculations, predictions, and simulations. The effective usage of this model is based logically on the way the factors of production and management interact (Al-Najjar, 2001). Moreover, Sherwin (2000) affirms that for better waste-cutting advantage, the model has an integrated information technology system supporting its implementation that guides the way factors of production and management interact. Sherwin (2000) observed that the model combines a TQM/terotechnology/LCP-supporting information technology system, and the essence of this merger is for effective information exchange, cooperation, and growth.

Hipkin and De Cock (2000) affirm that for the maintenance process, the information technology system as identified in the model will determine the way maintenance activities of the organisation are managed. The information technology system is intruded into the model to save time; the time saved is then translated into staff reductions. The existing staff must use the information technology system and adopt its underlying managerial theory. Nevertheless, a system for MM must be conducted within it, through mainly a module of the information technology system, through data-sharing, feedback links, and connecting with other modules (Coetzee, 1997).

Reliability-Centred Maintenance Model

In the sixty's, reliability-centred maintenance (RCM) came into being. However, it was originally oriented towards the maintenance of aircraft. The model concept directs maintenance efforts towards units and parts where reliability is critical. However, Gabbar et al. (2003) developed an improved RCM process integrated with computerised MM systems (CMMS). The major components of the enhanced RCM process are identified, and a prototype integrated with the various modules of the adopted CMMS is implemented. Also, Wessels (2003) developed a cost-optimised scheduled maintenance interval that uses costs as the constraint and overcomes quantitative complexity using computer/software technology. This interval enables maintenance organisations to implement a comprehensive RCM programme effectively. Eisinger and Rakowsky (2001) opined that in many practical applications leading to non-optimum maintenance strategies, the uncertainties in the decision-making of RCM might be unacceptable. Nevertheless, the main goal of RCM is to rank the consequence of each failure mode through maximisation of the reliability of the physical asset (Mungani & Visser, 2013).

Effectiveness-Centred Maintenance Model

According to Sherwin (2000: 139), the importance of effectiveness-centred maintenance (ECM) is that it stresses "doing the right things" instead of "doing things right." Mungani and Visser (2013) maintain that ECM focuses on customer services and system functions. According to Pun, Chin, Chow, and Lau (2002), ECM as a model comprehends core concepts of quality management, total production maintenance, and RCM. Moreover, Pun, Chin, Chow, and Lau (2002) emphasised that the ECM concentrates more on customer services and system functions and better MM practices based on continuous improvement in organisations regardless of their business size and nature. The ECM approach involves quality improvement, people participation, strategy development for maintenance, and performance measurement. That is why it is more comprehensive compared to total production and RCM (Mungani & Visser, 2013).

Total Productive Maintenance Model

As noted by Pun et al. (2002), one of the advantages of implementing total productive maintenance (TPM) in an organisation is a better understanding of the equipment performance. Venkatesh (2007) defines total production maintenance as a programme involving a newly defined concept for maintaining plants and equipment. The total production maintenance programme's main goal is to increase production significantly and, at the same time, increase employee morale and job satisfaction. Chen and Meng (2011) and Williamson (2006) suggest the best way to promote sustainable maintenance activities is through TPM. In addition, Ollila and Malmipuro (1999) state that TPM is an effective strategic improvement initiative for improving quality in maintenance engineering operations and activities. Nakajima (1989) and Willmott (1994) maintained that the three main goals of TPM are zero defects, zero accidents, and zero breakdowns. Finlow-Bates, Visser, and Finlow-Bates (2000: 286) suggest that to successfully implement the TPM model successfully, three strong tools that include "seven simple tools of TQM," "root cause analysis" and four thinking models of "Kepner-Tregoe" are to be navigated as all three are harmonise with each other.

Strategic Maintenance Model

According to Garg and Deshmukh (2006), in the strategic MM (SMM) approach, maintenance is viewed as a multidisciplinary activity. The SMM approach overcomes some of the deficiencies of both TPM and RCM approaches as it does not deal with issues such as outsourcing of maintenance, long-term strategic issues, operation load, and operating load on the equipment and its effect on the degradation process (Garg & Deshmukh, 2006). To a large extent, the SMM approach is qualitative or, at the most, semi-quantitative. Garg and Deshmukh (2006) further posit that the SMM approach is more quantitative, encompassing mathematical models that integrate commercial, technical, and operational aspects from a business perspective (Figure 3.3). This makes SMM perspective views of maintenance broader than the TPM and RCM approach. Murthy, Atrens, and Eccleston (2002) state that maintenance must be managed strategically through the SMM approach.

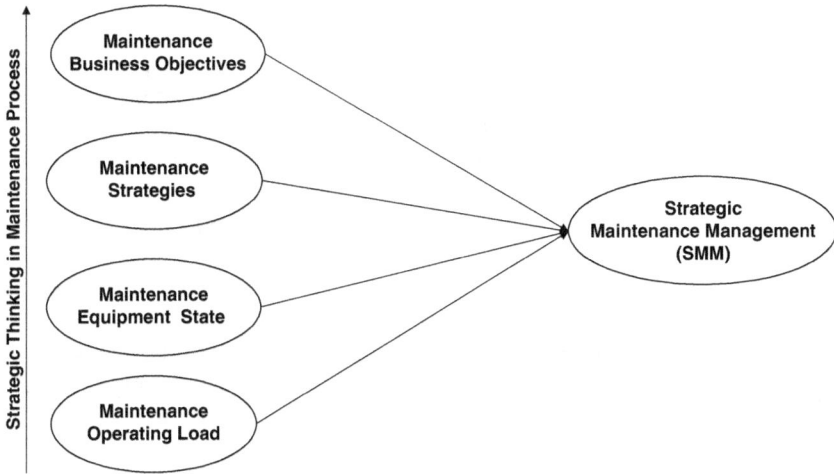

Figure 3.3 Strategic thinking in the maintenance process.

Source: Adapted from Murthy et al. (2002).

Policy and Planning in Maintenance Management of Buildings

The British Standards Institution defines maintenance policy as a strategy to make maintenance decisions (BS 3811, 1974). Pintelon and Muchiri (2009) state that maintenance policy is a strategy for decisions on maintenance tasks. It is a general guide that defines the common purpose and establishes the guidelines for units and individuals entrusted with implementing overall plans (Lind & Muyingo, 2012). A maintenance policy defines the ground rules for allocating resources (men, materials, and money) against the alternative types of maintenance activities available to management (Alexander, 1996). The main substance of the policy is to establish, in each instance, what is appropriate and how best it may be achieved (Pintelon & Muchiri, 2009).

Several factors influence the maintenance policy of an organisation. Nonetheless, Lind and Muyingo (2012) maintain that the use of buildings must be considered the prime factor in determining the requisite standard of care. According to Pintelon and Muchiri (2009), five main factors that could be regarded in maintenance policy formulation are the following:

- The aims of the parent organisation or property owner;
- The maintenance standards required to be kept;
- Legal liability and compliance with statutory regulations such as safety;
- Method of execution or maintenance system to be adopted; and
- Cost and method of financing maintenance work.

Maintenance Policy

Apart from these, the most important consideration for maintenance policy formulation must also include the objective of the maintenance, that is, what maintenance has to achieve (Atkin, 2003). An efficient maintenance policy attempts to achieve economic life cycle costs in the life of assets, i.e., low maintenance cost (Bradley, 2002). Furthermore, it is expedient to foresee the maintenance benefits, that is, what should be gained in the short term or the long run. Therefore, policies determine how to proceed and the structure of the maintenance organisation or department (Pintelon & Muchiri, 2009).

A policy may be in written form or simply understood from the guidelines laid down by top management (Alexander, 1996). It is not a once and for all decision but a continuing process that takes place at various stages, or at various times, before and during the life of a business undertaking (Bradley, 2002). It is expedient to know that for effective management, the policy must be formulated, interpreted, and implemented to provide the necessary platform for assessing results and producing other criteria for measuring the achievement of goals and operational efficiencies (Visser, 2002).

However, the contents of a maintenance policy may include issues such as the use of premises and availability, standards of care, accessibility life of a building, and its services (Visser, 2002). It could also include the privacy of maintenance, prevention of defects, categories of maintenance, means of effecting maintenance, maintenance programmes, maintenance budget, and information feedback (Pintelon & Muchiri, 2009).

In maintenance organisation, maintenance activity requires appropriate and workable building maintenance policies to be formulated to ensure an efficient maintenance programme (James, 1996. Also, maintenance principles involved in planning and execution must be carefully integrated into the maintenance policy (James, 1996). This, according to James (1996), generally involves the systematic implementation of the work entailed by the following activities anticipating, forecasting, and visualisation of future maintenance work by collating information from inspection reports, complaints, and requests from the users of the building, analysis of the history of the building, specification of the nature and details of the work required. It also involves the determination of the best way of executing that work, procurement, storage, issuing of the required materials, allocating tools of work to maintenance personnel, instructing labour about job schedules and procedures for carrying out maintenance work, and following up on, checking the progress of and evaluating maintenance work (Alexander, 2013; Pintelon & Muchiri, 2009).

An organisational maintenance policy can be based on knowledge of the maintenance needs of the organisation, corporate maintenance policies as specified by housing estate owners, the desired maintenance standards, frequency of inspection cycle, and the system of maintenance practiced (Visser, 2002). Fatokun (1997) posits that a maintenance policy and plan focus on clearly defining the parameters of the maintenance standard desired and broadly identifying categories of work to be classified under varying maintenance types. It also laid down broad approaches to executing maintenance works, whether by direct labour or through contractor or both, sourcing or method of funding maintenance works, and instituting of internal feedback system to indicate achievements, problems, defects, and changes to the planned maintenance programme (Visser, 2002).

However, in the organisation's maintenance policy, a proper record of all the items to be maintained will be outlined, a schedule of work for each item, and a programme showing how this schedule will be executed (Fatokun, 1997). It will also provide detailed work specifications for a complex task, work orders to provide instruction on when to carry out the work, and a system of recording work in progress (Fatokun, 1997; Visser, 2002). With a maintenance policy, maintenance organisations will organise, plan, control, and execute maintenance work effectively (Visser, 2002; Ogunbayo, Aigbavboa, Thwala, Opeoluwa, & Edwards, 2022).

Policy Formulation Significance in Maintenance Management of Buildings

According to Márquez (2007: 175), a policy is a definite course or method of action selected from among alternatives and in light of given conditions to guide and determine present and future decisions. It is also a guide for decision-making that links strategy formulation with implementation. Consequently, maintenance organisations can use a policy to guide their personnel in making decisions and to ensure building facilities are well maintained (Lateef, 2010). Thus, maintenance organisations could use policy to develop a maintenance strategy. Márquez (2007) states that the maintenance policy definition is important for an organisation.

Pintelon and Muchiri (2009) postulated that maintenance policies are important tactical-level decisions because they outline the rules for triggering maintenance actions. Several types of maintenance policies can be considered to trigger, in one way or another, either corrective or preventive interventions in maintenance activities. Nevertheless, an organisation's maintenance policy can also favour preventative maintenance against corrective maintenance or whether to use external contractors to carry out maintenance or to use in-house personnel. Olanrewaju and Abdul-Aziz (2015) hypothesised that in any MM system, there is a need for a maintenance policy that defines the ground rules and environment for the delivery of MM services. Accordingly, Olanrewaju and Abdul-Aziz (2015) further state that while prioritising maintenance services, a maintenance policy is a rule that guides the distribution of resources required to maintain a building.

Pintelon and Muchiri (2009) identified the significance of the maintenance policy as follows:

- It ensures that equipment is always in a ready and reliable condition. This ensures the company can respond to any sudden change in demand;
- It ensures that equipment is always calibrated to provide good-quality products. This ensures that there are no sudden and frequent break-downs and reduces the production of defective products;
- It ensures there is no loss of inventory or market share for companies;
- It ensures that costs are always controlled; and
- It is particularly important in capital-intensive industries.

Olanrewaju and Abdul-Aziz (2015) affirm that maintenance policy helps MM determine the types of maintenance strategies they choose to procure maintenance services. Pintelon and Muchiri (2009) states that parts of the maintenance policy of an organisation are usually detailed in the performance indicator of the organisation. Thus, Olanrewaju and Abdul-Aziz (2015) posited that a maintenance policy aims to do the following:

- Provide a clear statement of the organisation's objectives for the maintenance of its buildings;
- Explain how the identified objectives support and facilitate the delivery of the organisation's services; and
- Identify the unit, department, or section responsible for building maintenance management.

Furthermore, Pintelon and Muchiri (2009) posited that maintenance policies are either reactive, corrective, predictive, proactive, or passive. Márquez (2007) opines that the formation of maintenance policies is based on techno-economic considerations but not solely on technical considerations. Pintelon and Muchiri (2009) state that the policies adopted for maintenance operations greatly impact maintenance activities and productivity. Therefore, Olanrewaju and Abdul-Aziz (2015) maintained that maintenance policy forms a significant part of maintenance planning, ensuring that maintenance resources are used efficiently and effectively and that the organisations' objectives are achieved. Olanrewaju and Abdul-Aziz (2015) and Pintelon and Muchiri (2009) affirm that while formulating maintenance policy, the following factors must be considered:

- The aims of the organisation;
- The required maintenance standard;
- Legal and statutory liability;
- Procurement strategies for maintenance services;
- The consideration of cost/benefit of the analysis, particularly to the four above; and
- The condition of the buildings.

Nevertheless, for any maintenance organisation to fully develop an effective and efficient MM system, it must be based on a sound maintenance policy. Buttressing this statement, Olanrewaju and Abdul-Aziz (2015) posited that the organisation should state and develop its maintenance policy together with a procedure for its implementation, and the content of the policy should be made known to all the employees. Developing and implementing a logical maintenance policy and continuous monitoring apparatus minimises waste and error and further makes for smoother production and services operation.

However, regarding the continuous improvement of building maintenance within an organisation, Olanrewaju and Abdul-Aziz (2015) suggested that the following ideas should be set out in a maintenance policy to achieve effective maintenance operations:

- Identify the end-user needs and wants;
- Assess the ability of the organisation to meet the needs and wants economically;
- Ensure that materials, components, and labour meet the required standards of performance and efficiency;
- Ensure that contractors and suppliers share the organisation's value and process goals;
- Focus on proactive and holistic rather than a preventive and/or corrective philosophy;
- Educate and train staff and operatives for quality improvement at all levels, even among the contractors, subcontractors, and suppliers;
- Monitor performance and measure the users' satisfaction; and
- Review the maintenance policy periodically when and where the need arises.

Thus, a maintenance policy concerns all employees and its principles. Its objectives must be communicated to all within the maintenance organisation as widely as possible. This will help better understand employee maintenance procedures and the subcontracting supply chain. However, in ensuring the effectiveness of maintenance policy within an organisation, practical assistance and training should be given where necessary to acquire the relevant knowledge and experiences for the successful implementation of the policy.

Evolution of Maintenance Policy Framework

Over time, maintenance has evolved from a non-issue into a strategic concern. During this period, maintenance's role in managing buildings and other axillary structures has drastically been transformed. Maintenance, at first, was seen as nothing more than a mere inevitable part of the production process; however, maintenance is needed as an essential strategic element to accomplish business objectives.

According to Ding and Kamaruddin (2015: 1267), the MM of structures has always been cast in stone, "where all it needs is the sculptor's chisel of changing organisational requirements of accommodation to reveal its lineaments." Thus, Spedding and Holmes (1994) state that in the last part of the previous century, value concepts in building and construction economics have been progressively refined through viable policies for effectiveness. More recently, maintenance policies have begun to focus more on the connection between the building and the people. Although MM has existed as long as buildings have, its recorded history is a nanosecond in time, and in recent years it has received much-needed worldwide recognition (Becker, 1990).

Van Horenbeek and Pintelon (2015) stated that in the early 1940s, maintenance was viewed as nothing more than an unavoidable part of production; it was simply a necessary evil (see Figure 3.4). Maintenance policy in this era was based on replacing and repairing failed facilities, which were tackled when needed. At this early stage, optimisation maintenance questions were raised. Moreover, maintenance was characterised by maintenance managers working in isolation from the rest of the organisation, and maintenance was perceived as an overhead expense (Becker, 1990). Alexander (1996: 7) observed that despite the benefit of "leaving maintenance by itself," countless numbers of managers mistakenly view facilities as a necessary evil rather than a strategic asset. Price and Akhlaghi (1999) noted that building facilities were meant to manage for minimum cost rather than optimum value.

By the 1960s, maintenance was conceived as a technical matter (see Figure 3.4). The maintenance policy during this period was developed on maintenance processes and their management. The policy in this era seeks to promote the process focus between the organisation's businesses and the MM organisation by making maintenance activities continuous. Thus, the policy around this era was to move maintenance operations towards better management of facilities, especially the buildings and their auxiliary services, as both buildings and their occupants became more important and sophisticated (Amaratunga, 2001). It is noteworthy that the maintenance policy around this time makes the organisation concentrate on developing accountable, integrated, and value-adding services. However, management of maintenance operations at the business and corporate level is missing, as maintenance policy around this time also attempts to identify needs and design and specify the

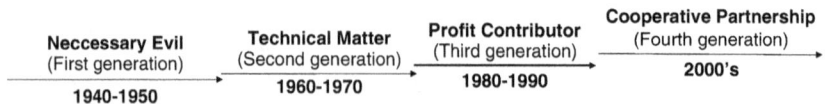

Neccessary Evil (First generation)	Technical Matter (Second generation)	Profit Contributor (Third generation)	Cooperative Partnership (Fourth generation)
1940-1950	1960-1970	1980-1990	2000's

Figure 3.4 Maintenance policy evolution.

Source: Adapted from Van Horenbeek and Pintelon (2015).

service to satisfy the maintenance services. Nevertheless, the policy around this era helps balance the maintenance processes that continuously match the provision of buildings, services, and systems to changing needs (Alexander, 2013).

However, in the 1980s, the maintenance policy developed for buildings and other facilities moved away from the "isolated MM function" towards an effective maintenance policy that is more concerned with "resource management (see Figure 3.4). As observed by Van Horenbeek and Pintelon (2015), a maintenance policy in this era shows clearly that the management of maintenance operations has become a complex function, encompassing management and technical skills while still requiring flexibility to cope with the dynamic business environment. Then (1999) postulated that MM during this period shifted towards resource integration with the emphasis on the provision of an enabling working environment where the issues of processes, people, and buildings are elements of the same problem seeking a common solution.

During this period, top management of the maintenance organisations recognised that having a well-thought-out maintenance strategy together with careful implementation of that strategy could have a significant financial impact on maintenance operations. The maintenance policy during this period led to organisational downsizing, which collectively imposed a burden on many maintenance organisations to seriously review their internal competencies necessary for managing the new era of choices and flexibility (Then, 2003). Conversely, because of the maintenance policy around this period, the maintenance of buildings continued to grow in importance owing to the flexibility it brought to maintenance organisation with a closer integration of facilities and a more appropriate focus on the user together with strategic needs that brings important business advantages.

By the early 2000s, maintenance had become a full-blown function against a production subfunction. Because of this, maintenance organisations started thinking more strategically about effective management systems within their maintenance organisations (see Figure 3.4). As posited by Van Horenbeek and Pintelon (2015), this has led to treating maintenance as a mature partner in business strategy development with consideration towards establishing external outsourcing and partnership of the maintenance function. Then (2003) posited that maintenance organisations started developing maintenance policies toward achieving the much-needed alignment between organisational structure, work processes, and the enabling physical environment. Moreover, this effort was in line with the assertion by Endrenyi et al. (2001) that a maintenance policy is one of the operating policies, and, in a given setting, it is designated to satisfy both financial constraints and technical requirements in the management of buildings.

Nowadays, the fact that maintenance has become more critical implies that a thorough insight into the impact of maintenance interventions is indispensable. A good maintenance policy is used for the right allocation of resources (personnel, spares, and tools) to guarantee higher reliability and availability of the installations by deciding on the suitable combination of maintenance actions (Van Horenbeek & Pintelon, 2015).

Globally, a maintenance policy has been used to develop the maintenance concept within the organisation and set up solid foundations for excellence in MM. Furthermore, maintenance policies have been used to anticipate and avoid the consequence of operation and facility failure within the maintenance organisation (Márquez, 2007). Moreover, the Inter-Agency and Expert Group on SDG Indicators developed a global indicator framework and a framework agreed on for developing countries. The essence of the framework was that by 2030, developing countries need to upgrade infrastructure and retrofit industries to make them sustainable (Kapto, 2019). The development of this framework will include developing an effective and viable maintenance policy toward the effective MM of both existing and future infrastructure for sustainability and good living conditions for the users. The report of the Commission, which included the global indicator framework, was then noted by ECOSOC (2016) at its 70th session in June 2016. This further strengthened and developed quality, reliable, sustainable, and resilient infrastructure, including regional and transborder infrastructure, to support economic development and human well-being, focusing on affordable and equitable access for all.

Nevertheless, Olanrewaju and Abdul-Aziz (2015) observed that in most developing countries, a maintenance budget constitutes a problem for an effective maintenance policy framework. He noted further that this always creates an insufficient allocation to maintenance. This is because when the maintenance budget or the maintenance allocation is cut or reduced, maintenance operations are often deferred to cut costs. Therefore, as noted by Ogunbayo et al., (2022), maintenance policy forms an important aspect of maintenance planning which ensures that resources are used efficiently and effectively and that the organisation's objectives are achieved.

Forms of Maintenance Policy

The essence of any form of maintenance policy is to reduce the frequency of service interruptions and the many undesirable consequences of such interruptions. As noted by Van Horenbeek and Pintelon (2015), as new techniques happen in the maintenance operation system, the economic implications of maintenance are realised with a direct impact on the maintenance policies, which is predictable. Over the years, many maintenance policies have been developed to guide the optimisation of the policy setting. This has given credence to the maintenance policy settings over the years. Understanding its efficiency and effectiveness continues to be fine-tuned as any other management science. Nevertheless, several maintenance policies can be considered preventive or corrective maintenance interventions. Van Horenbeek and Pintelon (2015: 391) provide an overview of the main form of maintenance policy available, and they include the following:

- Failure-based maintenance policy;
- Time/used-based maintenance policy;
- Condition-based maintenance policy;
- Opportunity-based maintenance policy;
- Design-out maintenance policy; and
- E- maintenance policy.

The Failure-Based Maintenance Policy

Failure-based maintenance (FBM) is a reactive maintenance policy where unscheduled maintenance or repair returns the machine or maintainable facility to a defined state. In this type of policy, there are no interventions to machines or facilities until failure has occurred. Repairing activities take place when the condition shows that failure may be imminent. Lewis (2000) described FBM as reactive maintenance, which, if there is any emergency breakdown, will create a bigger impact on the operation with a heavy cost implication. Buttressing this point, Burhanuddin et al. (2011) observed that since the FBM policy is based on unplanned maintenance, using this type of policy might lead to a big loss to maintenance organisations in terms of time and cost. The simplest method of the FBM involves the four human senses of sound, sight, smell, and touch to predict failure.

Time / Used - Based Maintenance Policy

Time/used-based maintenance (TBM/UBM) is a maintenance policy where the component is always maintained after usage or failure, whichever occurs first. In other words, the TBM/UBM policy is maintenance performed based on a calendar schedule. This simply means that time is the maintenance trigger for this form of maintenance policy. With the TBM/UBM in place within an organisation, maintenance is performed each time the calendar rolls over the specified number of days. Thus, the TBM/UBM policy is carried out on a fixed interval regularly over the service life of an asset, irrespective of its age. De Jonge, Teunter, and Tinga (2017) postulated that the TBM/UBM policy is easy to implement, and no condition monitoring is needed. The TBM/UBM policy, as observed by Albrice and Branch (2015), works perfectively with some assets, especially those highly regulated for safety reasons, but for other assets that behave differently over time, it is not appropriate.

Condition-Based Maintenance Policy

Flores-Colen and de Brito (2010) define the condition-based maintenance (CBM) predictive maintenance policy as a maintenance policy initiated based on knowledge of the deterioration level of an item through routine inspection and continuous monitoring, thus reducing the total cost of repairs. Nevertheless, Van Horenbeek and Pintelon (2015) posit that a CBM policy is mainly applied when the investment in condition monitory equipment was justified due to high risks. They observed further that a CBM policy is the potential savings in spare parts replacements for MM operations thanks to the accurate and timely forecasts on demand. This, in turn, enables better spare parts management through well-coordinated logistics. As noted again by Ugechi, Ogbonnaya, Lilly, Ogaji, and Probert (2009), CBM policy, compared to other forms of policies, is the most advantageous policy for maintenance operations and management. They argued further that CBM implementation allows for sufficient lead time to schedule, organise, and carry out necessary repairs before any failure occurs.

Consequently, costly downtime and major breakdowns can be avoided. However, Ellis and Byron (2008) contended that CBM does not apply to all services and assets. Buttressing this argument further, Horner, El-Haram, and Munns (1997) maintained that to attain the full advantage of a CBM policy, facility conditions must be monitored. This showed that before the implementation of the CBM policy, condition monitoring techniques are required. Van Horenbeek and Pintelon (2015) noted that this might enable better spare parts and facility management through coordinated logistics support.

Opportunity-Based Maintenance policy

Opportunity-based maintenance (OBM) is a passive maintenance policy. However, Muchiri (2014) states that this type of maintenance policy is only carried out if an occasion or need arises since no maintenance plan is developed. Muchiri (2014) observed that an OBM intervention for a failed or deteriorated facility could be carried out only when the asset risk is negligible. However, Van Horenbeek and Pintelon (2015) state that OBM uses non-critical components with a relatively long life. Amin (2016) postulated that under an OBM policy, maintenance could only be carried out on certain assets, components of a building, or facility when the opportunity arises while maintaining other more critical components. However, Amin (2016) concluded that using an OBM policy requires understanding and analysis of asset or facility life, usage, and cost considerations. From the foregoing literature, it appears that an OBM policy is largely applied to assets with a relatively long life that is considered non-critical.

Design-Out Maintenance Policy

Design-out maintenance (DOM) is a proactive policy used to maintain physical assets. The DOM policy is designed for the possible design changes that may avoid maintenance in the first place. As noted by Muchiri (2014), a DOM policy indicates that maintenance managers were proactively involved in the design stage to solve potential problems with maintenance and safety. A DOM policy's main aim is to avoid maintenance throughout a facility's operating life. Amin (2016) posited that at the earlier stage of the product life of a facility, a DOM policy focuses on the economic and technical reliability aspects of improving the asset design. The DOM policy makes maintenance simpler by enhancing availability and safety. Van Horenbeek and Pintelon (2015) postulated that one of the core goals of DOM is to eradicate the requirement to maintain the asset throughout its operational life, which can be practically unrealistic.

E-Maintenance Policy

The growth of e-business as the standard of business communication in the 1990s, with the emergence of the Internet as an enabling technology, makes e-maintenance also appear on the radar of maintenance policies (Van Horenbeek & Pintelon, 2015). However, e-maintenance can be considered as a means or enabler to some, if not all, the previous policies rather than a policy. Nevertheless, the e-maintenance policy is a platform for a fully integrated maintenance technique without boundaries (Han & Young, 2006). As postulated by Lee (2001), e-maintenance on its own can support other policies. An e-maintenance policy develops with globalisation and the fast growth of communication and information technologies. Nevertheless, Han and Young (2006) submitted that proper implementation of the e-maintenance policy would benefit both the maintenance manager and users of buildings to have process reliability with optimal asset performance and seamless integration.

Objectives of Maintenance Policy

Maintenance policies vary from organisation to organisation and from country to country. There is no one-fits-all solution or framework for maintenance policy formulation (Daragh Naughton & Tiernan, 2012). Basic steps for formulating policy for maintenance operation as stated by Bagadia (2006) are request, approval, plan, schedule, performing work, recording data accounting for costs, developing management information, updating equipment history, and providing management control reports. Nevertheless, it will be completely wrong to design a new policy, impose changes to an existing policy or implement a maintenance policy without a clear view of the purpose of the maintenance policy (new or reviewed). This should logically be examined with the framework of the overall purpose of maintenance policy. Hence, the primary objective of any maintenance policy should be to ensure the maximum efficiency and availability of production equipment, utilities, and other related facilities within the building environment at optimal cost and under satisfactory conditions of quality, safety, and protection.

The objective of maintenance policy, as affirmed by Van Horenbeek and Pintelon (2015), provides a strategic platform for keeping key facility components within the maintained building under satisfactory conditions of quality, safety, and protection for the users. Akasah, Shamsuddin, Rahman, and Alias (2009) hypothesised that a maintenance policy helps designate the maintenance responsibility for facilities, equipment, and infrastructure when maintenance is required and how it is performed or carried out. Further, Blissett (2004) asserts that in policies developed for maintenance activities or operations, one of the key objectives will be to define the terms of maintenance operation and describe the decision-making process governing the assignment of maintenance priorities and the selection of cost analysis, and quality assurance. Kelly (2006) asserts that maintenance policy objectives involve resourcing and executing repair, replacement, replacement, and inspections. Kelly (2006) states further that it is concerned with the following:

- Creating the organisation to enable the scheduled and unscheduled maintenance work to be resourced;
- Formulating a maximum best life plan for each maintenance unit; and
- Stating the maximum best life plan for each maintenance unit.

Han and Young (2006) postulated that the objectives of maintenance policies vary from organisation to organisation and from one professional body to another. One common purpose is that it guides and helps determine the present and future decisions on maintenance activities within the organisation (Han & Young, 2006). Accordingly, based on this statement, it is clear that no single ideal approach or strategy exists and that centralised external agencies cannot decide on priorities. This indicates that maintenance managers or organisations can determine their maintenance priorities. Nevertheless, the better the situation on maintenance formulation, the more the range of choices available to people and the more merit and assistance received.

Consequently, it is noteworthy that the maintenance policy objective requires a framework that locates the MM of buildings issue within its broader issues, which should inform the directions of housing policy in the specific content of any country.

Maintenance Policy Instruments

The policy instrument refers to tools that include economic tools (taxes, spending, incentives) and regulations (legal) used by governments to pursue outcomes. Over time, to respond to the policy constraints of the MM of facilities, different maintenance policy instruments have been developed (Akasah, Shamsuddin, Abd Rahman, & Alias, 2009). Though Howlett (2019) states that no matter how loosely defined or publicly beneficial policy (maintenance policy) goals might be, policymaking is all about creating and implementing mixes of policy instruments expected to attain policy goals. As noted by Salamon (2002), some characteristics are particular to each policy instrument tool that exists as a bundle of attributes utilising one or more governing resources. Aigbavboa (2014) postulated that a policy instrument is not projected to be obsolete, but when it can no longer perform the function, it was designed to terminate or modify it.

As noted by Akasah et al. (2009), most existing instruments for maintenance policy are not meeting the set maintenance goals. They are sometimes supplemented through additional instruments to provide a flexible solution to the needs and expectations of users and national interests on the sustainability of public facilities. An essential mix of maintenance policy instruments is targeted in developing a working maintenance framework that will include the development of effective and viable MM of existing and future infrastructure for economic sustainability and to improve the living conditions of the citizenry. Nevertheless, the new maintenance policy instrument tends to focus attention on sectors that have been previously ignored. These instruments are inclined to place greater emphasis on flexibility and reaction to local situations, particularly by the level of deterioration of the to-be-maintained facility. According to Olanrewaju and Abdul-Aziz (2015), Malaysia's government's maintenance strategy was often reviewed to pave the way for another with fewer workable years of initially approved ones. In Nigeria, for example, the government has tried different maintenance policy instruments to maintain public facilities.

In Nigeria, in 2019, there has been a greater emphasis on developing maintenance policy instruments that will institutionalise a maintenance culture and provide an inventory of government assets and job opportunities. In making this new maintenance policy work well, a national pilot plan (maintenance policy instruments) for the inventory of the Nigerian national assets and a maintenance procurement manual were made available to all the Ministries, Departments, and Agencies (MDAs) of government nationwide for better implementation. The new maintenance policy instruments it was believed will help to assess the conditions and value of both existing and future facilities. This will aid the development of a maintenance framework about what needs to be done after assessment, and then a maintenance procurement manual can then be developed for the country.

Thus, no single policy instrument has been found to fulfil the maintenance policy objectives and set goals of an organisation and, subsequently, the national maintenance policy of a country. However, there is no clear "best" maintenance policy instrument in all circumstances. Every policy instrument has its relative advantage to a degree as determined by the MM context of a particular country and by the set goals of the specific maintenance policy.

Performance Measurement of Maintenance Management of Buildings

According to Weber and Thomas (2005), performance measurement is an essential management principle. Weber and Thomas (2005) further posited that measurement in the MM of buildings' performance is vital because current performance gaps will be identified between desired and existing buildings' performance. Additionally, Weber and Thomas (2005) postulate that performance measurement indicates progress towards closing the identified gaps. Nonetheless, to identify and take action toward improving building performance based on the identified gaps, carefully selected key performance indicators (KPIs) will help identify where to take action and improve the building performance through an effective MM system (Amaratunga, 2001).

Performance Concept

In simple terms, BS 8210 (1986) defines performance as the behaviour of a product in use. In relating it to buildings, Williams and Williams (1994) posited that a building can contribute to fulfilling the functions of its anticipated use. Based on the concept of management which is the accomplishment of pre-selected objectives and missions, performance is concerned with achieving quality (Bartlett & Simpson, 1998). Performance, as defined by Williams and Williams (1994), points to three main aspects of in-use in a building, namely physical efficiency of the building, functional efficiency of the building, and financial efficiency of the building. The study of Amaratunga (2001) shows that an extensive percentage of most organisations' operating costs and assets is represented by their buildings' physical, functional, and financial aspects (see Figure 3.5). Today building performance assessment is becoming a formal and regular part of MM activities and processes in order to determine the effectiveness and efficiency of the maintained buildings.

Preiser (1995) and Obiegbu (2005) postulated that performance evaluation is an analytical tool that allows managers to identify and value critical aspects of a

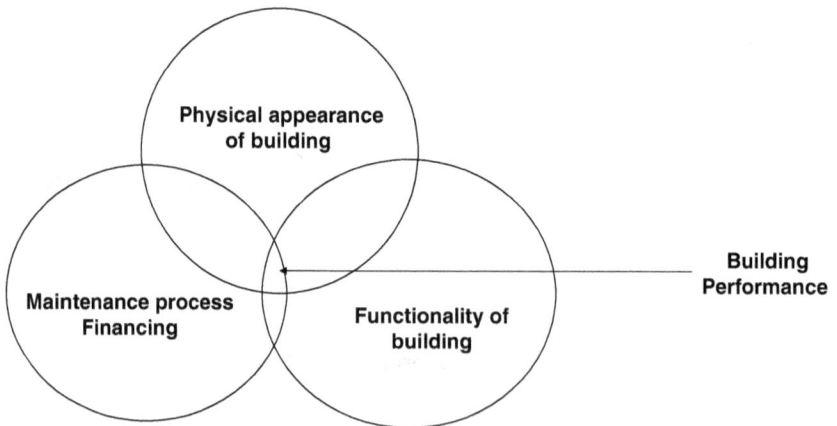

Figure 3.5 Building performance interrelatedness three facets.

Source: Adapted from Amaratunga (2001).

building to develop design guidelines and measures for future installations. In order to identify and appreciate essential elements of a building, Palm (2008) suggested that there is a need for proper documentation of building performance. Lackney (2001) and Zimring and Rashidi (2008) maintained that performance assessment of buildings established a clear link between users of the building and the physical environment. Also, the non-assessment of the performance of the building is attributed to a lack of budget for assessment, lack of cooperation among personnel, absence of required skill, and the complex and uncertain nature of building design (Zimring & Rashidi, 2008).

However, Parker (2000) postulated that performance measurement is an important management aspect that helps an organisation understand its operational processes, resources, products, resources, and environments. Also, Barrett (1992) opines that the performance measurement helps organisations develop a reference point for comparisons for future assessments, predict, evaluate, and gather the information that helps identify problems and track improvement efforts. Nevertheless, Amaratunga (2001) noted that measuring the performance of facilities (buildings) in use must be guided by some key performance criteria. These include the purpose of evaluation, key performance criteria definition, post-occupancy evaluation planning, criteria measurement, data evaluation/making an assessment, and feedback/stating the lessons learned. Again, Dixon, Nanni, and Vollmann (1990) submit that measurement provides the link between strategies and actions. Equally, Dixon et al. (1990) posit that appropriate performance measures will provide and strengthen this link and lead to the fulfilment of the strategic goals of the organisation.

Nevertheless, the significance of performance measurement in an organisation has been emphasised by various scholars. Oakland (1993) posits that the maintenance organisation is important in improving MM activities. The significance of performance measurement in a maintenance organisation, as noted by Oakland (1993), includes the following:

- The provision of standards,
- To provide an organisation scoreboard for monitoring organisation performance levels,
- To determine priority areas, to assess the quality of work carried out,
- To justify the resource input, to provide a platform for continuous improvement, and
- To drive the improvement effort through feedback.

Subsequently, Parker (2000), in his study, stated that performance measurement needs to be aligned with the organisational objectives, strategy, and policy. The concept of the performance measurement of building MM guides the different functions within the maintenance activities and controls the management process. Apart from this, the performance measurement of building MM also simplifies overall maintenance objectives through effective management channels (planning, assessment, and feedback) to other lower maintenance personnel.

Performance Measurement of Buildings

Many building organisations do not perform above expectations owing to poor operations and management strategies. This might be because most key players have little interest in its performance during the procurement stage of buildings (Ogunbayo & Aigbavboa, 2019). Mayaki (2005) observed that when the building is in-used, its performance is often never assessed. Leaman (2004) postulated that a survey of buildings in the United Kingdom shows a loss of an average of 2% in users' productivity due largely to building-related problems caused by the absence of performance evaluation based on usage. The study concluded that the performance of the majority of buildings does not meet the users' expectations.

Knirk (1993) insists that the performance of buildings should address a broad range of user-access needs and purposes for which it was built. Mbamali (2003) observed that the general maintenance of facilities by professionals in the management of buildings had remained practically the same. However, there has been a gradual departure from traditional building maintenance methods (Park, 1998). With the creation of all-embracing professionals (maintenance manager/facility manager) who harness and complement functions of other major professionals towards building performance evaluation when in use (Park, 1998). Nonetheless, Amaratunga (2001) opines that the main goal for either maintenance managers or facility managers is to relieve individual homeowners of the burden of ensuring the buildings run efficiently. Li, Qu, and Peng (2017) posit that the performance measurement function of buildings (PMB) is performed at three levels: tactical, operational, and strategic. Preiser (1995), as cited by Obiegbu (2005), states that the performance measurement of buildings is an analytical tool that allows maintenance managers to assess and identify critical aspects of a building, to develop a strategic maintenance approach, or system for future maintenance activities. Okolie (2011) noticed that the performance measurement of buildings is an extension of what was previously referred to as post-occupancy evaluation. Zimring (2001) postulated that performance measurement of buildings deals with the continuous process of evaluating the effectiveness and performance of one or more aspects of buildings in terms of safety, functionality, accessibility, aesthetics, cost-effectiveness, productivity, security, and sustainability.

Nonetheless, Douglas (1996), in analysing the relevance of building performance measurement to MM, emphasises that buildings and their auxiliary facilities are key functions as well as economic resources and should therefore be regarded as assets rather than liabilities. Douglas (1996) states that the performance measurement of buildings is a basic tool for realising set objectives in the MM of the buildings. According to Then (2003), the performance measurement of buildings provides a platform for strategic plans in the MM of buildings (Then, 2003). According to Barrett and Baldry (2003), performance measurement of buildings is broadly divided into user- and expert-based measurement systems. Barrett and Baldry (2003) postulated that the user-based measurement system uses building users based on their needs and productivity to evaluate the performance of a building. The user-based measurement system is also known as post-occupant evaluation. However, expert-based measurement systems rely on relevant expert assessments of the performance of the building together with organisational involvement. The expert-based measurement systems typically cover far more areas such as organisational growth, organisational objective, policy, information technology provision, and operational strategy changes in work style (Barrett & Baldry, 2003).

Thus, Ornstein et al. (2009) state that the objective of performance measurement of buildings is to measure the extent to which a building, after construction, meets its conception and design purpose. Also, Mayaki (2005) postulated that the objective of performance measurement of buildings is to create a more functional facility by improving design practices that better support service delivery. Amaratunga and Baldry (2000) maintain that the performance measurement of buildings is essential in ensuring the effective implementation of organisational strategy in management buildings. Additionally, Amaratunga and Baldry (2000) postulated that performance measurement of buildings allows organisations to assess their building performance through continuously stimulating action utilising what is to be done, who is required to act, and in what manner. Ornstein, Moreira, Ono, França, and Nogueira (2009) maintain that the objectives of performance measurement of buildings are not about running costs of buildings only but cover other strategic issues in the management of the building.

Equally, Amaratunga and Baldry (2002) suggested that, in general, construction organisations' performance measurement of buildings incorporates the design and management of space and other attached facilities for users and maintenance processes based on organisational set goals. Also, Madeley (1996) and Mayaki (2005) conjecture that in building maintenance activities, performance measurement focuses attention on feedback loops, which influences maintenance managers' reactions to users' maintenance requests. However, Lackney (2001) postulated that performance measurement creates an avenue to learn from past, present, and future contemporary trends in the building and its axillary facilities. Consequently, collection, interpretation, and information analysis about performance measurement of buildings will provide the strategy for better planning, policy, and design for the MM of buildings (Figure 3.6).

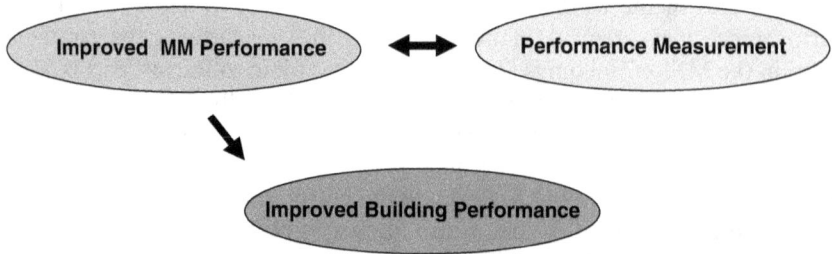

Figure 3.6 Maintenance management's relationship to building performance.
Source: Adapted from Amaratunga (2001).

KPIs in Maintenance Management of Buildings

According to Allen (1993), using resources to ensure that the building process and its auxiliary facilities comply with standards requires KPIs. Weber and Thomas (2005) observed that maintenance of the in-use building is gaining momentum worldwide. Nevertheless, Wireman (2005) argued that the maintenance of the building has an uncoordinated and fragmentary practical technique. Nonetheless, Wardhaugh (2004) posits that the efficiency and quality of MM operations depend on information on building conditions, users' needs and expectations, and the records of operation maintenance strategies used in executing maintenance activities, which can be measured based on different performance indicators.

Nevertheless, Okolie (2011) postulated that there is a need for KPIs for educational buildings, which will provide a nexus between the building performance and the productivity of its users. Omar, Ibrahim, and Omar (2017) hypothesised that KPIs help maintenance organisations reflect on operational achievement and progress as stipulated in the operational maintenance benchmark in the MM of buildings. Conversely, in measuring the performance of buildings, KPIs are concerned not only with doing good maintenance work but also with removing the risk of failure from maintenance work and its management. Wardhaugh (2004) states that KPIs guide maintenance choices through driving reliability growth for improving maintenance effectiveness and efficiency. Wardhaugh (2004) further observed that KPIs help maintenance managers to identify issues instigating the maintenance effects. Additionally, Omar et al. (2017) state that KPIs support the maintenance manager in understanding and selecting the right strategy for effective and efficient maintenance results. Omar et al. (2017) postulated that KPIs help to create a suitable base for MM activities for effective performance.

Furthermore, Alzaben (2015) postulated that in measuring the MM of buildings, some of the performance indicators commonly used include resources, budgetary information, response time, and recurrent problems, among others. Hence, Omar et al. (2017) assert that using KPIs is the best approach to measuring building performance through a maintainable system. Irajpour, Anajafabadi, Mahbod, and Karimi (2014) opined that KPIs for the maintenance process are proposed to measure whether the requirements of each maintenance process are satisfied. To this end, Ali and Mohamad Nasbi Bin Wan Mohamad (2009) contended that KPIs are fundamental principles in the MM of buildings. Omar et al. (2017) suggested that KPIs for institutional buildings MM could be subdivided into three factors (see Figure 3.7), and they include the following:

- **Individual factors:** Monitory and supervision, task planning and scheduling, Information management, and computerised MM systems;
- **Technical aspects:** Maintenance approach, spare part management, and outsourced strategy; and
- **Administrative and organisation factors:** Policy deployment and organisation, human resources management, finance aspect, and continuous improvement.

Performance Independent Variables in Maintenance Management

According to Wireman (2005), Omar et al. (2017), and Mekasha (2018), to identify the effectiveness of MM factors in measuring the performance of the building, the following key performance independent variables will be required. This includes monitoring and supervision, task planning and scheduling, computerised MM system, maintenance method/approach, spare part management, human resources, outsourcing strategy, policy deployment, maintenance budget, continuous improvement, and training.

Monitoring and Supervision

Continuous monitoring and supervision are some of the strategic ways of assessing and ascertaining the performance of maintenance organisations towards implementing maintenance plans, objectives, policies, and procedures related to MM (Duffuaa, 2000; Wordsworth & Lee, 2001; Campbell & Jardine, 2001). Nevertheless, Ahzahar, Karim, Hassan, and Eman (2011) and Manaf (2005) opined that monitoring and supervision help the maintenance managers to identify gaps between present building performances and expected performance that could guide the organisation to identify the prospect of maintenance process improvement (Ahzahar et al., 2011). Vaisnys et al. (2006) sustained that effective maintenance plans and schedules will help assign the resource for each work toward efficient operational monitoring and supervision. Nonetheless, to meet the maintenance buildings' performance expectations, appropriate supervision must be implemented to ensure that the maintenance work and its processes are performed through the organisation's maintenance policy and procedure. Moreover, Hassanain, Froese,

EFFECTIVENESS OF MAINTENANCE MANAGEMENT FACTORS		
INDIVIDUAL FACTORS	**TECHNICAL FACTORS**	**ADMINISTRATIVE FACTORS**
Monitoring & supervision Maintenance tools & planning Maintenance activities Scheduling Maintenance information system Computerised Maintenance management system	Maintenance outsource strategy Spare part management Maintenance approach Personnel training Resource allocation	Maintenance outsource strategy Spare part management Maintenance approach Personnel training Resource allocation

Figure 3.7 Factors for effectiveness of MM of building.

Source: Adapted from Omar et al. (2017).

and Vanier (2001) and Takata et al. (2004) identified variables such as suitable replacement material usage, constant rescheduling of maintenance activities, monitoring of maintenance operations, monitoring of safety procedure, observing faults' trigger components, periodic maintenance planning, and appropriate maintenance strategy as some of the factors that influence the monitoring and supervision factors of MM. Similarly, Crespo Márquez et al. (2009) and Márquez (2007) identified factors such as risk reduction diagnosis of maintenance task, meeting maintenance targets, material wastage reduction, maintenance personnel involvement, value improvement, and zero-error tolerance. Likewise, Pintelon et al. (1992) identified factors such as the determination of facilities performance, maintenance operation inspection, maintenance process monitoring, and performance reporting.

Task Planning and Scheduling

Another key performance indicator is task planning and scheduling. According to Omar et al. (2017), the need for organisational maintenance success has compelled routine monitoring and supervision to be supported by a well-guided task planning and working schedule. Consequently, Weber and Thomas (2005) affirm that for effective maintenance operation, maintenance supervisors/line managers should provide task planning and scheduling to ensure that the proposed maintenance plans are relevant, efficient, and effective. Equally, Wireman (2005) noted that task planning and scheduling for maintenance activities would lead to great responsiveness, while unplanned maintenance activities will lead to the abandonment of maintenance work with a severe cost implication. Also, Ogunbayo et al. (2022) maintain that task planning and scheduling will help provide feedback on maintenance activities. Moreover, Okolie (2011) noted that the lack of planning and scheduling by the maintenance managers of buildings would restrict the maintenance operations within the building.

Computerised MM System

According to Wireman (2005) and Omar et al. (2017), computerised MM systems (CMMS) are one of the vital key performance indicator variables for MM of buildings. Wireman (2005) stated that the CMMS provides the link between the current status of maintenance function to the objectives of maintenance. Furthermore, Omar et al. (2017) state that the maintenance function of an organisation can be supported through CMMS. Also, Wireman (2005) maintained that the CMMS could be used as an aid tool to manage and control the maintenance operations, purchase materials inventories, analyse, record, and coordinate maintenance activities. Hence, Manaf and Alias (2005) posits that the CMMS are more accurate and quicker than manual techniques for maintenance activities, and they are an efficient tool that has proficiencies of reporting the analysis related to MM of building precisely. Additionally, Irajpour et al. (2014) postulated that the CMMS provide more reliable and accurate information for the maintenance process and its administration.

Maintenance Method / Approach

The maintenance method used for maintenance activities and its process is acknowledged as a significant factor that highly determines the effectiveness of an organisation. Wireman (2005) postulated that for maintenance effectiveness, the proactive maintenance philosophies were adopted by many maintenance organisations such as RCM and TPM, which are committed to the long-term maintenance process and improvement. Nevertheless, Omar et al. (2017) observed that the reactive preventive planned maintenance approach adopted by some maintenance managers for maintenance work planning programmes drastically reduces errors, especially during maintenance operations and overall MM preparation.

Spare Part Management

One variable that is of utmost importance in KPIs in building performance is the replaceable spare part for the maintainable component of a building. The selection of this spare part management is often affected by the maintenance approach used for maintenance work by an institution or organisation (Vaisnys et al., 2006; Omar et al., 2017). Often, spare parts required for replacing a building component are not readily available when routine management requires, this prolonged maintenance time and cost implications. Vaisnys et al. (2006) observed that spare part management is critical during inventory. Most maintenance managers stockpile their maintenance parts because this utility absorbs excessive costs and expenses. Wireman (2005) posits that to keep the building and its components in a stable condition, a maintenance manager should ensure that the rights spare parts are in stock.

Outsourced Strategy

Contracting of maintenance work is a significant maintenance approach conducted through an in-house or outsourcing strategy for the sustainability of maintenance activities. The study of Ali et al. (2009) shows that most world-class maintenance organisations or municipalities outsourced 30% of their jobs. However, outsourcing becomes an issue in maintenance when the right strategy is not adopted. Ali et al. (2009) postulated that the most outsourced strategy used for contracting in maintenance work is based on contracting out the task and how much benefit parties to the contract will gain. Omar et al. (2017) state that outsourcing strategy depends on how maintenance organisations manage the contract and task and how much profit is achieved through the use of the outsourcing strategy.

Organisational Maintenance Policy

Another variable that needs special attention in the MM of the building is organisational maintenance policy drafting and applications. Vaisnys et al. (2016) posit that maintenance policy deployment preparation is an acknowledged factor that affects the MM effectiveness of an organisation. Ali et al. (2009) opine that the organisational maintenance policy contains a written plan, policies, and procedures which describe how the organisation will manage each specific component of buildings and their auxiliary facilities and services. Moreover, Wireman (2005) suggested that maintenance organisations should consider a maintenance policy as a business strategy to improve the overall performance of their maintenance activities and procedure.

Maintenance Budget

For successful MM operations in an organisation, budgeting is a key factor of progression. Ahzahar et al. (2011) state that as an important key performance indicator variable like other KPIs, it supports all maintenance planning and executions. Omar et al. (2017) observed that budget control and allocation play a critical role in encouraging the overall MM procedures and success. Wireman (2005) suggested that MM's finance elements include cost control for labour, cost of monitoring the contractor, and maintenance budget control, among others. Ogunbayo, Ohis Aigbavboa, Thwala, and Akinradewo (2022) postulated that in a maintenance organisation, the maintenance budget is a key element of progression that impacts the financial planning of maintenance operations.

Human Resources Management

Human resources management is an important key performance indicator variable for MM of buildings. As posited by Omar et al. (2017), there is a need for organisations to have effective human resource management to run the business efficiently as well as to certify the optimum performance of personnel. Ahzahar et al. (2011) assert that human resources management is one of the factors of MM because qualified personnel is needed to manage, supervise, plan, and execute the maintenance works. Also, Wireman (2005) observed that staffing is an important part of the effective maintenance of human resource management.

Continuous Improvement

For continuous improvement in the maintained buildings, management aspects, maintenance policy, and strategy must be constantly assessed (Omar et al., 2017). Moreover, Manaf (2005) postulated that continuous improvement is part of the condition for the quality management system as well as the benchmark for the effectiveness of the current practice of MM. Weber and Thomas (2005) claimed that continuous improvement helps maintenance organisations to determine better ways to conduct their maintenance operations and activities with a commitment to a long-term improvement to maintenance practice. Nevertheless, Omar et al. (2017) state that adopting proactive maintenance, use of performance measurement, and management commitment were factors that principally led to the continuous improvement of the MM process and operations.

Training

Training, as defined by the International Labour Organisation (ILO) (2009), is the process of developing skills, abilities, and knowledge. Ferraz and Gallardo-Vazquez (2016) assert that with constant global changes in the organisational environment and competitiveness, organisations must be continuously prepared through regular personnel training. Training in the MM process deals with on-job knowledge improvement for maintenance personnel toward continuous improvement and maintenance task efficiency (Ferraz & Gallardo-Vazquez, 2016). Accordingly, Fatoni and Nurcahyo (2018), citing De Groote (1995), contend that the competency of the maintenance personnel is a significant factor that affects maintenance operation output. This statement shows clearly that effective MM operations depend on the skill and knowledge of the maintenance personnel.

Bhatti and Kaur (2010) postulated that in achieving organisational maintenance tasks and improving personnel performance, there is a need to design training programmes toward creating a win-win situation for maintenance organisations and maintenance personnel. Schreiber (2007) opines that errors will be better managed with enhanced training among personnel maintenance. Usanmaz (2011) submitted that all the basic training should be given to maintenance personnel before and during employment. Nikandrou, Brinia, and Bereri (2009) postulated that the important factors related to training in the MM process include the training objectives and extent of training, training equipment, and the training methods and means. Nonetheless, Kempton (1996) noted that the main goal of training within the maintenance organisation is to apply knowledge gained in work practice and equip personnel with additional skills to enable them to undertake new tasks. Kempton (1996) affirms that personnel training will provide maintenance organisations and their teams with the skills and knowledge to effectively organise and motivate a multi-skilled workforce. Training enhances personnel capabilities (Khan et al., 2011). Also, Noe (2010) states that the maintenance personnel training will add to the personnel's skills and knowledge that they can apply to their day-to-day maintenance operations.

Moreover, Brinia and Efstathiou (2012) assert that nine factors guide effective training within a maintenance organisation. These include motivation to learn, transfer training, opportunity to use training, personal career goals, motivation from work content of training, organisational commitment, colleagues' support, and superiors' support. Additionally, Wilson (2005) and Fatoni and Nurcahyo (2018) identified reliability of personnel, safety of personnel, priorities of the organisation objectives, operation objectives improvement, current maintenance knowledge, training of personnel on maintenance skills, maintenance training opportunities, on-job training on maintenance, and understanding organisation maintenance policy as the factors that influence training in the MM process. Furthermore, Wireman (2005) and Mekasha (2018) also affirmed that maintenance personnel training is important in MM. It is influenced by a better understanding of organisation policy, technology advancement, proper procedures, appropriate tools, personnel training, problem-solving skills development, time management, interpersonal skills development, and integration of new techniques.

Building Performance and Users' Productivity

Building users, as described by Lateef (2009) and Ogunbayo and Mhlanga (2022), are groups of individuals, entities, or organisations who are interested in the usage and adequate functioning of the building. Additionally, Jones and Sharp (2007) posit that the activities of users of these buildings affect the building performance while the users' productivity is also affected by the performance of the building. According to Jones and Sharp (2007), it is not the physical condition or appearance of the building that is critical to users but the performance of the building to support activities around the attached facilities efficiently and effectively. Dewulf and Van Meel (2007) submitted that the users' needs are required to formulate the requirements for building in the design stage for better performance. Okolie and Shakantu (2009) suggested that in determining the productivity of users based on the building performance with consideration of many aspects of building components and its environment, the experts that know most about it were the end-users. Atkin and Brooks (2000) noted that there has been very little in the area of communication between the end-user of the building and the other stakeholders during maintenance activities, especially during the replacement of damaged or deterioration components. Hence, Atkin and Brooks (2000) posit that users of the buildings find fault in the process, and a costly alteration is made to the building to meet their needs and productivity.

Buildings (public or private) are designed, constructed, and operated for various purposes and uses. Dewulf and Van Meel (2007) postulated that the users' needs are one of the factors that will determine the usage of the building for better productivity. Moreover, Atkin and Brooks (2000) assert that in respect of the building types, the requirements of the various building end-users invariably need to be met to boost their productivity. Hence, Lomash (1997) and Elsevier (2008) list five important issues that designers must consider when designing buildings for the efficient performance of the building in meeting users' needs for better productivity. These include the following:

- Who are the users?
- What are the needs of the users?
- Where do the needs exist?
- When are the needs to be fulfilled? and
- How long will the needs exist?

Kelly and Male (2003) observed that buildings were designed to provide suitable space for any activities to be carried out within them. Moreover, the building can be said to be performing if it adds value to the activities within its environment (Kelly and Male, 2003).

However, in educational buildings, a research laboratory that is not conducive to the researchers or a learning classroom that is not conducive to teaching and learning is of little value. Thus, Bateman and Snell (2009) postulated that using an effective strategy in the MM of educational buildings will achieve a suitable building performance and boost users' productivity. Bateman and Snell (2009) further state that using this strategy will deal with the total resources and their availability for effective and efficient building functions to meet users' expectations and improve their productivity. In their study, Häkkinen and Nuutinen (2007) maintain that not considering basic users' needs in the buildings' design and maintenance operations might lead to an unsuitable working environment that could affect their productivity.

From the foregoing views, it can be said that the more the users' needs (in terms of reliable building performance, safety, comfort, and the like) and perception at less cost are met, the more value will be added to the users in boosting their productivity. Then (2003) theorises that the critical MM of buildings is typically about fitness for purpose for the users, especially towards their productivity. Elsevier (2008) postulated that the significant driving force for building performance evaluation and users' productivity is the prospect of the users of the building to articulate real needs concerning different functionalities of a building. Therefore, Barrett and Baldry (2003), Larssen and Bjørberg (2004), and Zimring and Rashidi (2008), in their studies, concluded that users' satisfaction with building and productivity come from enhancing the performance of the building and environmental comfort.

Summary

This chapter reviewed the literature concerning maintenance policy, planning, and performance measurement for MM of buildings. The chapter highlighted that the MM of buildings is dependent upon a good management context of planning, coordinating, and controlling operations within the buildings' structure and other auxiliary facilities. The chapter reveals that the benefit of MM of buildings is to provide a support platform for users of the building. This is done by addressing a broad spectrum of occupant-related issues such as creating personal user needs, technology, equipment, temperature and noise control, and a physically comfortable environment with adequate lighting. It was further revealed that building performance assessment is becoming a formal and regular part of MM activities and processes. The chapter further showed that KPIs for institutional building MM could be subdivided into three factors that include individual factors, technical aspects, and administrative and organisational factors. The chapter pointed out that buildings designed without adequate consideration for the basic end-users needs might provide an unsuitable working environment that could affect their productivity. Finally, the chapter revealed that the performance measurement of a building is vital because it identifies current building performance. Furthermore, it shows that measuring the performance of a building in use provides the decision-makers with information relating to a series of key performance criteria to guide their operations.

References

Abdul Lateef Olanrewaju, A. (2012). Quantitative analysis of defects in university buildings: User perspective. *Built Environment Project and Asset Management*, 2(2), 167–181.

Ahzahar, N., Karim, N. A., Hassan, S. H., & Eman, J. (2011). A study of contribution factors to building failures and defects in construction industry. *Procedia Engineering, 20*, 249–255.

Aigbavboa, C. O. (2014). An integrated beneficiary centred satisfaction model for publicly funded housing schemes in South Africa, Published doctoral dissertation, University of Johannesburg: Johannesburg, South Africa.

Akasah, Z. A., Shamsuddin, S. H., Rahman, I. A., & Alias, M. (2009). School building maintenance strategy. A new management approach. *Malaysian Technical Universities Conference on Engineering and Technology* (MUCEET 2009), Pahang, Universiti Malaysia Pahang, 20–22 June 2009, pp. 1–5.

Albrice, D., & Branch, M. (2015, April). A deterioration model for establishing an optimal mix of time-based maintenance (TBM) and condition-based maintenance (CBM) for the enclosure system. In *Fourth Building Enclosure Science & Technology Conference (BEST4)*, Kansas City, Missouri, 12–15 April 2015, pp. 231–237.

Alexander, K. (2013). *Facilities management: theory and practice*. London: Routledge (Taylor & Francis).

Ali, M., & Mohamad Nasbi Bin Wan Mohamad, W. (2009). Audit assessment of the facilities maintenance management in a public hospital in Malaysia. *Journal of Facilities Management, 7*(2), 142–158.

Allen, D. (1993). What is building maintenance? *Facilities – London Then Bradford, 11*, 7.

Al-Najjar, B. (2001). A concept for detecting quality deviation earlier than when using traditional diagram in automotive: A case study. *International Journal of Quality & Reliability Management, 18*(9), 917–940.

Alzaben, H. (2015). *Development of a MM framework to facilitate the delivery of healthcare provisions in the Kingdom of Saudia Arabia* (Published doctoral dissertation, Nottingham Trent University).

Amaratunga, D., & Baldry, D. (2002). Performance measurement in facilities management and its relationships with management theory and motivation. *Facilities, 20*(10), 327–336.

Amaratunga, D., & Baldry, D. (2000). Assessment of facilities management performance in higher education properties. *Facilities, 18*(7/8), 293–301.

Amaratunga, R. G. (2001). *Theory building in facilities management performance measurement: Application of some core performance measurement and management principles* (Published doctoral dissertation, University of Salford).

Amin, R. A. (2016). *Condition-based maintenance: Innovation in building maintenance management* (Published doctoral dissertation, University College London).

Atkin, B. (2003). Contracting out or managing services in-house. *Nordic Journal of Surveying and Real Estate Research, 1.*

Atkin, B., & Brooks, A. (2000). *Total facilities management.* London: Blackwell Science.

Bagadia, K. (2006). *Computerized maintenance management systems made easy.* New York: McGraw-Hill Professional.

Barrett, P. (1992). Development of a post-occupancy building appraisal model. In P. Barrett (Ed.), *Facilities management: Research directions.* London: RICS Books.

Barrett, P., & Baldry, D. (2003). *Facilities management: Towards best practice.* Oxford: Blackwell.

Bartlett, E. V., & Simpson, S. (1998). Durability and reliability, alternative approaches to assessment of component performance over time. In *World building congress* (pp. 35–42). Gävle Sweden, 7–12 June 1998.

Bateman, T. S., & Snell, S. (2009). *Administración: Liderazgo y colaboración en un mundom competitivo* No. Sirsi) a 458252.

Bhatti, M. A., & Kaur, S. (2010). The role of individual and training design factors on training transfer. *Journal of European Industrial Training, 34*(7, 656–672.

Becker, F. (1990). Facility management: A cutting-edge field? *Property Management, 8*(2), 108–116.

Blissett, R. (2004). Effective leadership in managing proactive maintenance and reliability programs. *The Maintenance Journal.* Retrieved from www.maintenancejournal.com

Borkowski, P., Pawlowski, M., & Makowiecki, T. (2011). Economical aspects of building management systems implementation. In *2011 IEEE Trondheim Power Tech* (pp. 1–6). IEEE.

Bradley, P. S. (2002). *Designing the best maintenance.* Retrieved from http://www.samicorp.com, 1–5.

Brinia, V., & Efstathiou, M. (2012). Evaluation of factors affecting training transfer on safety in the workplace: A case study in a big factory in Greece. *Industrial and Commercial Training, 44*(4), 223–231.

British Standards Institution (1974). *Glossary of general terms used in maintenance organization.* BS 3811: 1974. London, BSI.

British Standards Institution [BSI] (1986). *Guide to building maintenance management.* BS 8210: 1986. London, BSI.

Burhanuddin, M. A., Halawani, S. M., Ahmad, A. R., & Tahir, Z. (2011). Failure-based maintenance decision support system using analytical hierarchical process. *International Journal of Advanced Computer Science, 1*(1), 1–9.

Campbell, J. D. (1995). Outsourcing in maintenance management: A valid alternative to self provision. *Journal of Quality in Maintenance Engineering, 1*(3), 18–24.

Campbell, J. D., & Jardine, A. K. (2001). *Maintenance excellence: Optimizing equipment life-cycle decisions.* New York: Marcel Dekker Inc.

Chelson, V. J., Payne, C. A., & Reavill, R. P. (2005). *Management for engineers scientists and technologists* (2nd ed.). Chichester England: Wiley.

Chen, L. X., & Meng, B. (2011). How to apply TPM in equipment management for Chinese enterprises. *Chinese Business Review, 10*(2), 137–145.

Cholasuke, C., Bhardwa, R., & Antony, J. (2004). The status of maintenance management in UK manufacturing organisations: Results from a pilot survey. *Journal of Quality in Maintenance Engineering, 10*(1), 5–15.

Coetzee, J. L. (1997). Towards a general maintenance model. In: *Proceedings of International Foundation for Research in Maintenance*, Hong Kong. 97, paper 12, pp. 1–9.

Crespo Márquez, A., de León, P. M., Gomez Fernandez, J. F., Parra Marquez, C., & López Campos, M. (2009). The maintenance management framework: A practical view to maintenance management. *Journal of Quality in Maintenance Engineering, 15*(2), 167–178.

Daft, L. R. (1989). *Organization theory and design*, 3rd ed. New York: West Publishing Company.

Daragh Naughton, M., & Tiernan, P. (2012). Individualising maintenance management: A proposed framework and case study. *Journal of Quality in Maintenance Engineering*, 18(3), 267–281.

De Groote, P. (1995). Maintenance performance analysis: A practical approach. *Journal of Quality in Maintenance Engineering, 1*(2), 4–24.

De Jonge, B., Teunter, R., & Tinga, T. (2017). The influence of practical factors on the benefits of condition-based maintenance over time-based maintenance. *Reliability Engineering & System Safety, 158*, 21–30.

Dewulf, G. P., & Van Meel, J. (2003). Democracy in design? *Workplace Strategies and Facilities Management*, 281–291. London, Routledge.

Ding, S. H., & Kamaruddin, S. (2015). Maintenance policy optimization—Literature review and directions. *The International Journal of Advanced Manufacturing Technology, 76*(5–8), 1263–1283.

Dixon, J. R., Nanni, A. J., & Vollmann, T. E. (1990). *New performance challenge: Measuring operations for world-class competition (Irwin/Apics series in production management)*. Homewood: McGraw-Hill Professional Publishing.

Douglas, J. (1996). Building performance and its relevance to facilities management. *Facilities, 14*(3/4), 23–32.

Duffuaa, S. O. (2000). Mathematical models in maintenance planning and scheduling. In *Maintenance, modeling, and optimization* (pp. 39–53). Boston, MA: Springer.

ECOSOC, U., & Union, A. (2016). *Facing the Challenges of Land Monitoring in the Framework and Guidelines on Land Policy in Africa* –Towards Agenda 2063 and the 2030 Agenda for Sustainable Development. New York: A Working paper submitted at the ECOSOC Dialogue Workshop No. 8, 22–23 June 2016.

Eisinger, S., & Rakowsky, U. K. (2001). Modeling of uncertainties in reliability centered maintenance—A probabilistic approach. *Reliability Engineering & System Safety, 71*(2), 159–164.

Ellis, B. A., & Byron, A. (2008). Condition-based maintenance. *The Jethro Project, 10*, 1–5.

Elsevier, B. (2008). User driven innovation in building process. *Tsinghua Science and Technology, 13*(1), 248–254.

Endrenyi, J., Aboresheid, S., Allan, R. N., Anders, G. J., Asgarpoor, S., Billinton, R., & Grigg, C. (2001). The present status of maintenance strategies and the impact of maintenance on reliability. *IEEE Transactions on Power Systems, 16*(4), 638–646.

Fagbenle, B. J. (1998). Provision and maintenance of engineering infrastructure technological development in Nigeria: Kaduna. *Kaduna Publication Ltd.* Nigeria

Fagbenle, O. I., & Oluwunmi, A. O. (2010). Building failure and collapse in Nigeria: The influence of the informal sector. *Journal of Sustainable Development, 3*(4), 268–276.

Fatokun, T. O. (1997). Prospect control of schemes of maintenance, repairs and improvement works. Paper presented at a 2-Day Workshop on *Residential Property Maintenance Planning and Management*, organized by Estate Surveyors Registration Board of Nigeria in with First State Communication.

Fatoni, Z. Z. Z., & Nurcahyo, R. (2018). Impact of training on maintenance performance effectiveness. In *Proceedings of the International Conference on Industrial Engineering and Operations Management*, Paris, 26–27 July 2018, pp. 619–628.

Ferraz, F. A. D., & Gallardo-Vazquez, D. (2016). Measurement tool to assess the relationship between corporate social responsibility, training practices, and business performance. *Journal of Cleaner Production, 129*, 659–672.

Finlow-Bates, T., Visser, B., & Finlow-Bates, C. (2000). An integrated approach to problem solving: Linking K-T, TQM, and RCA to TPM. *The TQM Magazine, 12*(4), 284–289.

Flores-Colen, I., & de Brito, J. (2010). A systematic approach for maintenance budgeting of buildings façades based on predictive and preventive strategies. *Construction and Building Materials, 24*(9), 1718–1729.

Gabbar, H. A., Yamashita, H., Suzuki, K., & Shimada, Y. (2003). Computer-aided RCM-based plant maintenance management system. *Robotics and Computer-Integrated Manufacturing, 19*(5), 449–458.

Garg, A., & Deshmukh, S. G. (2006), Maintenance management: Literature review and direction. *Journal of Quality in Maintenance Engineering, 12*(3), 205–238.

HajShirmohammadi, A., & Wedley, W. C. (2004). Maintenance management – An AHP application for centralization/decentralization. *Journal of Quality in Maintenance Engineering, 10*(1), 16–25.

Häkkinen, T., & Nuutinen, M. (2007). Seeking sustainable solutions for office buildings. *Facilities, 25*(11/12), 437–451.

Halevi, G. (2001). *Handbook of production management methods.* Woburn: Reed Elsevier Plc group.

Hallberg, D. (2009). *System for predictive life cycle management of buildings and infrastructures* (Published doctoral thesis, KTH Research School—HIG, University of Gävle, Gävle, Sweden).

Han, T., & Young, B. (2006). Development of an e-maintenance system integrating advanced techniques. *Computers in Industry, 57*(6), 569–580.

Haroun, A. E., & Duffuaa, S. O. (2009). Maintenance organization. In *Handbook of maintenance management and engineering* (pp. 3–15). London: Springer.

Hassanain, M. A., Froese, T. M., & Vanier, D. J. (2001). Development of a maintenance management model based on IAI standards. *Artificial Intelligence in Engineering, 15*(2), 177–193.

Hipkin, I. B., & De Cock, C. (2000). TQM and BPR: Lessons for maintenance management. *Omega, 28*(3), 277–292.

Horner, R. M. W., El-Haram, M. A., & Munns, A. K. (1997). Building maintenance strategy: A new management approach. *Journal of Quality in Maintenance Engineering, 3*(4), 273–280.

Howlett, M. (2019). *Designing public policies: Principles and instruments*. Oxfordshire: Routledge.

International Labour Organisation (ILO) (2009). *Background studies on infrastructure sector in Ghana*. Cape Coast, South Africa: Directorate of Research, Innovation and Consultancy University of Cape Coast.

Irajpour, A., Fallahian-Najafabadi, A., Mahbod, M. A., & Karimi, M. (2014). A framework to determine the effectiveness of maintenance strategies lean thinking approach. *Mathematical Problems in Engineering, 2014*, 1–11.

James, P. T. (1996). *Total quality management: An introductory text*. New Jersey: Prentice Hall.

John, R. S. (1996). *Management and organizational behavior essentials*. New Jersey: Wiley.

Jones, K., & Sharp, M. (2007). A new performance-based process model for built asset maintenance. *Facilities, 25* (13/14), 525–535.

Kapto, S. (2019). Layers of politics and power struggles in the SDG indicators process. *Global Policy, 10*, 134–136.

Karia, N., Asaari, M. H. A. H., & Saleh, H. (2014). Exploring maintenance management in service sector: A case study. In Proceedings of *International Conference on Industrial Engineering and Operation Management*, Bali, 7–9 January 2014, pp. 3119–3128.

Kelly, A. (2006). *Strategic maintenance planning* (Vol. 1). Oxford: Elsevier.

Kelly, J., & Male, S. (2003). *Value management in design and construction*. London: Routledge.

Kempton, G. E. (1996). Training for organizational success. *Health Manpower Management, 22*(6), 25–30.

Khan, R. A. G., Khan, F. A., & Khan, M. A. (2011). Impact of training and development on organizational performance. *Global Journal of Management and Business Research, 11*(7), 1–7.

Knirk, F. G. (1993). Facility requirements for integrated learning systems. *Educational Facility Planner, 31*(3), 13–18.

Lackney, J. A. (2001). The state of post-occupancy evaluation in the practice of educational design. Paper presented at 32nd Annual Meeting of the Environmental Design Research Association Edinburgh, Scotland, 3–6 July 2001.

Larssen, A. K., & Bjørberg, S. (2004). User need/demands (functionality) and adaptability of buildings – A model and tool for evaluation of buildings. In *Conference Proceedings, 12th CIB Triennial World Building Congres*, Toronto, Canada, 7–8 May 2004.

Lateef, O. A. (2009). Building maintenance management in Malaysia. *Journal of Building Appraisal, 4*(3), 207–214.

Lateef, O. A. (2010). Case for alternative approach to building maintenance management of public universities. *Journal of Building Appraisal, 5*(3), 201–212.

Leaman, A. (2004). *Post-occupancy evaluation: Building use studies* [Online]. Retrieved from: www.usablebuildings.co.uk (Accessed 7 October 2020).

Lee, H. H. Y., & Scott, D. (2009). Overview of maintenance strategy, acceptable maintenance standard and resources from a building maintenance operation perspective. *Journal of Building Appraisal, 4*(4), 269–278.

Lee, J. (2001). A framework for web-enabled e-maintenance systems. In *Proceedings Second International Symposium on Environmentally Conscious Design and Inverse Manufacturing*, Tokyo, Japan, 11–15 December 2001, pp. 450–459. IEEE.

Lewis, L. (2000). *Condition of America's public school facilities, 1999.* US Department of Education, Office of Educational Research and Improvement: National Center for Education Statistics.

Li, L., Qu, M., & Peng, S. (2017). Performance evaluation of building integrated solar thermal shading system: Active solar energy usage. *Renewable Energy, 109*, 576–585.

Lind, H., & Muyingo, H. (2012). Building maintenance strategies: Planning under uncertainty. *Property Management, 30*(1), 14–28.

Lomash, S. (1997). *Value management* (pp. 167–173). New Delhi: Sterling publishers.

Madeley, A. (1996). *The performance of organisations and facilities management: An exploration into the opportunities afforded for facilities management to contribute toward corporate performance'* (Unpublished M.Sc dissertation, University of Strathclyde).

Manaf, Z., & Alias, A. (2005). Training needs in facilities management a study among local authority offices in Malaysia. In *International Real Estate Research Symposium (IRERS)*, Kuala Lumpur, Malaysia, 11–13 April, pp. 1–14.

Márquez, A. C. (2007). *The maintenance management framework: models and methods for complex systems maintenance.* London: Springer-Verlag.

Mayaki, S. S. (2005). Facility performance evaluation. In *International Conference Organized by the Nigerian Institute of Building.* Nasarawa, 20–22 April 2005, pp. 34–47.

Mbamali, I. (2003). The impact of accumulation deferred maintenance on selected buildings of two federal universities in the Northwest zone of Nigeria. *Journal of Environmental Science, 5*(1), 77–83.

Mekasha, E. (2018). *Maintenance management framework development for competitiveness of food and beverage industry*: A case study on Asku Plc (Thesis Draft, Addis Ababa Institute of Technology, Addis Ababa University).

Muchiri, P. N. (2014). Strategic approach in critical asset maintenance. In *The First Dekut International Conference on Science, Technology and Innovation*, Nairobi, 4th–7th November 2014. pp. 15.

Mungani, D. S., & Visser, J. K. (2013). Maintenance approaches for different production methods. *South African Journal of Industrial Engineering, 24*(3), 1–13.

Murthy, D. N. P., Atrens, A., & Eccleston, J. A. (2002). Strategic maintenance management. *Journal of Quality in Maintenance Engineering, 8*(4), 287–305.

Nakajima, S. (1989). *TPM development program: Implementing total productive maintenance.* Florida: Productivity press-Routledge.

Niebel, B. W. (1994). *Engineering maintenance management.* New York: CRC Press. Taylor & Francis group.

Nikandrou, I., Brinia, V., & Bereri, E. (2009). Trainee perceptions of training transfer: An empirical analysis. *Journal of European Industrial Training, 33*(3), 255–270.

Noe, R. A. (2010). *Employee training and development* (5th ed.). New York: McGraw-Hill.

Oakland, J. S. (1993). *Total quality management: Theroute to improving performance.* Oxford: Butterworth-Heinemann.

Obiegbu, M. E. (2005). Overview of total performance concept of buildings: Focusing on quality, safety, and durability. In *34th Annual Conference of the Nigerian Institute of Building*, Abeokuta, 27–28 April 2005, pp. 2–10.

Ogunbayo, B. F., Aigbavboa, C. O., Thwala, D. W., & Oguntona, O. A. (2022). Assessing Critical Factors Affecting The Development Of A National Maintenance Policy. *Proceedings of the Fourth European and Mediterranean Structural Engineering and Construction Conference*, Leipzig, Germany, June 20 - June 25, 2022, *9*(1), ISSN 2644-108X.

Ogunbayo, B. F., & Aigbavboa, O. C. (2019). Maintenance requirements of students' residential facility in higher educational institution (HEI) in Nigeria. In *IOP Conference Series: Materials Science and Engineering* (Vol. 640, No. 1, p. 012014). IOP Publishing.

Ogunbayo, B. F., Aigbavboa, C. O., Thwala, W., Akinradewo, O., Ikuabe, M., & Adekunle, S. A. (2022). Review of culture in maintenance management of public buildings in developing countries. *Buildings, 12*(5), 677.

Ogunbayo, B. F., Aigbavboa, C. O., Thwala, W. D., Opeoluwa, O. I., & Edwards, D. (2022). Validating elements of organisational maintenance policy for maintenance management of public buildings in Nigeria. *Journal of Quality in Maintenance Engineering.* https://doi.org/10.1108/JQME-05-2021-0039

Ogunbayo, B. F., Ohis Aigbavboa, C., Thwala, W. D., & Akinradewo, O. I. (2022). Assessing maintenance budget elements for building maintenance management in Nigerian built environment: A Delphi study. *Built Environment Project and Asset Management, 12*(4), 649–666.

Ogunbayo, S. B., & Mhlanga, N. (2022). Effects of training on teachers' job performance in Nigeria's public secondary schools. *Asian Journal of Assessment in Teaching and Learning, 12*(1), 44–51.

Okolie, K. C. (2011). *Performance evaluation of buildings in educational institutions: A case of Universities in South-East Nigeria* (Published doctoral dissertation, Nelson Mandela Metropolitan University, Port Elizabeth, South Africa).

Okolie, K. C., & Shakantu, W. M. (2009). Design and user/occupier needs in building performance. In *Conference Proceedings of RIC COBRA Construction and Building Research Conference of the Royal Institution of Chartered Surveyors*, University of Cape, 10th–11th September 2009, pp. 9–11.

Olanrewaju, A. L., & Abdul-Aziz, A. R. (2015). Building maintenance processes, principles, procedures, practices, and strategies. In *Building maintenance processes and practices* (pp. 79–129). Singapore: Springer.

Ollila, A., & Malmipuro, M. (1999). Maintenance has a role in quality. *The TQM Magazine, 11*(1), 17–21.

Omar, M. F., Ibrahim, F. A., & Omar, W. M. S. W. (2017). Key performance indicators for maintenance management effectiveness of public hospital building. In *MATEC Web of Conferences*, Ho Chi Minh City, Vietnam, 5–6 August 2016, pp. 01056. (Vol. 7). EDP Sciences.

Ornstein, S. W., Moreira, N. S., Ono, R., França, A. J., & Nogueira, R. A. (2009). Improving the quality of school facilities through building performance assessment: Educational reform and school building quality in São Paulo, Brazil. *Journal of Educational Administration, 47*(3), 350–367.

Palm, P. (2008). *Closing the loop: The use of post occupancy evaluations in real estate management* (Published doctoral dissertation, Avdelningen för Bygg-och fastighetsekonomi KTH Stockholm).

Park, A. (1998). *Facilities management: An explanation.* London: Macmillan Press Ltd.

Parker, C. (2000). Performance measurement. *Work Study, 49*(2), 63–66.

Pinjala, S. K., Pintelon, L., & Vereecke, A. (2006). An empirical investigation of the relationship between business and maintenance strategies. *International Journal of Production Economics*, *104*(1), 214–229.

Pintelon, L. M., & Gelders, L. F. (1992). Maintenance management decision making. *European Journal of Operational Research*, *58*(3), 301–317.

Pintelon, L., & Muchiri, P. N. (2009). Safety and maintenance. In *Handbook of maintenance management and engineering* (pp. 613–648). London: Springer.

Preiser, W. F. E. (1995), Post-occupancy evaluation: How to make buildings work better. *Facilities*, *13*(11), 19–28.

Price, I., & Akhlaghi, F. (1999). New patterns in facilities management: Industry best practice and new organisational theory. *Facilities*, *17*(5/6), 159–166.

Pun, K., Chin, K., Chow, M., & Lau, H. C. W. (2002). An effectiveness-centered approach to maintenance management: A case study. *Journal of Quality in Maintenance Engineering*, *8*(4), 346–368.

Salamon, L. M. (2002). Introduction:The new governance and the tools of publicaction. *The tools of government: A guide to the new governance*, *28*(4), 1–46.

Schermerhorn, J. R. (2007). *Management*, 9th ed. New York: John Wiley.

Schermerhorn, J. R. (1996). *Essentials of management and organizational behavior*. New York: John Wiley and Sons.

Schreiber, F. (2007). *Maintenance briefing notes human performance error management*. Toulouse, France: Airbus Customer Service.

Sherwin, D. (2000). A review of overall models for maintenance management. *Journal of Quality in Maintenance Engineering*, *6*(3), 138–164.

Spedding, A., & Holmes, R. (1994). Facilities management. CIOB Book of Facilities Management, London: Longman Scientific & Technical.

Takata, S., Kirnura, F., van Houten, F. J., Westkamper, E., Shpitalni, M., Ceglarek, D., & Lee, J. (2004). Maintenance: Changing role in life cycle management. *CIRP Annals*, *53*(2), 643–655.

Then, D. S. (2003). Strategic management. In: R. Best, C. Langston, and G. De Valence (Eds.), *Workplace strategies and facilities management: Building in value*. London: Butterworth.

Tsang, A. H. (2002). Strategic dimensions of maintenance management. *Journal of Quality in Maintenance Engineering*, *8*(1), 7–39.

Ugechi, C. I., Ogbonnaya, E. A., Lilly, M. T., Ogaji, S. O. T., & Probert, S. D. (2009). Condition based diagnostic approach for predicting the maintenance requirements of machinery. *Engineering*, *1*(03), 177–187.

Usanmaz, O. (2011). Training of the maintenance personnel to prevent failures in aircraft systems. *Engineering Failure Analysis*, *18*(7), 1683–1688.

Vaisnys, P., Contri, P., Rieg, C., & Bieth, M. (2006). *Monitoring the effectiveness of maintenance programs through the use of performance indicators* (p. 22602). Pattern Netherland: European Commisison Report.

Van Horenbeek, A., & Pintelon, L. (2015). A joint predictive maintenance and inventory policy. In *Engineering asset management-systems, professional practices, and certification* (pp. 387–399). Cham: Springer.

Venkatesh, V. (2007) An introduction to total productive maintenance (TPM). Retrieved from www.plant-maintenance.com. 3–20.

Visser, J. K. (2002). Maintenance management – a neglected dimension of engineering management. In: *Proceedings IEEE Africon, Vols 1 and 2: Electro technological Services for Africa, IEEE*, New York, 2–4 October 2002. pp. 479–484.

Vosloo, M. M., & Visser, J. K. (1999). The development of a maintenance philosophy. *R & D Journal, 15*(2), 27–34.

Wardhaugh, J. (2004). Useful key performance indicators for maintenance. In *Singapore IQPC Reliability and Maintenance Congress*, Singapore, 2004, Lifetime Reliability Maintenance – The best practices, www.lifetime-reliability.com

Weber, A., & Thomas, R. (2005). *Key performance indicators. Measuring and managing the maintenance function*. Burlington: Ivara Corporation.

Wessels, W. R. (2003). Cost-optimized scheduled maintenance interval for reliability-centered maintenance. In *Proceedings of Annual Reliability and Maintainability: The International Symposium on product quality & integrity*, Tampa, Florida, USA, 27–30 January 2003, pp. 412–416, IEEE.

Williams, M. F., & Williams, B. L. (1994). *Building enclosure assemblies: U.S. Patent No. 5,279,091*. Washington, D.C.: U.S. Patent and Trademark Office.

Williamson, R. M. (2006). *Using overall equipment effectiveness: The metric and the measures*. Columbus: Strategic Work Systems.

Willmott, P. (1994). Total quality with teeth. *The TQM Magazine, 6*(4), 48–50.

Wilson, J. P. (Ed.). (2005). *Human resource development: learning & training for individuals & organizations*. New York: Kogan Page Publishers.

Wilson, D. C. (1992). *A strategy of change: Concepts and controversies in the management of change*. London: Routledge.

Wireman, T. (2005). *Developing performance indicators for managing maintenance*. New York: Industrial Press Inc.

Wordsworth, P., & Lee, R. (2001). *Lee's building maintenance management*. London: Blackwell Science.

Zimring, C. (2001). Post-occupancy evaluation and organizational learning. In: F. F. Council (Ed.), *Learning from our buildings: A state-of-the-practice summary of post-occupancy evaluation*. Washington: National Academy Press.

Zimring, C., & Rashidi, M. (2008). *Facility performance evaluation* [Online]. Retrieved from: www.wbdg.org/Resources/Fpe.Php (Accessed 7 October 2020).

4 Theories, Models, and Concepts in Maintenance Management Studies

Introduction

This chapter reviews the literature in relation to the theoretical and conceptual perspective of maintenance management (MM) research. Also, it includes a discussion on previous MM theories and models, subjective definition of MM, the relative nature of MM study, the complexity of MM studies, and methodological issues in the study of MM. Moreover, various models of MM of buildings are highlighted. Lastly, based on the reviewed models, the researcher's conceptualisation has been done before giving a chapter conclusion.

Issues Raised in Maintenance Management Study

Attempts by scholars to empirically prove models for MM studies have encountered several problems. The problems are usually grouped into three, namely issues relating to the definition of the term "MM" (the meaning of MM is subjective and has multiple definitions, the complexity of MM – problems relating to the attributes that determine MM), and their relative contribution towards the understanding of maintenance dimensions. Lastly, issues relating to the process and relative nature of MM studies (Wireman, 1990; Jonsson, 2000; Mc Kone & Schroeder, 2001). These problems have resulted in difficulties in formulating attributes or variables that holistically determine MM in a related given organisation or industry.

Subjective Definition of Maintenance Management

Several bodies, including professionals and researchers, have defined MM based on their understanding of its concept. This, in turn, has influenced the characteristics or attributes that impacted or determined MM in their respective studies or reports. Hence, the unpredictable views are advanced by researchers in MM studies (Marquez & Gupta, 2006). However, three views are generally held in literature in relation to MM definitions. There is one group of researchers that define MM based on technical attributes, while

DOI: 10.1201/9781003344681-4

another group of researchers defines MM based on administrative and managerial attributes in an organisation or an institution, and finally, researchers that integrate technical, administrative, and managerial attributes in defining MM.

Moreover, studies that defined MM based on technical attributes include those of Wireman (1990), Campbell (1995), and Marquez and Gupta (2006), in which MM is defined as the combination of all technical actions during the life cycle of a building, aimed at retaining it in or restoring it to, a condition in which it can perform the required function. Here, the emphasis is on technical attributes. According to an administrative and managerial-based view, researchers generally inform that MM comprises all activities of the management that determine maintenance priorities or objectives, responsibilities, and strategies (Wordsworth & Lee, 2001, Ogunbayo et al., 2022). That could be implemented through means such as maintenance planning and organisation, maintenance control and supervision, and various other improving procedures, including economic and business aspects in the organisation (Duffuaa, 2000; Wordsworth & Lee, 2001; Campbell & Jardine, 2001). Researchers who have defined MM based on technical, administrative, and managerial attributes influencing MM include the studies by Duffuaa (2000), Campbell and Jardine (2001), and Wireman (2005).

Regarding what defines the attributes that help organisations attain effective MM, authors have differed in terms of definitions. For instance, Karia, Asaari, and Saleh (2014) inform that MM is principally a preventive management philosophy that is considered a business function that provides opportunities to retain the value, life, and quality of buildings and improve risk, cost, and productivity concerns in organisations. This suggests that the responsibilities for managing the maintenance and operations of all the physical facilities, including an organisation's buildings, fall with the maintenance organisation, which includes the maintenance department, the facilities department, and the property division, among others. According to the definition of MM by Campbell and Jardine (2001), it is the management of all assets, including buildings owned by an organisation, based on maximising the return on investment in the asset.

This suggests that the MM of an organisation is characterised by a return on investment in the asset maintained. However, according to the Royal Institution of Chartered Surveyors (1993), MM consists of three distinctive but interconnected areas: the management of support services, the management of the building, and the management of the MIS. This suggests that a combination process exists between the management and personnel of the maintenance organisation. Blessing, Richard, and Emmanuel (2015) argue that an organisation's MM is defined by its orderly and systematic approach to planning, organising, monitoring, and evaluating maintenance activities and their costs. This shows that the MM of an organisation is defined by systematic planning, organising, monitoring, and supervising maintenance activities for cost-effectiveness.

Similarly, Puķīte and Geipele (2017) defined MM as a legal and technical set of operations required to maintain the buildings and preserve their usable condition and attached components. This suggests that the MM of an organisation, apart from the administrative and managerial aspects, needs a technical aspect to execute its maintenance operations. Moreover, Bagadia (2006) defined MM as a system that makes equipment and facilities available, which helps maintenance organisations' minimal downtime. This suggests that where equipment is available for maintenance operations, MM will achieve minimal downtime and reduced wastage in terms of time and money. Also, the Centre for Facilities Management (1992) defined MM as a procedure through which a maintenance organisation maintains structured quality support services to meet the organisation's objectives at the best cost. This suggests that MM provides guidelines for maintenance processes and procedures for operations toward meeting the organisation's set goals at the best operational cost. Lateef (2009) defined maintenance as a system of planning, directing, controlling, and organising maintenance processes through technical, administrative, and managerial strategies to obtain maximum returns for the investment. This suggests that MM deals with the management of both technical and administrative aspects of the main-tenance organisation. From the various definitions of MM, it could be de-duced that MM has no generic definition.

Over the years, definitions for MM have varied among researchers and have been fairly limited (Bagadia, 2006; Goyal & Maheshwari, 2012). Furthermore, the meaning of MM is dependent on the expectation of a maintenance organisation under investigation (Zhu, Gelders, & Pintelon, 2002; Daragh Naughton & Tiernan, 2012). Throughout the literature, the researchers' definitions of MM influence the constructs or attributes that determine MM. The diverse definitions of MM are greatly influenced by the research method adopted for the study. Some definitions focused mainly on the technical attributes, while others focused on the administrative and managerial attributes that influence MM. However, a more comprehensive definition of MM must consider technical, administrative, and managerial attributes that influence MM in a given maintenance organisation. Thus, Marquez and Gupta (2006: 314) defined MM as "... the combination of all technical, administrative, and managerial actions during the life cycle of a structure (building) envisioned to restore it to, or retain it in, a state in which it can perform the required function of meeting users' requirements". This aligns with a view earlier advanced by British Standard (EN 13306: 2001).

Thus, this study assumes that the definition of MM is not limited to technical attributes (aspects) only or administrative and managerial attri-butes (aspects) in the maintenance organisation, but a collection of both (technical, administrative, and managerial aspects) defines MM in a given maintenance organisation. These technical, administrative, and managerial aspects could be improved for continuous maintenance improvement (Omar, Ibrahim & Omar, 2017). Thus, this study supports the definition of

MM given by British Standard (EN 13306: 2001), Marquez and Gupta (2006), and Omar et al. (2017) as all embrace the technical, administrative, and managerial attributes that determine MM.

The Relative Nature of Maintenance Management Study

The variables that determine MM differ from one organisation to another and from one nation to another, owing to the relative nature of MM studies (Zhu et al., 2002; Goyal & Maheshwari, 2012; Daragh Naughton & Tiernan, 2012). Hence, MM is measured in relation to a particular organisation in a specific nation within which the maintenance organisation or agency operates (Dhillon & Subramanian, 2001). On the other hand, the relative nature of variables that determine MM change with time from one particular maintenance set of objectives to the other within an organisation or nation due mostly to changes in economic situations (Ahuja & Khamba, 2007). Thus, studies on MM have been considered and focused on different organisations and industries with different variables within a nation (Mc Kone & Schroeder, 2001; Daragh Naughton & Tiernan, 2012). However, a plethora of existing literature affirms that a relatively large number of studies on MM have focused on organisations in developed countries and their findings are only largely applicable to the industries they studied (Parida, Kumar, Galar & Stenström, 2015; Mekasha, 2018).

Additionally, among the relative nature of MM is that professionals and scholars use different words to describe similar or the same MM concepts (Parida et al., 2015; Kym, 2015). The concept of MM as described by authors with different terms include maintenance model, maintenance systems, maintenance strategies, maintenance philosophies, maintenance types, maintenance methods, maintenance framework, and maintenance techniques which are often used to describe MM models in most literature on the same notion on maintenance (Parida et al., 2015; Kym, 2015; Mekasha, 2018; Ogunbayo, Ohis Aigbavboa, Thwala, & Akinradewo, 2022). Nonetheless, in respect of the existing nature of MM studies, fewer studies focus on the MM of organisations, companies, and institutions in developing countries. It is worth emphasising that since the organisational characteristics and policies, as well as the administrative and technical processes in relation to MM in developed countries, may not necessarily be the same, this current study seeks to develop a MM framework for effective maintenance of municipal buildings in the education sectors of developing economies, using South Africa as a case study.

The framework seeks to empirically ascertain and explain the attributes that will aid effective MM of educational buildings in the education sector and the relative contribution of the attributes in the education sectors of the developing economies, using South Africa as a case study. Hence, in addressing the issues surrounding the relativity of the concept of MM, this study focuses on municipal buildings in the South African education sector.

Complexity of Maintenance Management Study

As approaches to the study of MM have varied over the years, researchers have not yet reached a consensus on the variables that comprehensively influence MM in an organisation, company, country, or geographical block (Mekasha, 2018). Marquez and Gupta (2006) opined that this is possible because MM is often associated with a series of difficulties. For instance, the studies of Jonsson (2000), Wireman (1990), and Mc Kone et al. (2001) show that MM is to a certain degree "under-developed" with an absence of effective prevention methodology and the dearth of integration of these methods in organisations, companies, and institutions in most countries. Also, the study of Jonsson (2000) indicates that there is also a lack of MM models that show a better understanding of complex issues in maintenance. Therefore, Marquez and Gupta (2006) came up with a list of factors impacting complexity issues in MM, including lack of CMMS, low technical expertise of the maintenance staff, low level of operator knowledge and involvement in maintenance. They also listed the lack of maintenance procedures in place, automation and process integration level, variety of technologies used in the production process, complexity in the production process technology, and lack of historical data.

Several causes have been identified by different studies to have influenced effective MM. Nevertheless, few of these studies organised them into models to explain how maintenance organisations could attain effective MM. Moreover, Alzaben (2015), Daragh Naughton and Tiernan (2012) and Ogunbayo, Aigbavboa and Thwala (2022) posit that previous models that seek to explain how effective MM could be attained have focused on different tools and techniques such as:

- Balanced scorecard,
- Reliability blocked diagram,
- Fault tree analysis,
- Current reality tree,
- Life cycle cost analysis,
- Critical path method,
- Reliability analysis,
- Reliability-centred maintenance,
- Overall equipment effectiveness and
- Critical success factors.

For example, the models by Vanneste and Van Wassenhove (1995) and Waeyenbergh and Pintelon (2009), which measure the way equipment is operated and failures incurred during their operation, integrated the failure mode and effect analysis (FMEA) to give the constructs that measure MM a more comprehensive approach. Similar approaches were used by Daragh Naughton and Tiernan (2012), Alzaben (2015), and Mekasha (2018).

However, there is a dearth of studies on organisation MM that use all available tools or techniques to give it all-inclusive approaches (Ogunbayo & Aigbavboa, 2019). This is because maintenance concepts vary from organisation to organisation and one-fits-all solution, and literature is absent around MM model studies (Daragh Naughton & Tiernan, 2012; Mekasha, 2018). Hence, this present study contributes to addressing these gaps by developing a holistic MM framework that considers different maintenance performance management approaches to the study of MM. These approaches include the strategic asset performance approach, value-driven performance measures, balanced scorecard approach, TQM approach, user requirement approach, and organisation change management approach. Hence, the attributes of MM that this present study conceptualises cover relevant aspects of approaches in MM studies.

The variables that determine MM are subjective. They depend on the organisation in question and the stakeholders' definition of MM in that particular organisation (Bagadia, 2006; Alzaben, 2015). Hence, they could be influenced by stakeholders' socio-cultural differences, as well as the prevailing cultural, economic, social, political conditions and user requirements in a particular nation (Cholasuke, Bhardwa, & Antony, 2004; Mekasha, 2018). According to Nakajima, Conforti, and Rubbia (1997) and Cholasuke et al. (2004), many studies on MM concepts have modelled MM using econometric, statistical, and/or correlation analysis. For example, Ben-Daya, Duffuaa, Raouf, Knezevic, and Ait-Kadi (2009), Okolie (2011), Alzaben (2015), Blessing et al. (2015), Olanrewaju and Abdul-Aziz (2015) and Mekasha (2018).

Nevertheless, little is known about the use of the Delphi technique in modelling MM. Hence, this study models an educational-centred integrated framework for attaining MM in the developing economies' education sector using a Delphi survey (DS). It is important to note that a dearth of studies used the Delphi techniques in developing economies, and South Africa is no exception. Hence, this study employs the Delphi survey in determining the relationship between exogenous variables and how they affect the MM of municipal buildings in the developing economies using the South African education sector. Aigbavboa (2014) asserted that Delphi survey usage is an effective, robust, and suitable tool for capturing vital data in qualitative studies.

Methodology Issues in Maintenance Management Study

Available literature has suggested diverse ways of measuring MM at the organisation, industry, and national levels. Also, existing literature informs that there is no generic measurement of MM as the variables of MM have varied according to definitions and concepts as developed by researchers (Goyal & Maheshwari, 2012; Waeyenbergh & Pintelon, 2002; Daragh Naughton & Tiernan, 2012; Mekasha, 2018). Some literature measured MM's outcomes by using an organisation maintenance strategy. Some of the strategic measures that have been used in alignment with the company's business plans include maintenance costs, reliability, maintenance quality, personnel management, inventory of spare parts, overall equipment effectiveness, safety/risk, capital replacement decisions, logistics, life-cycle optimisation and output quality (Van Horenbeek, Pintelon & Muchiri, 2010). According to Barberá, Crespo, Viveros, and Stegmaier (2012), MM in a continuous improvement process is better measured in alignment with an organisation's strategy because maintenance achievement depends on the organisation's business plan.

Nevertheless, Pintelon et al. (1992) argued that MM is not an isolated process; its measures should depend on all factors related to MM, which should include not only internal factors of the organisation but its external factors as well. Nonetheless, the advantage of MM strategy measures in alignment with organisational business needs includes minimising indirect maintenance costs associated with production losses (Vagliasindi, 1989; Van Horenbeek et al., 2010). On the other hand, some researchers have measured MM by productivity measures. Generally, maintenance productivity measures describe the organisation-specific need required to attain transparency and uniformity amongst stakeholders, including all employees of the organisation (Parida & Kumar, 2009). Specifically, the maintenance productivity measures take into account the life cycle of each organisation's physical assets. According to Barberá et al. (2012), it ensures the reduction of overall production costs, the correct performance of equipment, the reduction in the level of risk, and the negative impacts on the environment.

Additionally, it generates activities and processes that support these objectives. Parida and Kumar (2009) noted that maintenance productivity measures various factors and issues that need to be considered, including the value created by the maintenance process, resources allocations, health, safety, and environmental (HSE) issues, management of knowledge, adoption of new operating and maintenance strategies and changes in organisational management and structure. Moreover, maintenance productivity measures are related to a powerful competitive factor of an organisation in which businesses can grow (Barberá et al., 2012).

Another issue with MM's study is that some researchers measured MM through maintenance re-engineering measures. According to Alzaben (2015), the basic concept of measuring maintenance re-engineering is the continuous improvement of the MM process. Moreover, the maintenance re-engineering measures take into account strategies for asset and human resources, monitoring and control of individual assets, maintenance performance measurement systems, planning and scheduling of maintenance activities, maintenance tactics, and the application of TPM and RCM for continuous improvement (Campbell, 1998; Alzaben, 2015). In measuring the maintenance re-engineering, the MM is incorporated with knowledge, intelligence, and analysis, which support maintenance decision-making for the continuous improvement of the MM process (Barberá et al., 2012).

Similarly, researchers also measured MM through maintenance performance measures. With fast changes in the organisation's business, maintenance performance has caught the imagination and involvement of researchers and managers alike from the maintenance organisation (Murthy, Atrens, & Eccleston, 2002). Andersen and Fagerhaug (2002) posit that maintenance performance measures can be viewed along three dimensions that include effectiveness (satisfaction of user needs), efficiency (economic and optimum use of organisational resources), and changeability (strategic plan to handle organisational changes). Parida (2006) states that maintenance performance measures are concerned with the multi-disciplinary process of justifying and measuring values through maintenance investment and meeting the organisation's stakeholder requirements based strategically on the overall business perceptions. Various issues that are considered in maintenance performance measures include PM and equipment history, information systems, management training, organisational policy and staffing, labour productivity, education, and technical training, staff motivation, management control and budgeting, work order planning, stores, material and tool control, engineering and condition monitoring, work measurement and incentives, and facilities (Raouf, 1994). Moreover, maintenance performance measures are important for a successful organisation to remain competitive and cost-effective in business (Parida & Kumar, 2009).

Also, some other researchers measure MM through value-driven maintenance (VDM) measures. According to Stenström, Parida, Kumar, and Galar (2013), the philosophy behind value-driven maintenance is to understand a delicate balance between the value that improved reliability brings and the cost of maintenance. Moreover, Olanrewaju and Abdul-Aziz (2015) assert that the philosophy behind value-driven maintenance measures is the alignment of building performance with organisational corporate strategy and maintenance resources with users' satisfaction. Additionally, value-driven maintenance measures ensure holistic consideration of MM processes, procedures, practices, and their implementation within the organisation (Olanrewaju & Abdul-Aziz, 2015). However, various issues in value-driven maintenance measures are based on the methodology developed on four value drivers in the maintenance process: resource allocation, asset utilisation, cost control, and health, safety, and environment (Kumar et al., 2013). Moreover, VDM measures' significance is optimising the value derived from maintenance at any particular point in time (Kumar et al., 2013).

Consequently, the question of how to measure MM efficiently has been an issue in establishing effective measuring tools for maintenance. Researchers' opinions differ on establishing methods or concepts that will combine measuring tools such as organisational maintenance strategy, maintenance productivity, maintenance re-engineering, maintenance performance, and value-driven maintenance. Nonetheless, Olanrewaju and Abdul-Aziz (2015) opined that MM requires a multi-disciplinary approach for effectiveness from technological, engineering, business, economic, social, and management perspectives. This also appears suitable for maintenance organisations as it seeks to capture the organisation's present, past, and future performance in meeting its business plans, maintenance objectives, equipment prioritisation, cost/benefit analysis, and users' requirements, among others.

Nevertheless, an integrated approach to measuring MM gives a more holistic framework for measuring MM. Thus, this study adopts an integrated approach. Consequently, the next section reviews and discusses previous MM models and theoretical frameworks to inform the integrated MM this present study seeks to develop for Municipal buildings in developing economies' education sector.

Maintenance Management Conceptual Models

This section focuses on the vital MM models for the study. The philosophy behind the model is explained and supported by its specific details. A MM model is a conceptual structure that allows different processes containing tasks/activities to be organised, managed, coordinated, and monitored systematically to facilitate the maintenance work to be completed satisfactorily. This section focuses on the vital theoretical models for the study.

Supporting Structure Models

Marquez and Gupta (2006) present a MM process and framework that suggest the alignment of the MM process with the three levels of business activities that include strategic, operational, and tactical levels. The model is backed by three basic supporting structural pillars: information technology (IT), maintenance, and organisational techniques. The framework at the strategic level focuses on transmuting business priorities into maintenance priorities. The tactical level of the framework emphasises the assignment of maintenance resources such as the test equipment, materials, and skills. Marquez and Gupta (2006) state further that the tactical level of the process requires a maintenance plan and task scheduling. At the operational level, emphasis is more on the maintenance tasks that will be executed by the skilled maintenance technicians based on scheduled time, correct procedure, and proper tool usage. Marquez and Gupta (2006) state that this level will require data to document maintenance activities through MIS. This will help provide information on the history of work done on each piece of equipment. However, the MM framework shows that the IT pillars would allow the maintenance manager and personnel to access all equipment data through CMMS. The data provided through the CMMS will be transformed into information that would be used to make decisions at the levels of the business activities and prioritise actions. It will allow control of assets and proper monitoring. The MM framework identified a set of key techniques that constitute the maintenance technique pillar. They include RCM, TPM, quantitative tools, tactical activity-oriented stochastic tools, and management techniques that focus on optimising maintenance resources management.

Decision-Making Model

Pintelon and Gelders (1992) suggest a MM model that contains three simple building blocks. The first block places the MM within the wider business perspective, where finance, marketing, and operation are the bases for key maintenance decisions. MM under this block is considered sub-functions of the maintenance operations within the scope of resources in maintenance. The second block in this model identified planning and control as the core elements, with sub-variables that include maintenance managers' decisions on major business functions (operations, marketing, and finance), performance reporting, and management of resources. The importance of training was also highlighted, such as training on repair, monitoring techniques, or techniques related to better maintenance design. This was emphasised to improve the maintenance personnel's knowledge for them to operate in a safe environment. Finally, the last block, the third block in this MM framework, is called the MM toolkit, which is the core element in this block. It consists of statistical tools with other OR/MS (Operations Research/ Management Science) techniques that guard against the occurrence of failures in the system statistically. The blocks focused on optimising maintenance resources management using CMMS to help optimise the policies and actions of the maintenance process.

Generic Maintenance Management Model

Fernández and Márquez (2009) suggest a generic MM model. The suggested framework contains a total of eight blocks in sequence organised to cover four functions that represent the core element of the framework, namely effectiveness, efficiency, assessment, and continuous improvement. The effectiveness in the maintenance model covers maintenance objectives and related KPIs to be described, the appropriate maintenance strategy specified, the asset to be prioritised, and a weak point with high impact to be acted upon. The efficiency in the model shows the optimisation and design of preventive maintenance plans that include resources schedule and resources. Assessment in the model focuses on maintenance control, execution, and monitoring and controlling. It further focused on replacement optimisation and asset life cycle analysis. Improvement in the model focuses on issues relating to continuous improvement through the integration of new techniques where applicable. The framework highlighted the significance of integrating new techniques and engineering tools with management concepts.

Life Cycle Maintenance Management Model

Takata et al. (2004) proposed an MM model where there is a bridge between the developing phase of maintenance activities and the operation phase connected by maintenance strategy planning. As Takata et al. (2004)

recommended, the life cycle maintenance model involves three feedback loops or stages. The first loop of the maintenance process, as designed in the framework, uses Deming's PDCA cycle (Plan-Do-Control-Action) to plan the maintenance and evaluate the results of the outcome of the task. The model's second loop, analysing the deterioration process and evaluating failure, involves selecting a maintenance strategy. The effectiveness of maintenance technology during the product's life cycle may be revised after evaluating the maintenance approach used. The third loop seeks to improve the design of equipment based on the 'lessons' learned previously.

Maintenance Strategy Model

Márquez (2007) developed a model for maintenance strategy toward achieving effective MM. The model emphasised that the maintenance strategy-setting process should follow standard organisational planning based on the policies and objectives derived from corporate goals, which should be communicated to all maintenance personnel, including external parties. Márquez (2007) pointed out that the maintenance strategy's objectives include equipment availability, reliability, safety, personnel training, and a maintenance budget. However, the model pointed out that the building's performance and the target performance measure (KPIs) should be determined based on the maintenance objectives. The model suggests continuous improvement based on accepted operation, user, and MM performance indicators. The model further notes that planning, execution, assessment, analysis, and improvement will guide the strategy implementation to achieve effective MM.

Risk Assessment Maintenance Management

The model was developed by Crespo Márquez et al. (2009). Crespo Marquez et al. (2009) postulated that using this model for MM will reduce the substantial maintenance costs, the indirect maintenance costs, production losses, environmental risk, and risk associated with maintenance and ultimately reduce users' and customers' dissatisfaction. The model shows that to carry out an asset's criticality analysis, the following risk assessment should be considered. They include the purpose and scope of the analysis should be defined, the risk factors to take into account, and the overall procedure for identifying and prioritising the critical assets should be established. The model posits further that assessing criticality will be specific to each system, plant, or business unit. The model also shows that effective asset assessment, asset prioritisation, and effective maintenance operations may be obtained when maintenance action is aligned to the business at any time. However, asset priority will be defined with a setup of the strategy to be used for each asset category.

The "Smart" Pyramid Model

The strategic measurement analysis and reporting technique (SMART) was developed due to dissatisfaction with traditional performance measures by Wang Laboratories (Lynch & Cross, 1995). The main objectives of this model were to devise an effective management system with performance indicators designed to define and sustain success. The model is represented by a four-level pyramid of objectives and measures. The first level in the model shows that for efficient operation, corporate vision or organisation strategy is necessary, whereby management assigns a corporate portfolio role to each unit with resources to support them. The second level indicates that the objectives for each operations unit are defined in the market and financial terms. The third level illustrates that an organisation's operating objectives and priorities should be defined in terms of customer satisfaction, flexibility, and productivity, which is a very important component of MM of buildings. The fourth level elucidates that customer satisfaction, flexibility, and productivity at departmental levels are represented by specific operational criteria such as quality, delivery, process time, and cost. In this model, the operational measures at the fourth level representing the foundation of the performance pyramid that guides organisational strategy and operation are the key factors to achieving higher-level results and ensuring successful implementation of the organisation's strategy.

Management Process Model

The management process model for the MM of buildings was developed by Vanneste and Van Wassenhove (1995). The model suggested that a maintenance structure should include two management processes. The two management processes include analysis of process effectiveness and analysis of process efficiency. The model postulated that an effective management process seeks to identify the most important problems in maintenance activities and identify their potential solutions. The efficiency management process focuses on identifying suitable procedures for maintenance operations. The model identified eight stages loops for evaluating the MM process towards achieving effectiveness and efficiency. The first stage in the model is to determine the maintenance structure's current performance, which includes planning, supervising, and monitoring. The second stage identified by the model is to analyse the quality and downtime problems achieved through an organisational maintenance policy. However, the third stage of the model shows that there should be an effective analysis of a potential solution to maintenance problems through continuous improvement. The fourth stage of the model indicates that an efficient analysis of maintenance procedures should be achieved through a suitable maintenance approach. The fifth phase of the model also identifies planning and execution as a stage required for evaluating the MM process, which can be achieved through planning and scheduling. Stages six, seven, and eight of the models place more emphasis on the importance of data collection and implementation actions, data processing and monitoring, and effective information handling procedures, whereas all of this can be achieved through CMMS, which can be used for proper information gathering, processing, and sharing towards achieving effective and efficient MM.

Integrated Maintenance Management model

Hassanain, Froese, and Vanier (2001) suggest a generic framework for integrating the MM for built and in-use assets. According to the framework for effective and efficient MM, five sequential management steps are necessary. These include asset identification, identification of the performance requirements of the asset, assessment of the current performance of the assets, an asset maintenance plan, and maintenance operations management and control. However, the model shows the need to define the required maintenance objectives and effective development of MIS for information and effective maintenance operations. The model recognised asset identification, asset performance, performance expected of the asset identified, and a value-driven management system as vital for effective MM.

Wireman Maintenance Management

Wireman (2005) proposes a successive implementation of steps to ensure that all maintenance functions are in order. The model shows that a preventive maintenance programme should be in place before advancing to the next level of maintenance activities. The model posits further that before one considers the implementation of RCM and predictive maintenance programmes, there should be CMMS implementation with a suitable work order release system (to schedule appropriate prioritised tasks). There should also be a provision of spare parts and maintenance personnel training (maintenance resources management system). The model states further that there is a need for the implementation of TPM. As postulated by this model, TPM would considerably help achieve operator involvement and routinising optimisation techniques. TPM would also help guide configuring a necessary maintenance organisation structure and applying statistical tools for financial optimisation. However, the model signified continuous improvement in maintenance practices to achieve an effective MM system.

Campbell Maintenance Management

Campbell (1998) proposed a formal structure model for effective MM. The model starts by identifying maintenance strategies for the asset and stressing the associated human resources-related aspects required to produce the needed working culture. The model shows that the organisation needs monitoring and supervising to ensure the functionality of each asset throughout its life cycle. This can be achieved through implementing a CMMS, a maintenance function measurement system, and planning and scheduling maintenance activities. The model posits that depending on the value that assets represent and the risks they entail for the organisation, it is accomplished according to one or more of the following eight tactics: re-dundancy, run to failure, scheduled overhaul, scheduled replacement, ad-hoc maintenance, preventive maintenance (use-based or either age-based), condition-based maintenance, and redesign if necessary. To realise con-tinuous improvement, the model suggests using two highly successful maintenance methods, TPM and RCM. The model signifies the use of process re-engineering techniques to maintain the top level of improvements already achieved in the maintenance process (Table 4.1).

Table 4.1 Individual conceptualisation of maintenance management model

Authors	Models	Conceptualisation
Marquez and Gupta (2006)	Supporting Structure Pillars Model	The key elements identified in the model to achieve effective MM of buildings are MIS, outsource strategy, organisational maintenance policy, human resources management, monitoring and supervision, training, and maintenance budget.
Pintelon and Gelders (1992)	Decision-making Model	The core element identified by the model includes task planning and scheduling, monitoring and supervision, MIS, maintenance budget, human resources management, monitoring and supervision, maintenance strategy, and organisational maintenance policy.
Fernández and Márquez (2009)	Generic Maintenance Management Model.	The core elements in the model include organisational maintenance policy, monitory and supervision, maintenance budget, human resources management, and training
Takata et al. (2004)	Life Cycle Maintenance Management Model	The key element in this model includes spare part management, monitory and supervision, outsourcing strategy, training, human resources management, and maintenance budget
Márquez (2007)	Maintenance Strategy Model	The model dimensions are MIS, monitoring and supervision, organisational maintenance policy, continuous improvement, maintenance budget, and human resources management.

(Continued)

Table 4.1 (Continued)

Authors	Models	Conceptualisation
Marquez et al. (2009)	Risk Assessment Maintenance Model	In relation to effective MM, the key elements identified are organisational maintenance policy, monitoring, and supervision, maintenance budget, human resources management, user need, and satisfaction.
Lynch (1995)	The "Smart" Pyramid Model	For effective MM, the dimension of the model is organisational maintenance policy, maintenance strategy, human resources management, maintenance budget, training, users' need, and expectation.
Vanneste and Van Wassenhove (1995)	Management Process Model	The key dimension in this model concerning MM effectiveness and efficiency are task planning and scheduling, monitoring & supervision, organisational maintenance policy, continuous improvement, training, maintenance approach, and MIS.
Hassanain et al. (2001)	Integrated Maintenance Management Model	The key element in this model includes human resources management, organisational maintenance policy, maintenance budget, training, monitoring, and supervision.
Wireman (2005)	Wireman Maintenance Management Model	The key dimensions in this model include organisational maintenance policy, MIS, provision of spare parts; training; maintenance approach; human resources management; maintenance budget; monitoring, and supervision.
Campbell (1998)	Campbell Maintenance Management Model	The key elements in this model include organisational maintenance policy, human resources management, monitory and supervision, MIS, maintenance budget, and training.

Source: Researcher's compilation (2022) as reviewed from the literature.

Drivers of Effective Maintenance Management of Buildings

In an attempt to sustain the effective MM of buildings, several concepts were developed, as seen in all the models discussed above. According to Marques and Gupta (2006), the alignment of the MM process with the three levels of business activities (strategic, tactical, and operational) is backed by three basic supporting structure pillars (information technology [IT], maintenance technique, and organisational techniques) will improve the MM process. Pintelon et al. (1992) posit that key decisions should guide the maintenance planning, operation, and techniques to achieve effective and efficient buildings. Márquez and Gupta (2006) postulated that MM operations should be guided by key factors such as effectiveness, efficiency, assessment, and continuous improvement. Takata et al. (2004) state that for MM to be effective, there is a need for a link between the developing phase of maintenance activities and the operation phase connected by maintenance strategy and planning. In addition, Márquez (2007) informed that effective MM should be based on a maintenance strategy-setting process that should follow standard organisational planning methods, while Fernández et al. (2009) posited that effective MM maintenance actions should be aligned with business targets. Lynch and Cross (1995) asserted that performance indicators based on an organisation's maintenance plan and objectives would drive effective MM.

Similarly, Vanneste and Van Wassenhove (1995) opined that for the attainment of effective MM of buildings, the management process should identify the most important problems in maintenance activities as well as their potential solutions through planning and scheduling, supervising and monitoring, policy deployment, and organisation. While the efficiency management process should focus more on identifying suitable procedures for maintenance operations through using a suitable maintenance approach and CMMS for data gathering and processes and continuous improvement. Also, Hassanain et al. (2001) posited that asset identification, asset performance, the identified performance expected of the asset, maintenance planning, and operations management and control are vital for effective MM. Wireman (2005) postulated that for effective MM, there is a need for a preventive maintenance programme, CMMS with suitable work order, spare parts availability together with maintenance personnel training, and continuous improvement of the maintenance process. Campbell (1998) asserted the need for developing strategies for assets, human resources management, CMMS implementation, a measurement system for maintenance functions, planning and scheduling, and suitable continuous improvement arrangements within the maintenance operation and activities.

Hence, from all the models reviewed, no single model is holistically sufficient to contain all the key drivers of effective MM. The key drivers identified in all the models reviewed were related to a particular industry, system, or nation, which may have a different economic and cultural value that differs from that of developing economies. Equally, organising the various key drivers of effective MM in a single study to ascertain their relative contribution empirically is a novel approach. However, the study carried out a qualitative inquiry using an adoptive mode into the key drivers of effective MM of municipal buildings in the developing economies' educational sector as well as empirically ascertaining the relative contribution of the key drivers of MM of municipal buildings in the developing economies' educational system.

Summary

In an attempt to achieve the aim of the study, this chapter describes the problems posed in the research, raises theoretical and methodological queries that have affected MM research, and suggests possible solutions that guide this present study. From the literature review, one of the most outstanding findings in this chapter revealed that the Wireman MM model (2005) and the Campbell MM model (1998) were the most comprehensive conceptual models for effective MM of buildings, on which many studies on MM have been based. Similarly, findings from the reviewed literature on the existing conceptualised models on the MM of buildings identified various key dimensions in achieving an integrated MM of municipal buildings. These core elements or independent variables include an organisational maintenance policy, monitoring and supervising, maintenance budget, human resources management, training, MIS, planning and scheduling, a maintenance approach/strategy, users' needs, and continuous maintenance process improvement. Finally, each of the core elements identified through the review of the existing models has its indicators or constructs. These have provided the base for the next chapter, which builds on the established core element in the gap identified from the existing MM models, especially for the developing economies context, which this study is positioned to address.

References

Ahuja, I. P. S., & Khamba, J. S. (2007). An evaluation of TPM implementation initiatives in an Indian manufacturing enterprise. *Journal of Quality in Maintenance Engineering, 13*(4), 338–352.

Aigbavboa, C. O. (2014). *An integrated beneficiary-centered satisfaction model for publicly funded housing schemes in South Africa* (Published doctoral dissertation, University of Johannesburg, Johannesburg).

Alzaben, H. (2015). *Development of a MM framework to facilitate the delivery of healthcare provisions in the Kingdom of Saudia Arabia* (Published doctoral dissertation, Nottingham Trent University).

Andersen, B., & Fagerhaug, T. (2002). Eight steps to a new performance measurement system. *Quality Progress, 35*(2), 112.

Bagadia, K. (2006). *Computerized maintenance management systems made easy*. New York: McGraw-Hill Professional.

Barberá, L., Crespo, A., Viveros, P., & Stegmaier, R. (2012). Advanced model for maintenance management in a continuous improvement cycle: Integration into the business strategy. *International Journal of System Assurance Engineering and Management, 3*(1), 47–63.

Ben-Daya, M., Duffuaa, S. O., Raouf, A., Knezevic, J., & Ait-Kadi, D. (Eds.). (2009). *Handbook of maintenance management and engineering* (Vol. 7). London: Springer.

Blessing, O., Richard, J., & Emmanuel, A. (2015). Assessment of building MM practices of higher education institutions in Niger State – Nigeria. *Journal of Design and Built Environment, 15*(2).

British Standard [B.S.] (2001). *Maintenance terminology*. EN 13306.

Campbell, J. D. (1995). Outsourcing in maintenance management: A valid alternative to self-provision. *Journal of Quality in Maintenance Engineering, 1*(3), 18–24.

Campbell, J. D. (1998). *Uptime, strategies for excellence in maintenance management*. Portland, OR: Productivity Press.

Campbell, J. D., & Jardine, A. K. (2001). *Maintenance excellence: Optimizing equipment life-cycle decisions*. New York: Marcel Dekker Inc.

Centre for Facilities Management. (1992). *An overview of the facilities management industry* (Part 1). Strathclyde: Strathclyde Graduate Business School.

Cholasuke, C., Bhardwa, R., & Antony, J. (2004). The status of maintenance management in UK manufacturing organisations: Results from a pilot survey. *Journal of Quality in Maintenance Engineering, 10*(1), 5–15.

Crespo Márquez, A., Moreu de León, P., Gómez Fernández, J.F., Parra Márquez, C., & González, V. (2009). The maintenance management framework: A practical view to maintenance management, Safety, Reliability and Risk Analysis: Theory, Methods and Application. Taylor & Francis Group, London.

Daragh Naughton, M., & Tiernan, P. (2012). Individualising maintenance management: A proposed framework and case study. *Journal of Quality in Maintenance Engineering, 18*(3), 267–281.

Dhillon, B. S., & Subramanian, P. (2001). Reliability analysis of triple modular computer systems with redundant voters and restricted maintenance. *Journal of Quality in Maintenance Engineering, 7*(2), 151–164.

Duffuaa, S. O. (2000). Mathematical models in maintenance planning and scheduling. *In Maintenance, modeling, and optimization* (pp. 39–53). Boston, MA: Springer.

Fernández, J. F. G., & Márquez, A. C. (2009). Framework for implementation of maintenance management in distribution network service providers. *Reliability Engineering & System, 94*(10), 1639–1649.

Goyal, R. K., & Maheshwari, K. (2012). Maintenance management practices: A retrospective and literature review. *International Journal of Advances in Engineering Research, 3*(2), 1–18.

Hassanain, M. A., Froese, T. M., & Vanier, D. J. (2001). Development of a maintenance management model based on IAI standards. *Artificial Intelligence in Engineering, 15*(2), 177–193.

Jonsson, P. (2000). Towards a holistic understanding of disruptions in operations management. *Journal of Operations Management, 18*(6), 701–718.

Karia, N., Asaari, M. H. A. H., & Saleh, H. (2014). Exploring maintenance management in service sector: A case study. In: *Proceedings of International Conference on Industrial Engineering and Operation Management*, Bali, 7–9 January 2014, pp. 3119–3128.

Kumar, U., Galar, D., Parida, A., Stenström, C., & Berges, L. (2013). Maintenance performance metrics: A state-of-the-art review. *Journal of Quality in Maintenance Engineering, 19*(3), 233–277.

Kym, F. (2015). Maintenance management models – A study of the published literature to identify empirical evidence. *International Journal of Quality and Reliability Management, 32*(6), 635–661.

Lateef, O. A. (2009). Building maintenance management in Malaysia. *Journal of Building Appraisal, 4*(3), 207–214.

Lynch, R. L., & Cross, K. F. (1995). *Measure up: How to measure corporate performance.* Malden, MA: Blackwell.

Márquez, A. C. (2007). *The maintenance management framework: Models and methods for complex systems maintenance.* London: Springer-Verlag.

Márquez, A. C., de León, P. M., Fernandez, J. F. G., Marquez, C. P., & Campos, L. M. (2009). The maintenance management framework: A practical view to maintenance management. *Journal of Quality in Maintenance Engineering, 15*(2), 167.

Marquez, A. C., & Gupta, J. N. (2006). Contemporary maintenance management: Process, framework, and supporting pillars. *Omega, 34*(3), 313–326.

Mc Kone, K. E., Schroeder, R. G., & Cua, K. O. (2001). The impact of total productive maintenance practices on manufacturing performance. *Journal of Operations Management, 19*(1), 39–58.

Mekasha, E. (2018). *Maintenance management framework development for competitiveness of food and beverage industry: A case study on Asku Plc.* (Thesis Draft, Addis Ababa Institute of Technology, Addis Ababa University).

Murthy, D. N. P., Atrens, A., & Eccleston, J. A. (2002). Strategic maintenance management. *Journal of Quality in Maintenance Engineering, 8*(4), 287–305.

Nakajima, S., Conforti, M., & Rubbia, S. (1997). *Introduction to TPM: Total productive maintenance.* Cambridge, MA: Productivity Press Inc.

Ogunbayo, B. F., & Aigbavboa, O. C. (2019). Maintenance requirements of students' residential facility in higher educational institution (HEI) in Nigeria. In *IOP conference series: Materials science and engineering* (Vol. 640, No. 1, p. 012014). IOP Publishing.

Ogunbayo, B. F., Aigbavboa, C. O., Thwala, W., Akinradewo, O., Ikuabe, M., & Adekunle, S. A. (2022). Review of culture in maintenance management of public buildings in developing countries. *Buildings, 12*(5), 677.

Ogunbayo, B. F., Ohis Aigbavboa, C., Thwala, W. D., & Akinradewo, O. I. (2022). Assessing maintenance budget elements for building maintenance management in Nigerian built environment: A Delphi study. *Built Environment Project and Asset Management, 12*(4), 649–666.

Ogunbayo, B., Aigbavboa, C., & Thwala, D. (2022). Occupants' maintenance requests in higher educational institution (HEI) residential buildings. In AIP Conference Proceedings (Vol. 2437, No. 1, p. 020146). AIP Publishing LLC.

Okolie, K. C. (2011). *Performance evaluation of buildings in educational institutions: A case of Universities in South-East Nigeria* (Published doctoral dissertation, Nelson Mandela Metropolitan University, Port Elizabeth, South Africa).

Olanrewaju, A. L., & Abdul-Aziz, A. R. (2015) Building maintenance processes, principles, procedures, practices, and strategies. In *Building maintenance processes and practices* (pp. 79–129). Singapore: Springer.

Omar, M. F., Ibrahim, F. A., & Omar, W. M. S. W. (2017). Key performance indicators for maintenance management effectiveness of public hospital building. In *MATEC Web of Conferences*, Ho Chi Minh City, Vietnam, 5–6 August 2016, pp. 01056. (Vol. 7). EDP Sciences.

Parida, A. (2006). *Development of a multi-criteria hierarchical framework for maintenance performance measurement: Concepts, issues, and challenges* (Doctoral dissertation, Luleå tekniska universitet).

Parida, A., & Kumar, U. (2009). Maintenance productivity and performance measurement. In *Handbook of maintenance management and engineering* (pp. 17–41). London: Springer.

Parida, A., Kumar, U., Galar, D., & Stenström, C. (2015). Performance measurement and management for maintenance: A literature review. *Journal of Quality in Maintenance Engineering, 21*(1), 2–33.

Pintelon, L. M., & Gelders, L. F. (1992). Maintenance management decision making. *European Journal of Operational Research, 58*(3), 301–317.

Puķīte, I., & Geipele, I. (2017). Different approaches to building management and maintenance meaning explanation. *Procedia Engineering, 172,* 905–912.

Raouf, A. (1994). Improving capital productivity through maintenance. *International Journal of Operations & Production Management,* 14(7), 44–55.

Royal Institution of Chartered Surveyors (RICS) (1993) *CPD – Review of Policy and Future Strategy.* London: RICS.

Stenström, C., Parida, A., Kumar, U., & Galar, D. (2013). Performance indicators and terminology for value driven maintenance. *Journal of Quality in Maintenance Engineering,* 19(3), 222–232.

Takata, S., Kimura, F., van Houten, F. J., Westkamper, E., Shpitalni, M., Ceglarek, D., & Lee, J. (2004). Maintenance: Changing role in life cycle management. *CIRP Annals, 53*(2), 643–655.

Vagliasindi, F. (1989). *Gestire la manutenzione: perché e come.* Milano: FrancoAngeli.

Van Horenbeek, A., Pintelon, L., & Muchiri, P. (2010). Maintenance optimization models and criteria. *International Journal of System Assurance Engineering and Management, 1*(3), 189–200.

Vanneste, S. G., & Van Wassenhove, L. N. (1995). An integrated and structured approach to improve maintenance. *European Journal of Operational Research, 82*(2), 241–257.

Waeyenbergh, G., & Pintelon, L. (2002). A framework for maintenance concept development. *International Journal of Production Economics, 77*(3), 299–313.

Waeyenbergh, G., & Pintelon, L. (2009). CIBOCOF: A framework for industrial maintenance concept development. *International Journal of Production Economics, 121*(2), 633–640.

Wireman, T. (1990). *World class maintenance management.* New York, NY: Industrial Press Inc.

Wireman, T. (2005). *Developing performance indicators for managing maintenance.* New York: Industrial Press Inc.

Wordsworth, P., & Lee, R. (2001). *Lee's building maintenance management.* London: Blackwell Science.

Zhu, G., Gelders, L., & Pintelon, L. (2002). Object/objective-oriented maintenance management. *Journal of Quality in Maintenance Engineering, 8*(4), 306–318.

5 Maintenance Management Research Theories

Introduction

This chapter presents a framing overview of maintenance management (MM) research, highlighting some theories with complementary perspectives on the MM of buildings, followed by a discussion on maintenance policy, the evolution of maintenance, legal policy framework, and forms of maintenance policy. Also, the objectives of the maintenance policy are assessed. The chapter finally closes with an outline of maintenance policy instruments that enable the intentions of a maintenance policy to be realised.

Theoretical Framework for Maintenance Management Research

An MM framework is essentially a conceptual structure that allows various processes containing activities to be organised, coordinated, monitored, and managed in a systematic way to enable the work of maintenance to be completed satisfactorily (Campbell 1998; Wireman, 2005). Lateef (2009) postulated that the MM of buildings involves planning, directing, controlling, and organising maintenance processes and services to obtain maximum returns for the investment and meet the maintenance needs of users of the building. Nevertheless, Mekasha (2018) states that the stimulus of maintenance operations and quality is moving from man to machine owing to the increased technological advancements. Phogat and Gupta (2017) opined that implementing a well-developed and organised MM framework may improve maintenance operation and quality. Regardless of the terminology used, the central theme for MM research is the study of a preventive management philosophy (Gökçe, Stack, Gökçe & Menzel, 2009). Karia, Asaari, and Saleh (2014) postulated that the preventive management philosophy is considered a business function that provides opportunities to retain the quality, life, and value of assets and improve cost, risk, and productivity concerns in organisations. Alzaben (2015) states that, unlike other management techniques focusing mainly on deploying resources to achieve defined business goals, MM seeks to ensure operations and associated equipment are properly functioning.

DOI: 10.1201/9781003344681-5

On the other hand, MM theoretical framework guides equipment availability and facility sustainability (Mukelasi, Zawawi, Kamaruzzaman, Ithnin & Zulkarnain, 2012; Ogunbayo & Aigbavboa, 2019). Availability in the context of maintenance operation means maintenance personnel can demand and receive any equipment required for maintenance when requested, while sustainability means an effective MM system in meeting organisational objectives and meeting users' maintenance needs. Based on this, Bagadia (2006) suggested that a viable MM theoretical framework for buildings will help realise minimal downtime.

Over the years, many new MM frameworks have been developed for effective building maintenance (Mukelasi et al., 2012). Mekasha (2018) posits that the MM framework as a management strategy is somewhat underdeveloped compared to other management disciplines. However, the concept of MM is largely traced to Wireman (2005), who emphasised the need for a preventive maintenance programme, CMMS with proper work order, spare parts availability together with maintenance personnel training which is vital for the implementation of the total productive maintenance. Other MM concepts, as suggested by Campbell (1998), Márquez (2007), and Waeyenbergh and Pintelon (2002), were all concerned with the business aspect of the MM organisation. Furthermore, Sharma (2013) stated that it is widely believed that a lack of funds causes MM problems, and the quality concept is a major difficulty in managing buildings. Bagadia (2006) noted that the basic steps common to most identified existing MM concepts are the plan, schedule, request, approval, performance work, data recording, cost accounting, developing management information, providing management control reports, and updating equipment history. Naughton and Tiernan (2012) and Zhu, Gelders, and Pintelon (2002), in their studies on the validation of an effective framework for maintenance operations, observed that there is no one-fits-all solution among the literature published around generic commercial frameworks for maintenance. However, Kolawole (2002) stated that viable MM requires the correct diagnosis of defects, current remedial measures, thorough technical knowledge of material usage of management resources, and the formulation and implementation of integrated plans and policies to sustain utility. The absence of these qualities has led to the decay of the nation's physical, social, aesthetic, and economic environment. The earlier attributes of MM focused on cost control, repair skill, scheduled overhaul, employment management techniques, and the business management of maintenance to bring flexibility to organisations in a continually changing climate with closer integration of facilities and a more appropriate focus on the user and strategic needs that lead to important business advantages.

However, in contemporary studies, the attributes of MM have taken several dimensions, which include the individual factor (such as monitory and supervision), technical aspects (such as maintenance approach and outsourced strategy), and administrative, organisational factors (such as policy deployment and organisation) as they have been established to have contributed to the MM of the building (Omar, Ibrahim, and Omar, 2017). Thus, there has been a gradual shift in the development of a MM concept based not only on cost and quality but also on human behaviour, attitude (culture), and interaction (communication), which might foster an effective MM concept. Around the world, each year, the lack of effective MM developed based on human behaviour and attitude has led to billions of dollars being spent on equipment and facility maintenance (Mekasha, 2018; Ogunbayo et al., 2022). However, no single theoretical framework adequately explains MM (Söderholm, 2005).

This present research further explores the theoretical framework of MM studies. Thus, this MM study considers the total quality management (TQM) theory, the balanced scorecard (BSC), the requirement management theory, and the change management theory.

Total Quality Management Theory

The word "total" in organisational management, according to Kermally (1996), means that everyone in the organisation, all processes, systems, levels of management, and employees, must be involved in satisfying the users of maintained facilities. In other words, "quality," as Garvin (1998) opined, is an unusually elusive concept, easy to visualise and yet exasperatingly difficult to define. The diverse conceptions of "quality" have brought several and sometimes unharmonious definitions. Thus, Wilkinson (1998) defines quality as follows:

• Conformance to standard;
• Excellence;
• Fitness for use;
• Meeting or exceeding users' needs or expectations;
• Zero defects; and
• Right the first time.

The International Standards Organization 8402 (ISO 1986) describes "quality" as the totality of features and characteristics of a product or service that bear on its ability to meet a stated or implied need. The ISO 8402 (1986) recognised that users' needs could be defined in terms of safety, availability, versatility, usability, reliability, maintainability, compatibility with other products, overall cost (including maintenance cost, product life, and purchasing price), environmental impact or other desired characteristics. Some key phrases about quality, according to Crosby (1992), mean conformance to requirements, while Juran (1992: 28) posited that "Quality is fitness for use. However, Deming (2018) states that quality should be aimed toward present and future users' needs and expectations. Taguchi (1995) observed that lack of quality in the maintenance of buildings would affect users' productivity and output.

Furthermore, Taguchi (1995) states that facilities for human habitation should follow a quality process, which should include the production process and the quality of the material used. This vision is related to today's concept of sustainable development as pronounced by the World Commission on Environment and Development (WCED, 1987).

Kast (1985) posited that the word "management" is an activity that has been practiced for as long as man existed. According to Cole and Kelly (1996), to manage is to plan, forecast, command, coordinate, organise, and control. Anastasiadou (2015) more recently posits that researchers have spread the word that management should have ways of measuring quality. Equally, for Macdonald and Piggott (1993), the term "management" recognises that total quality management (TQM) is not an accidental phenomenon of any orga- nisation's activities. It is a managed process that involves people, systems, and supporting tools and techniques (Deming, 2018). It further indicates that continuous quality improvement must be planned, measured, and controlled (Deming, 1994). TQM was well-thought-out as a management tool in the manufacturing and services industry and was widely used. In the construction industry, especially in the maintenance of built facilities, it is only considered an effective tool in recent times. It is a concept in which organisation, process, and people (users) are interwoven in the correct format so that the right thing is done accurately at a precise time (Macdonald Piggott, 1993).

In the same way, TQM emphasises innovation and the adoption of new technology to improve quality. It is an approach that seeks to enhance quality and performance to meet user needs and expectations (Deming, 2018). TQM is a theory-based option that allows managers to reward truly exceptional performance while increasing the capacity for cooperation and involvement of the entire system (Wilkinson, 1998). It deals with clients' satisfaction (users) at all levels of MM of an organisation.

Equally, Bamisile (2004) posited that TQM is a set of philosophies through which a management system can efficiently achieve the organisa- tion's objectives to ensure users' satisfaction and maximise stakeholder value. TQM looks at the overall quality measures used by an organisation, including managing quality design and development, quality control and maintenance, quality improvement, and quality assurance. As Harris (2001) described, TQM is based on the philosophy of continuously improving goods and services that everyone in an organisation should be involved in and committed to from top to bottom. Harris (2001) further posits that for successful TQM, an organisation should ensure that its services meet the following measures:

- Supply a quality product that is so much better than the competition that clients want regardless of the price;
- Be fit for purpose on a consistently reliable basis; and
- Delight the customer with the services which accompany the quality of production.

The level of quality in maintenance organisations involves understanding and implementing quality management principles and concepts in every aspect of the maintenance of buildings. According to Dahlgaard, Reyes, and Chen (2018), TQM demands that the principles of quality management must be applied at every level, every stage, and department of the maintenance process or operation of buildings. They further posited that applying sophisticated quality management techniques must enrich the idea of the TQM philosophy. The method of TQM should go beyond the internal organisation to develop close collaboration with the maintenance system and operations (Dahlgaard et al., 2018). In the MM of buildings, quality delivery of maintenance work is not a luxury but a critical implementation of the maintenance plans with adequate supervision and total quality of work to ensure the success of maintenance operations. Any deviation from a design or specification implies a reduction in quality (Söderholm, 2005). Garvin (1998) points out that it is beneficial to be aware of different approaches to quality, even though specific approaches are probably focused on various phases of developing a product or for different departments in an organisation.

Evolution of Total Quality Management

The concept of quality has been in existence since the First World War; nonetheless, its meaning has evolved and changed. In the early 20th century, quality management meant inspecting products to ensure they met specifications (Dahlgaard et al., 2018). Throughout the Second World War, quality became more statistical: statistical sampling techniques were used to monitor quality and quality control charts in the production process (Harris, 2001). However, in the 1960s, with the help of the so-called "quality gurus," the quality concept took on a comprehensive meaning. Quality began to be viewed as something that should comprise the entire organisation, not just the production process. Since all functions were responsible for product quality and all shared the costs of poor quality, quality was seen as a concept that affected the entire organisation. This changes the meaning of quality radically. On the other hand, quality management was still viewed as something that needed to be inspected and corrected (Dahlgaard et al., 2018). From this point, quality management began to have a strategic meaning.

Today, a successful organisation, including maintenance organisations, understands that quality provides a competitive advantage. Most successful organisations prioritise their customers or users of their products and define quality as meeting or exceeding users' needs and expectations. Organisations, especially property and facility management organisations, focus on improving quality to be more competitive. In many of these organisations, quality and excellence have become a standard for doing business. Organisations that do not meet this standard will not survive. However, the term used for today's concept of quality is TQM. According to Feigenbaum (1983), the evolution and concept of quality control started in 19th century and spanned through the 20th century.

Nevertheless, the early people of the 1900s who presented the fundamentals of the TQM philosophy suggested concepts such as "client satisfaction," "fairness," "value," "associate training," and reward for performance to be managerial bases for business (Fisher & Nair, 2009: 7). The historical evolution of quality management may be described in different ways. One common description, as posited by Garvin (1998) and Dale, Van der Wiele, and Van Iwaarden (1999), is made up of four stages (see Table 5.1) that follow each other. These stages can be categorised as follows:

- Quality inspection,
- Quality control,
- Quality assurance, and
- TQM.

Table 5.1 Characteristics of quality stages

Quality stages	Characteristic
Quality Inspection (1910s)	Identify the source of non-conformance
	Correction
	Sorting
	Salvage
Quality Control (1920s)	Third-party approval
	Use of statistics
	Quality training
	Product testing
	Self-inspection
	Performance data
	Quality manual
Quality Assurance (1950s)	Third-party approval
	Process control
	Quality cost
	Quality manual
	Quality planning
	System audits
Total Quality Management (1980s)	Management leadership
	Company-wide application
	Prevention
	Performance measure
	Internal customer
	Continuous improvement
	Focused vision

Source: Dahlgaard et al. (1998).

Quality Inspection

This was the first directly linked concept of TQM that was ever introduced. At this stage, quality inspection (QI), the focus was on inspecting some critical characteristics of finished products relative to stated requirements. Accordingly, Garvin (1998) opines that QI is a simple inspection-based system where traditionally, an organisation would employ people to examine and check the work of others. As Garvin (1998) postulated, the basics of this system were that any work below a certain standard would be found before reaching a customer. Consequently, necessary documents were proofread for errors, and mistakes would be corrected when found. According to McCabe and Wilkinson (1998), the problem with this quality check process is that the checkers often fail to identify poor quality work, leaving the end-users with the consequence.

Quality Control

This is the result of the first change order that came about at the time of the Second World War when technology was becoming more complex and costly in terms of people and equipment relied heavily on inspection. As stated by McCabe and Wilkinson (1998), under this system, documentation control and product testing become a greater process control, and failure is reduced. At this stage, the production process's quality control (QC) characteristics were inspected at some appropriate time interval and compared to the process's inherent variation (Söderholm, 2005). Typical of such processes were data collection on performance, information feedback, and the beginning of self-inspection. In this stage, QC was carried out on a sampling basis dictated by statistical methods. However, at this stage, the responsibility for quality was mainly located in the manufacturing and engineering departments.

Quality Assurance

Quality assurance (QA) is the third stage; however, as postulated by Söderholm (2005), the whole production chain in this stage, from design to market, and the contribution of all functional departments are considered to prevent failures. This was a move away from product quality to system quality. Quality assurance was developed to certify that specifications are constantly met. It also creates sufficient confidence that a product, service, or structure would satisfy and meet the users' needs and expectations. Dahlgaard et al. (2018) posit that at this stage, the emphasis was on preventing bad quality rather than detecting activities. This stage is closely connected to issues related to responsibilities, routines, and organisations established in standards such as QS 9000 and ISO 9000 (Kartha, 2004). However, top management only involves a limited degree (Kartha, 2004).

Total Quality Management

This level of quality involves understanding and implementing quality management principles and concepts in every aspect of the organisational process. TQM is the fourth stage of quality evolution where everyone in the organisation is considered responsible for the quality, and top management exercises strong and committed leadership. Garvin (1998) and Dale et al. (2007) postulated that TQM widens the focus and emphasises the market and end-users needs and expectations. Equally, Dahlgaard et al. (1998) pronounced that quality management demands that the principle of quality management must be applied at every level, every stage, and department of any organisation, especially facility and property provider firms. Additionally, Levy (2006) said that TQM encompasses elements that form the basis of an organisation's wide focus on quality-element that extends from office to the field. However, James (1996) emphasised that typical of an organisation going through a total quality process would be a clear and unequivocal vision, excellent customer relations, time spent on training, and the realisation that quality was not just product but the quality of the organisation as a whole. Accordingly, Harris and McCaffer (2001) state that TQM is an important step beyond the mere implementation of quality assurance, quality, or quality control procedures to the management-led philosophy of providing complete user satisfaction.

In later years, Kroslid (1999) described another evolution of quality management. He stated that there are two different schools of thought on quality management contrary to the single evolutionary path described by, for instance, Dale et al. (1999) and Garvin (1998). The first school of thought is the Deterministic School of Thought, while the other is the Continuous Improvement School of Thought. These two schools form two paths of the new evolution of quality management (QM). The Deterministic School of Thought may be related to QI and QA. This part of QM started with Frederick W. Taylor at the end of the 19th century. The idea was developed roughly via Crosby (1992) and was later succeeded by British and American military standards, which later became the base for ISO 9000. Kroslid (1999: 41) postulated that the Deterministic School of Thought is specified as "evolving ... around a deterministic view of reality with a belief in the existence of one best way" This means conformance by standards is the best way to meet customer requirements.

The Continuous Improvement School of Thought, which focuses on improvement and variation, may be seen as related to QA and TQM. This part of QM can be traced back to Walter A. Shewhart in the 1930s (Kroslid, 1999). Bhuiyan and Baghel (2005) observed that continuous improvement over the years had become a frequently used terminology in the industry. Caffyn (1999) posited that continuous improvement is being referred to as a by-product of existing quality conduct like TQM by some scholars, and some express it as a completely new approach to enhancing creativity and gaining competitiveness. Nevertheless, Sanchez and Blanco (2014: 992) postulated that William Edwards Deming is the pioneer who defined continuous improvement as "improvement initiatives that increase successes and reduce failures." Bessant, Caffyn, Gilbert, Harding, and Webb (1994: 19) described the continuous improvement concept as a process and defined it as "a company-wide process of focused and continuous incremental innovation." Kaye and Anderson (1999: 492) reformulated the above definition. They stated that "continuous improvement is the planned, organised and systematic process of ongoing, incremental and company-wide change of existing practices designed at improving company performance."

However, Maschek, Khazrei, Hempen, and Deuse (2011) described the continuous improvement process as a frame of thought and the proceeding actions for targeted, formalised, and progressive improvement of activities. Furthermore, Zollo and Winter (2002) state that continuous improvement is a learned and stable pattern of collective action through which the organisation systematically generates and modifies its operating routines in pursuit of improved effectiveness toward achieving set goals.

The view and naming of quality management also differ between different descriptions, which can probably be explained by the different schools of quality management, as stated earlier. However, researchers including Shiba, Graham, and Walden (1993), Dean and Bowen (1994), and Hellsten and Klefsjö (2000) have all suggested a system approach to QM. As Hellsten and Klefsjö (2000) stated, QM could be described as a management system aiming to increase both external and internal users' satisfaction with reduced resources. This management system consists of three interdependent elements, as shown in Figure 5.1. They include values, methodologies, and tools.

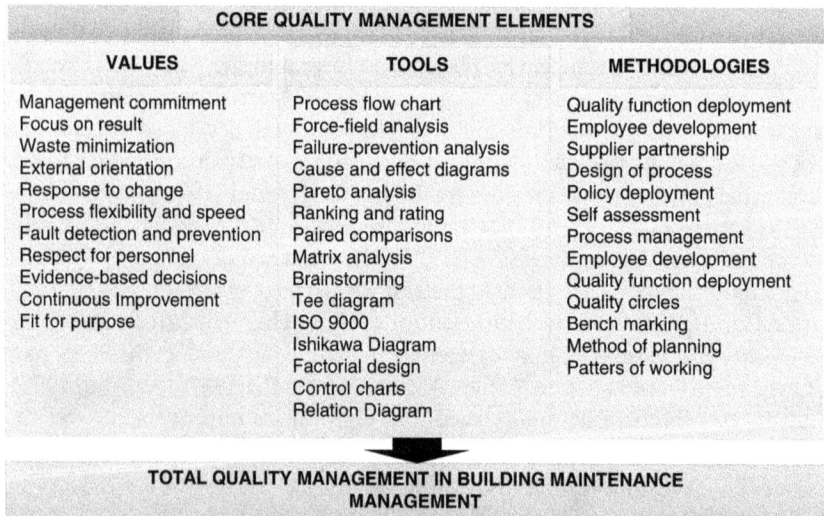

CORE QUALITY MANAGEMENT ELEMENTS		
VALUES	**TOOLS**	**METHODOLOGIES**
Management commitment	Process flow chart	Quality function deployment
Focus on result	Force-field analysis	Employee development
Waste minimization	Failure-prevention analysis	Supplier partnership
External orientation	Cause and effect diagrams	Design of process
Response to change	Pareto analysis	Policy deployment
Process flexibility and speed	Ranking and rating	Self assessment
Fault detection and prevention	Paired comparisons	Process management
Respect for personnel	Matrix analysis	Employee development
Evidence-based decisions	Brainstorming	Quality function deployment
Continuous Improvement	Tee diagram	Quality circles
Fit for purpose	ISO 9000	Bench marking
	Ishikawa Diagram	Method of planning
	Factorial design	Patters of working
	Control charts	
	Relation Diagram	

TOTAL QUALITY MANAGEMENT IN BUILDING MAINTENANCE MANAGEMENT

Figure 5.1 Core quality management elements for maintenance management of building.

Source: Adapted from Söderholm (2005).

Several researchers, including Oakland (1993), Lewis (1996), and Boaden (1997), stressed that the core values are fundamental to QM. Accordingly, Hellsten and Klefsjö (2000) state that the core values are the basis of goals set by the organisation. As stated by Hellsten and Klefsjö (2000), the core values constitute an important element as they represent an organisation's culture. Nevertheless, the formulation, name, and number of values differ among different authors. In their work, Dale et al. (1999: 22) discuss "eight key elements," Lamprecht (1991: 15) includes "eight management principles" while Söderholm (2005: 61) suggested "eleven values and concepts."

Values, which are the first element of the management system, can be summarised as those mentioned in Figure 5.1 and may be found in all the sources described above. Since these values are frequently mentioned in the literature describing quality management, they may be seen as core values of quality management (Hellsten & Klefsjö., 2000). The second element of the management system consists of tools that are well-defined and more concrete (Söderholm, 2005). These tools have a statistical basis to support decision-making or facilitate data analysis. Hellsten and Klefsjö (2000) and Akersten and Klefsjö (2003) identified some tools that support the methodologies element of management, including fault trees, the booklet of criteria to the Malcolm Baldrige National Quality Award, house of quality (HoQ), FMEA sheets, and factorial design sheets. Finally, the third element of the management of the system is the set of methodologies. The set of methodologies is how the organisation works toward meeting the objectives and goals. Söderholm (2005) identified some of these methodologies to include risk analysis, self-assessment, failure mode & effects analysis (FMEA), quality function deployment (QFD), and design of experiments (DoE).

However, TQM looks at the overall quality-related functions and processes throughout an organisation, including managing quality design and development, quality control and maintenance, quality improvement, and quality assurance. It considers all quality measures taken at all levels and involves all company employees in meeting users' needs and expectations. Against this backdrop, the TQM model will guide the development of an integrated MM framework for this study.

The Balanced Score Card

The balanced score card (BSC) was designed by Kaplan and Norton in 1996 in reaction to the increasing focus on purely financial measures for planning and managing a business. The BSC model provides strategic performance measurement and management for high-performance organisations (Kaplan & Norton, 1996; Amaratunga, 2001). Okolie (2011) states that the BSC integrates measures of process performance, product or service innovation, customer satisfaction, and finance in connecting short-term operational control to the long-term vision and strategy of the business. The BSC, as postulated by Kaplan and Norton (2007), addresses deficiencies in traditional management systems and further provides an instrument required by managers needed to navigate to future competitive success.

However, Okolie (2011) and Amaratunga (2001) opine that with BSC, an organisation's vision will translate into a set of performance indicators distributed among the following four perspectives, as shown in Table 5.2 below:

As shown in Table 5.2, Kaplan and Norton (1996) noted that the customer perspectives define customer value (users' value), how it can be satisfied, and why users will be willing to pay for it. The customer perspective should guide an organisation's internal processes and development efforts. Okolie (2011) posited that the internal business perspective encompasses all company processes through delivering the product/service and identifying required resources and capabilities for which the company needs to upgrade.

The learning and growth perspectives come from three sources: people, systems, and organisational procedures. Okolie (2011) observed that this perspective allows the organisation to ensure its capacity for long-term renewal, which is the basis for survival in the long run. Under this perspective, organisations consider what they need to develop, how to maintain the know-how, and the process's sustainability, efficiency, and productivity. As shown in Table 5.2, Kaplan and Norton (1996) affirm that the financial perspective help shows the results of strategic choices made in the other perspectives by establishing long-term goals as well as a large part of the general ground rules and premises for different perspectives. However, the chosen measures will therefore represent the relevant stage in the life cycle of the product or service.

However, Ashton (1998) hypothesised that the BSC provides the following for an organisational set:

- Dynamic communication and feedback;
- Easy identification of cause-and-effect relationships across operations;
- Both quantitative and qualitative information;
- The strategy that will set the foundation for management;
- A system that is capable of driving dramatic improvements in performance;
- A practical framework for implementing corporate strategy;

Table 5.2 Balanced scorecard perspectives

BSC perspectives	Performance indicator
Customer	To know what the existing and new customers value from the organisation
Internal process	To establish what internal processes must be excelled at to achieve the financial and customer perspective
Learning and growth	To establish whether the organisation can continue to improve and create future value
Financial	To establish how value can be created for shareholders

Source: Kaplan and Norton (1996).

- Management tool for linking business, team, and individual objectives and rewards to strategic goals;
- An effective mechanism for implementing change management;
- Good fit with the organisation's move away from a command and control culture to one of empowerment and coaching; and
- The ability to understand the drivers of business success.

Nonetheless, Amaratunga and Baldry (2000) state that the main aim of a BSC is to provide the management of an organisation with a concise summary of the key success factors of a business and to facilitate the alignment of business operations with the overall organisational strategy. BSC increases the economic value of an organisation's revenue growth and productivity (Okolie, 2011). The BSC is a management framework that measures economic and operational performance with an emphasis on non-financial measures (Amaratunga, 2001). The BSC makes a compelling case for including non-financial measures in an organisation's overall measurement system. The BSC framework tells the story of an organisation's strategy through a cause-and-effect model that eventually links all the measures to stakeholder value. Amaratunga and Baldry (2000) affirm that through the BSC, organisations monitor their current performance (finances, customer satisfaction, and business process results) and their efforts to improve processes, motivate and educate employees, and enhance information systems - their ability to learn and improve. It does include the hard financial indicators, but it balances these with other so-called "soft" measures, such as customer satisfaction and organisational learning. Okolie (2011) identified that the BSC model could be devised into a performance evaluation framework, especially for buildings. However, Kaplan and Norton (1996) posit that the BSC model advocates that no one measure can offer a clear view of the organisation. This is because the strength of the BSC falls within the fact that organisations do not have to choose between financial and non-financial measures. As noted by Okolie (2011), the BSC model at all levels of an organisation might not be used only as a strategy but also as a management tool for running organisations.

Requirements Management Theory

Mohd-Noor, Hamid, Abdul-Ghani, and Haron (2011) observed that despite the substantial resources for building MM, there is a dearth of the operational procedure due to a lack of understanding of the maintenance requirements of stakeholders. Subsequently, maintenance managers struggled with the lack of knowledge to implement MM operational systems. According to Straub (2010), to fulfil the users' maintenance needs, most maintenance organisations rely on the maintenance operators/personnel to provide maintenance services without recourse to stakeholders' requirements. Au-Yong, Ali, and Ahmad (2014) posited that the MM of buildings requires the involvement of stakeholders, including the maintenance organisation, building users, the maintenance manager, personnel, and maintenance service providers.

However, there are different definitions of requirements. According to Hull, Jackson, and Dick (2005: 112), requirements are "a condition or capability to which a system must conform," which could be any of the following:

- A capability must be met or possessed by a system to satisfy a contract, standard, specification, regulation, or other formally imposed documents;
- A capability needed by a customer or user to solve a problem or achieve an objective; or
- A restriction imposed by stakeholders.

Davis et al. (1993: 149) define a requirement as "a user need or necessary feature, function, or attribute of a system that may be sensed from a position external to that system." Furthermore, the ISO/IEC 15288 (2002: 38) states that "stakeholder requirements are expressed in terms of the wants, needs, desires, perceived constraints and expectations of identified stakeholders." Stakeholder requirements include, but are not limited to, the needs and requirements imposed by society, the constraints imposed by an acquiring organisation, and the capabilities and limiting characteristics of the operator staff of the organisation (Söderholm, 2005). The IEEE 610.12 (1990: 132) defines a requirement as:

- A condition or capability must be met or possessed by a system or system element to satisfy a contract, standard, specification, or other formally imposed documents;
- A condition or capability needed by a user to solve a problem or achieve an objective; or
- A documented representation of a condition or capability, as in one of the two former statements above.

Furthermore, Leong (2004) states that the absence of key stakeholders in the development or maintenance planning stage leaves gaps in analyzing or evaluating the maintenance requirements and strategies. The concept of stakeholder cannot be overemphasised in achieving a stable requirements management system within a maintenance activity. Hull et al. (2005) define a stakeholder as an individual affected by the system being developed or run within an organisation. Au-Yong, Ali, Ahmad, and Chua (2017) posit that a stakeholder is someone who has a vested interest in a system. However, Foley (2001) postulated that stakeholders were individuals or groups (e.g., staff, shareholders, customers, and users) capable of acting in a way that could impose unacceptable costs and threaten the organisation's enterprise viability if their requirements are not satisfied. Söderholm (2005) described requirements in this context as needs, expectations, and interests.

However, Leong (2004) noted that stakeholders have the means of bringing their requirements to the attention of MM or taking action if their requirements are not satisfied. The participation of key stakeholders in MM operations will provide additional and different comments, opinions, viewpoints, suggestions, effectiveness, and solutions concerning the maintenance activities (Newig et al., 2008). Nonetheless, this involvement requires the organisation's willingness, maintenance manager and personnel, building users, and service providers to understand the maintenance requirements of other stakeholders (Au-Yong et al., 2017). Jiao and Tseng (1999) posits that the organisation and management would benefit by accepting valuable and unexpected individual contributions through a viable requirement management system. Nevertheless, Jansson, Schade, and Olofsson (2013) single out shareholders and customers as having special characteristics for enterprise management. Accordingly, Söderholm (2005) states that the quality of goods and services mostly fulfils customer requirements.

The customers are not necessarily stakeholders; however, they provide the revenue necessary to fulfil other stakeholders' requirements, e.g., shareholders. MM, like business success, depends on a deep knowledge of stakeholders (customers/users) and the requirements that shape stakeholder behaviour and their dynamic interaction (Söderholm 2005). Furthermore, Au-Yong et al. (2014) postulated that effective MM requires the involvement of key stakeholders, including the organisation, maintenance manager, personnel, and building users.

Hull et al. (2005) identify people that can be considered stakeholders in a system (maintenance activities) as follows:

• Individuals participating in the development of the system (i.e., maintenance managers, maintenance personnel, quality assurance officers);
• Any individual contributing knowledge to the system (author of maintenance manual and operation manual or feedback notes used for requirement elicitation);
• Executives (head of maintenance operations or units);
• Individuals involved in maintenance and support (outsourcing, suppliers, contractors); and
• Providers of rules and regulations (maintenance organisation/government).

Söderholm (2005) stated that requirements could be classified in different ways. As Mylopoulos, Chung, and Yu (1999) affirmed, one traditional way is to classify the requirement as functional or non-functional. According to Söderholm (2005), functional requirements may be described as what the system does or performs, such as describing a specific function, whereas a non-functional requirement is how the system works with the organisation's procedures or constraints on the functional requirements of the stakeholders. Hull et al. (2005) argued that depending on the required format, common characteristics, and source, the requirements can be divided into different requirements. Some requirement types that are often used in projects that he listed are as follows:

- Stakeholders need a requirement from a stakeholder;
- Feature: a service provided by the system, usually formulated by a business analyst; the purpose of a feature is to fulfil a stakeholder need;
- Use cases: a description of system behaviour in terms of sequences of actions;
- Scenario: a specific sequence of actions; a specific path through a use case;
- Supplementary requirement: another requirement (usually non-functional) that cannot be captured in use cases; and
- Test case: a specification of test inputs, execution conditions, and expected results.

However, as shown in Figure 5.2, Hull et al. (2005) posited that these requirement types can be presented in a pyramid form. At the top level are the stakeholder needs, while features, use cases, and supplementary requirements are at the bottom. Nevertheless, the requirement pyramid shows that different details are captured at different levels of these requirements. The lower the level, the more detailed the requirements are (Davis et al., 1993).

One important best practice of requirements management is to have at least two different levels of requirement abstraction. One vision encompasses high-level requirements (features), and the other in the lower levels in the pyramid expresses the requirements at a detailed level. As opined by Hull et al. (2005), the main difference between needs and features is in the source of the requirement, while needs come from stakeholders (users) and features are formulated by business analysts (maintenance objectives and policy). This clearly shows that the requirement in maintenance is based on users, while method execution lies with the maintenance personnel. Leffingwell and Widrig (2000) state that the role of test cases is to check whether use cases and supplementary requirements are implemented correctly, while scenarios help derive use cases from test cases and facilitate the design and implementation of specific paths through use cases. Grimshaw and Draper (2001) postulated that requirements management might be considered an approach developed within the information technology industry in response to increasing costly failures and demands. Alexander (1996) and Davis et al. (1993) postulated that requirements management is a systematic approach to documenting, eliciting, organising, and managing a technical system's initial and changing requirements. Furthermore, Davis et al. (1993) consider requirements management partly an engineering discipline and partly a management process that can effectively manage both technical complexity and requirements of the system or an organisation.

Kotonya and Sommerville (1998) postulated that requirements management could be described as managing changes in the requirements and the relationship between requirements on a system or an organisation. One important aspect of requirements management is that it is economically and technically possible to perform the proposed changes (Söderholm, 2005). It is important to check what other requirements may be affected if the change applies to a specific requirement. This entails that links between

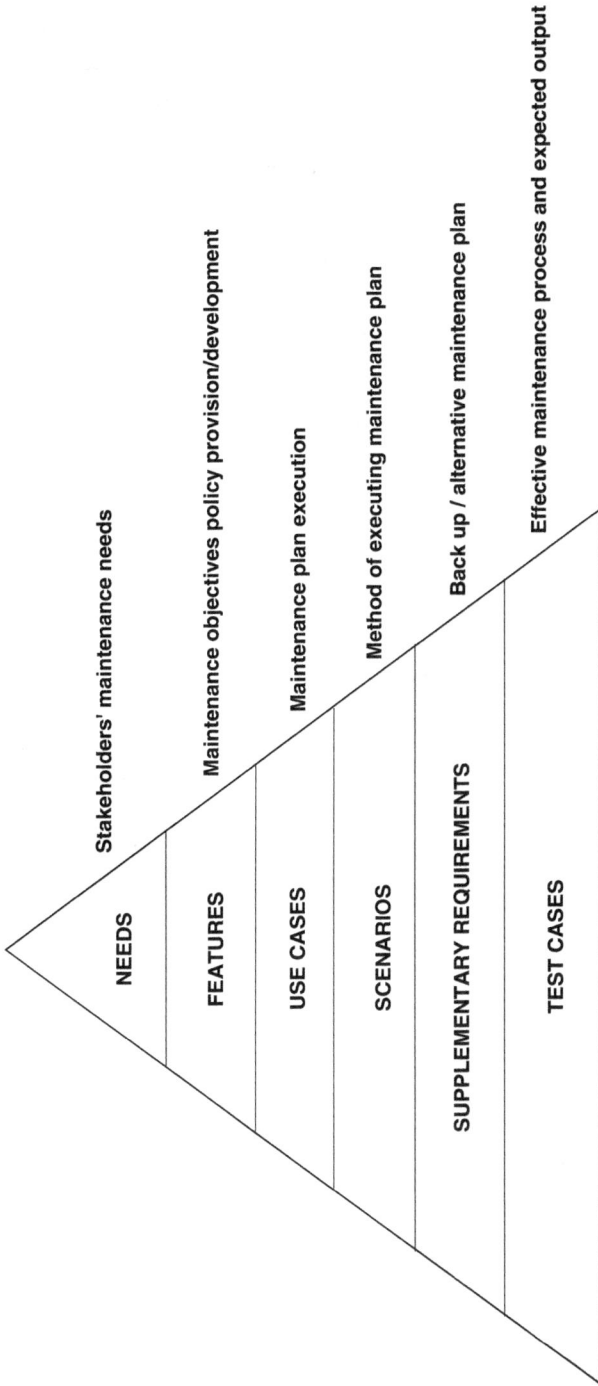

Figure 5.2 The requirements pyramid for the maintenance process.

Source: Adapted from Hull et al. (2005).

requirements, the system design, and the sources of requirements must be recorded, i.e., traceability information. The concept of traceability can be described as a technique that provides a relationship between different levels of requirements in the system organisation, which helps to determine the origin of any requirement (Hull et al., 2005). Söderholm (2005) affirms that the traceability concept points to one characteristic of a good requirement.

As shown in Table 5.3, typical characteristics of good requirements are that they should be unambiguous, consistent, complete, correct, traceable, modifiable, and ranked (Young, 2001).

Kotonya and Sommerville (1998) state that requirements management comprises tools and methodologies for executing and establishing a formal procedure for verifying, collecting, considering, and studying how requirements changes affect the system. From the feedback from these activities, requirement management may thus be seen as a way to manage four key activities (Bohner & Arnold, 1996). According to Söderholm (2005), the first key step is to study the proposed new requirements or adjustments to existing requirements and make appropriate decisions about necessary changes based on the wanted and unwanted effects of the proposal. This shows that the maintenance needs of users of the facility need to be identified. Subsequently, the changes must be specified and designed in the second key step, meaning there should be a plan and material for the replacement. Thirdly, based on the specification, the changes must then be executed, which means that the identified portion of the facility should be repaired properly. Lastly, the performed changes must be studied to see whether they meet the new requirements and whether the system meets the other existing requirements. This means that the maintenance carried out should meet the needs and expectations of the users together with the set goals of the maintenance organisation. Söderholm (2005) observed that one major universal component in various requirement management descriptions is the requirement of stakeholders in the system.

Table 5.3 Characteristics and descriptions of good requirements

Characteristics	Description
Correct	It should state something that the system shall meet.
Unambiguous	It should have only one interpretation by using a single unique term.
Complete	It should be both determined and significant.
Ranked	It should be ranked for importance and/or stability
Consistent	It should not conflict with other requirements.
Verifiable	It should have some cost-effective methodology.
Modifiable	It should be possible to change easily, completely, and consistently
Traceable	It should have a clear development from its origin to its furthest development.

Source: Young (2001) as adapted from IEEE 830 pp. 4–8.

Change Management Theory

Kanter, Stein, and Jick (1992) described the change as the crystallisation of new possibilities such as new behaviour, policies, methodologies, patterns, products, or new market ideas based on re-conceptualised patterns in the institution. This description means the architecture of change involves designing and constructing new patterns or re-conceptualising old ones to make new and hopefully more productive actions possible. According to Kanji and Moura e Sá (2003), changes include changes to structures, procedures, rules, and regulations. Kanter et al. (1992) state that change could be considered a shift in the behaviour of the whole organisation. Senge et al. (1999) categorically stated that organisations are influenced by environmental changes that require the adaptation of internal processes. Robbins (1990) maintained that in adding a new dimension to the change definition, all change initiatives must be planned in consultation with employees.

Similarly, Dunphy (1996) states that all change initiatives must be planned with relevant stakeholders. D'ortenzio (2012) suggests that planned change must have a specific purpose for the organisation to remain in a viable state. Lawrence, Dyck, Maitlis, and Mauws (2006) postulated that planned changes should be a continuous and adaptive process to sway employees to buy into new ideas consistent with the organisational strategic direction. However, D'ortenzio (2012) concluded that any failure on the part of the organisation in introducing continuous and adaptive change means they will experience adaptation without the energy that comes from employees' buying into different approaches and perspectives and reformulating their identities in ways that match the new direction. Buttressing this statement, Zimmerman (1995) argued that because employees are directly involved in the process of change in some form or other, they should always be the key players in implementing, facilitating, and managing change.

Nevertheless, Dervitsiotis (2003) posited that for organisational effectiveness, it is of importance that organisations give serious thought to the depth, kind, and complexity of the necessary changes before implementing them. As viewed by Harvey and Brown (1996), organisations that avoid changes must be able to endure a stable identity and achieve operational goals. Ford and Saren (1996) hypothesised that for successful change to occur in an organisation, it is necessary to foster good coordination, clear communication, and strong leadership to exploit and develop the organisational resources. For this to happen, D'ortenzio (2012) theories that an organisation must acknowledge that relationships are symbolic of living structures with intrinsic dynamic features and are characterised by the continuous change processes. He further stated that organisational change should consider fears, values, behaviours, and the aspirations of all stakeholders involved in the change process. De Jager (2001: 24) described the change as:

a simple process. At least, change is simple to describe. It occurs whenever we replace the old with the new. Change is about traveling from the old to the new, leaving yesterday behind in exchange for the new tomorrow. But implementing change is incredibly difficult. Most people are reluctant to leave the familiar behind. We are all suspicious about the unfamiliar; we are naturally concerned about how we get from the old to the new, especially if it involves learning something new and risking failure.

Based on the above description, D'ortenzio (2012) hypothesised that change involves moving from the known into the unknown. It is noteworthy that most employees do not support change unless convincing reasons are shown to them. Hence, for change to be effective in an organisation, it must involve, amongst other important factors, vision, mission, communication, strong leadership, participation, and culture. Accordingly, Hamel and Prahalad (1994) posit that vision in organisational change involves developing a future picture of the organisation. Senge and Roberts (1994) state that the vision helps set the organisational change scene. Handy (1996) claimed that communication and strong leadership are essential in preparing the organisation for change because they guide it through turbulent stages. Participation in the change process involves giving stakeholders a fair says (Zand, 1997). Moreover, McAdams et al. (2006) state that organisational culture is a collective understanding of the organisation's workings that influences change initiatives.

As a result, Harvey and Brown (1996) stated that to foster the development of positive relationships within an organisation. It is important to have an integrated approach to any change programme that combines technological, structural, and behavioural approaches. According to Pryor, Taneja, Humphreys, Anderson, and Singleton (2008), change management is an integral part of life and is constant in most organisations (see Figure 5.3). Szamosi and Duxbury (2002) sustain that it is a process that will handle

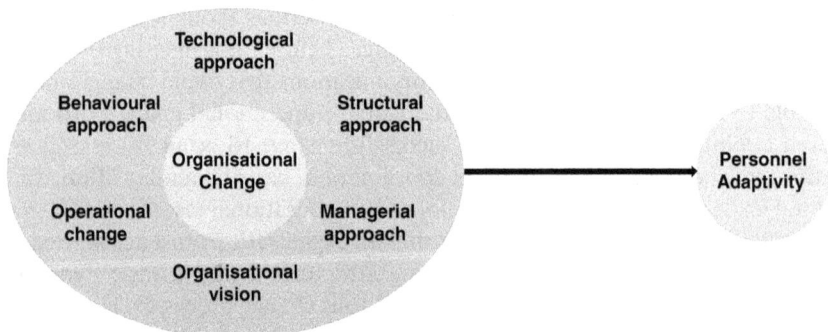

Figure 5.3 Adapting personnel response to organisation change.

Source: Adapted from Pryor et al. (2008).

unexpected situations. Thus, George and Jones (2002) postulated that change management is the organisational movement from the existing plateau toward a desired future state to increase organisational efficiency and effectiveness. Changes in an organisation might be ongoing or continuous improvement initiatives resulting from the reaction to external forces for changes (Pryor et al., 2008). Such changes could be part of improvement initiatives such as TQM.

Jansson (2008) postulated that change management could be described as making changes in a planned and managed or systematic fashion with the response over which the organisation exercises little or no control. Nonetheless, Burke et al. (2000) maintained that effective change and change management processes must be in place for an organisation to gain a competitive edge over its rivals.

Change management can involve elements of organisational structure and culture which might include technological developments, transformational relationships, organisational control, organisational structure, organisational culture, organisational locations, balance sheets, and others (Mead, 2005; Jansson, 2008). Nevertheless, Sætren and Laumann (2015) state that change within the organisation will depend on the degree and nature of transformation within the organisation. However, maintaining lasting changes in an organisation is no easy task due to cultural diversity, making change management difficult (Robertson, Roberts, & Porras, 1992; Ogunbayo, Ohis Aigbavboa, Thwala, Akinradewo, 2022). Therefore, for viable change, an organisation must strive to change the rites, rituals, behaviour, and values of individuals in the organisation (Jansson, 2008).

However, managing and implementing change can be ambiguous (Bolman & Deal, 2018). This simply means in embracing transformation, organisations must have effective processes and practices in place to manage those changes. Therefore, developing efficient and effective ways of promoting change in the organisation while at the same time encouraging all employees to accept the change is the primary function of change management. Additionally, change management is a process of ensuring that organisational goals are met through the proper use of organisational resources. Hence, to maintain a competitive advantage in public and private institutions, the organisation must embrace change as a way of life.

Thus, to better understand the essence of change management in an organisation, especially maintenance organisations, Kurt Lewin's and Kotter's model of change were engaged in buttressing it more holistically. These theories cover a wide scope of thought and appear close to containing all the essential vital skills required for effective management in an organisation (maintenance organisation) with a strong focus on the behavioural change of stakeholders that can influence their attitude toward the maintenance process.

Kurt Lewin's Model of Change

Kurt Lewin's change model involves three steps: unfreezing, moving, and refreezing (Burnes, 2004). According to Branch (2002), the three-step model is associated with an intentional change in the organisation (see Figure 5.4). To implement this change, change initiators may choose to use various strategies.

Burnes (2004) states that the first step, "unfreezing," is to neglect the old behaviour before adapting to new behaviour. Gilley, Gilley, and McMillan (2009) state that preparing the workers and organisation for change is a step. However, in this step, employees break away from how things have been done. In maintenance organisations, for example, for change to occur toward effective MM, maintenance personnel must embrace new work practices with a sense of urgency. Maintenance personnel could be encouraged to distance themselves from comfort zones that they were familiar with for them to acclimatise to new ideals or work practices. Harper (2001: 121) maintained that organisations (maintenance organisations) implementing change management should encourage employees (personnel) to abide by a plan that allows for the "sloughing off of yesterday" because "it will force thinking and action ... make available men and money for new things ... create a willingness to act."

The second step, as opined by Burnes (2004), is "moving." It is a process that helps employees learn acceptable behaviour. It is also the stage in which employees are involved in change processes (Gilley et al., 2009). At this stage, employees (maintenance personnel) can engage in activities that implement and identify new ways or activities of doing things to bring about change. Based on this, Harper (2001) suggests that for change to take place in an organisation (maintenance organisation), management must ensure that all relevant stakeholders (maintenance personnel, users, and suppliers) are allowed to be involved in decision-making and problem-solving collaboratively. Employees involved in decision-making are more

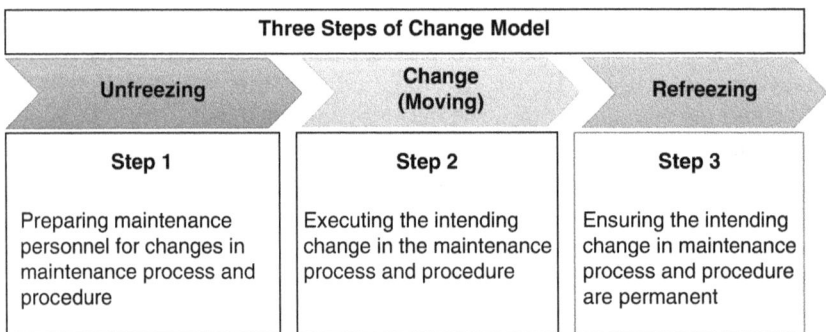

Three Steps of Change Model		
Unfreezing	**Change (Moving)**	**Refreezing**
Step 1	**Step 2**	**Step 3**
Preparing maintenance personnel for changes in maintenance process and procedure	Executing the intending change in the maintenance process and procedure	Ensuring the intending change in maintenance process and procedure are permanent

Figure 5.4 Lewin's three-step change application for the maintenance process.
Source: Adapted from D'ortenzio (2012).

likely to accept change and are committed to making change a success within the maintenance organisation. A better understanding of the needs and benefits of change among employees may lead to little or no resistance on their part.

Finally, the last step in this model is "refreezing." According to Burnes (2004), this stage is where the new behaviours are learned and maintained permanently. He states further that the new behaviour must be congruent with the employees. At this stage, the employer emphasises strengthening the maintenance organisation's new process (maintenance process) and task (maintenance objectives). For this step to be effective and successful, employees (maintenance personnel) must be acknowledged, as a reward is an important consideration. D'ortenzio (2012) opines that for effective behaviour modification, the reward is crucial. This simply means that employees should receive appropriate recognition for change in behaviour if they accept the change. In this regard, Harper (2001) concluded that reward recognises the new behaviour is valued and avoids earlier behaviour reoccurring.

Thus, Branch (2002) states that Lewin's model of change can be illustrated as follows:

- Changing various organisational structures and systems – reward systems, reporting relationships, work designs;
- Changing the individuals who work in the organisation (their skills, values, attitudes, and eventually behaviour) – with an eye to instrumental organisational change; or
- Directly changing the organisational climate or interpersonal style – how often people are with each other, how conflict is managed, and how decisions are made.

Kotter's Model of Change

Kotter's model of change views changes as driven from the bottom up rather than from the top down and further stresses that change is a continuous process of adaptation to changing conditions and circumstances. The model presumed change to be so rapid that it is impossible for change initiators to successfully identify, plan, and implement the necessary organisational changes (Kanter et al., 1992). Accordingly, Wilson (1992) states that this is why responsibility for organisational change has become increasingly devolved. The model approach also shows that change should be perceived as a continuous, open-ended process of adaptation to changing circumstances and conditions (Dawson, 1994).

Burnes (2004: 24) postulated that this model's concept promotes an "extensive and in-depth understanding of people, style, structure strategy and culture," and how these can function as sources of action that can block change or as devices to encourage the change process. However, Mabey et al. (1993) hypothesised that a change approach involves a learning process and is not just a method of changing organisational structures and practices. This change management approach shows that maintenance personnel must be continuously trained and educated on their job, especially on innovation, technology, techniques, material selection, job behaviours, and values. Their ability to learn and adapt within the organisation may influence the success or failure of the change management programme.

According to Burnes (2004), Kotter's model of change advocates eight steps in the change process. They include establishing a sense of urgency, creating the guiding coalition, developing a vision and strategy, communicating the change vision, empowering employees for broad-based action, generating short-term wins, consolidating gains, and producing more.

Summary

This chapter outlined a number of theories upon which MM studies should rest. It highlighted theories that have complementary perspectives on the MM of building. It showed that theories such as TQM, BSC, requirement need, and change management cover a broad scope of thought. The theories appear close to containing all the essential vital skills required for effective MM with a strong focus on quality, performance indicators, users' requirements, and behaviour change of stakeholders in the maintenance process and the maintained buildings.

References

Akersten, P. A., & Klefsjö, B. (2003). Total dependability management. In *Handbook of reliability engineering* (pp. 559–565). London: Springer.

Alexander, K. (1996). *Facilities Management: Theory and Practice*. London: Routledge (Taylor & Francis group).

Alzaben, H. (2015). *Development of a MM framework to facilitate the delivery of healthcare provisions in the Kingdom of Saudia Arabia* (Published doctoral dissertation, Nottingham Trent University).

Amaratunga, D., & Baldry, D. (2000). Assessment of facilities management performance in higher education properties. *Facilities, 18*(7/8), 293–301.

Amaratunga, R. G. (2001). *Theory building in facilities management performance measurement: Application of some core performance measurement and management principles* (Published doctoral dissertation, University of Salford).

Anastasiadou, S. D. (2015). The roadmaps of total quality management in the Greek education system according to Deming, Juran, and Crosby in light of the EFQM model. *Procedia Economics and Finance, 33*, 562–572.

Ashton, M. C. (1998). Personality and job performance: The importance of narrow traits. *Journal of Organizational Behavior: The International Journal of Industrial, Occupational and Organizational Psychology and Behavior, 19*(3), 289–303.

Au-Yong, C. P., Ali, A. S., & Ahmad, F. (2014). Improving occupants' satisfaction with effective maintenance management of HVAC system in office buildings. *Automation in Construction, 43*, 31–37.

Au-Yong, C. P., Ali, A.-S., Ahmad, F., & Chua, S. J. L. (2017). Influences of key stakeholders' involvement in maintenance management. *Property Management, 35*(2), 217–231.

Bagadia, K. (2006). *Computerized maintenance management systems made easy.* New York: McGraw-Hill Professional.

Bamisile, A. (2004). *Building production management.* Lagos: Foresight Press.

Bessant, J., Caffyn, S., Gilbert, J., Harding, R., & Webb, S. (1994). Rediscovering continuous improvement. *Technovation, 14*(1), 17–29.

Bhuiyan, N., & Baghel, A. (2005). An overview of continuous improvement: From the past to the present. *Management Decision, 43*(5), 761–771.

Boaden, R. J. (1997). What is TQM ... and does it matter? *TQM, 8*(4), 153–171.

Bohner, S. A. (1996). Impact analysis in the software change process: A year 2000 perspective. In *Proceedings of International Conference on Software Maintenance.* Monterey, CA, USA, 4–8 November 1996, pp. 42–51.

Bolman, L. G., & Deal, T. E. (2018). *Reframing the path to school leadership: A guide for teachers and principals.* California: Corwin Press.

Branch, K. M. (2002). Participative management and employee and stakeholder involvement. Management Benchmarking Study by Washington Research Evaluation Network, Chapter 10, 1–27.

Burke, W. W., Trahant, W. J., & Koonce, R. (2000). *Business climate shifts: Profiles of change makers.* London: Routledge (Taylor and Francis Group).

Burnes, B. (2004). *Managing change: A strategic approach to organisational dynamics.* (3rd ed.). Harlow: Pearson Education.

Caffyn, S. (1999). Development of a continuous improvement self-assessment tool. *International Journal of Operations & Production Management, 19*(11), 1138–1153.

Campbell, J. D. (1998). *Uptime, strategies for excellence in maintenance management.* Portland, OR: Productivity Press.

Cole, G. A., & Kelly, P. (1996). *Management: Theory and practice.* New York, NY: Continuum.

Crosby, P. B. (1992). *Completeness: Quality for the 21st century.* Dutton, New York, NY: Penguin.

D'ortenzio, C. (2012). *Understanding change and change management processes.* University of Canberra.

Dahlgaard-Park, S. M., Reyes, L., & Chen, C. K. (2018). The evolution and convergence of TQM and management theories. *TQM & Business Excellence, 29*(9-10), 1108–1128.

Dale, B.G., Van Der Wiele, T., & Van Iwaarden, J. (2007). *Managing quality.* Malden, MA, USA: Blackwell Publishing Limited.

Dale, B. G., Van der Wiele, A., & Van Iwaarden, J. D. (1999). TQM: An overview. *Managing quality,* Malden MA: Black Publishing Limited, 3–33.

Daragh Naughton, M., & Tiernan, P. (2012). Individualising maintenance management: A proposed framework and case study. *Journal of Quality in Maintenance Engineering, 18*(3), 267–281.

Davis, A., Overmyer, S., Jordan, K., Caruso, J., Dandashi, F., Dinh, A., & Ta, A. (1993). Identifying and measuring quality in a software requirements specification. In *Proceedings First International Software Metrics Symposium* MD, USA, 21–22 May 1993. pp. 141–152. IEEE.

Dawson, P. M. (1994). *Organizational change: A processual approach.* London, Greater London: Paul Chapman Publishing.

De Jager, P. (2001). Resistance to change: A new view of an old problem. *The Futurist, 35*(3), 24.

Dean Jr, J. W., & Bowen, D. E. (1994). Management theory and total quality: Improving research and practice through theory development. *Academy of Management Review, 19*(3), 392–418.

Deming, W. E. (2018). *The new economics for industry, government, education.* Cambridge: MIT Press.

Dervitsiotis, K. (2003). The pursuit of sustainable business excellence: Guiding transformation for effective organizational change. *TQM & Business Excellence, 14*(3), 251–267.

Dunphy, D. (1996). Organizational change in corporate settings. *SAGE Social Science Collections, 49*(5), 541–552.

Feigenbaum, A. V. (1983). *Total quality control.* New York: McGraw-Hill.

Fisher, N. I., & Nair, V. N. (2009). Quality management and quality practice: Perspectives on their history and their future. *Applied Stochastic Models in Business and Industry, 25*(1), 1–28.

Foley, K. J. (2001). *Meta management: A stakeholder/quality management approach to whole-of enterprise management.* Sydney: Standards Australia.

Ford, D., & Saren, M. (1996). *Technology strategy for business.* International Thompson Business Press.

Garvin, J. (1998). Managing with total quality management – Theory and practice. *International Journal of Manpower, 19*(5), 358–360.

George, J. M., & Jones, G. R. (2002). *Understanding and managing organizational behavior.* Upper Saddle River, NJ: Prentice Hall.

Gilley, A., Gilley, J. W., & McMillan, H. S. (2009). Organizational change: Motivation, communication, and leadership effectiveness. *Performance Improvement Quarterly, 21*(4), 75–94.

Gökçe, H. U., Stack, P., Gökçe, K. U., & Menzel, K. (2009, October). Maintenance scheduling based on the analysis of building performance data. In *Proceedings, of 26th W78 Conference on Information Technology in Construction*, Istanbul, Turkey, 15–17 July 2009, pp. 1–3.

Grimshaw, D. J., & Draper, G. W. (2001). Non-functional requirements analysis: Deficiencies in structured methods. *Information and Software Technology, 43*(11), 629–634.

Hamel, G., & Prahalad, C. K. (1994). Competing for the future. *Harvard Business Review, 72*(4), 122–128.

Handy, C. B. (1996). *Gods of management: The changing work of organizations.* Oxford University Press, USA.

Harper, J. T. (2001). Short-term effects of privatization on operating performance in the Czech Republic. *Journal of Financial Research, 24*(1), 119–131.

Harris, F., & McCaffer, R. (2001). *Modern construction management.* USA: Black well Scientific.

Harris, L. C. (2001). Market orientation and performance: Objective and subjective empirical evidence from UK companies. *Journal of Management Studies, 38*(1), 17–43.

Harvey, D., & Brown, D. (1996). Organization transformation: Strategy interventions. *An Experimental Approach to Organizational Development*, Hoboken New Jersey: Prentice-Hall.

Hellsten, U., & Klefsjö, B. (2000). TQM as a management system consisting of values, techniques, and tools. *The TQM Magazine, 12*(4), 238–244.

Hull, E., Jackson, K., & Dick, J. (2005). Requirements engineering in the solution domain. In *Requirements Engineering*, London: Springer.

IEEE Standards Coordinating Committee (1990). IEEE standard glossary of software engineering terminology (IEEE Std 610.12-1990). Los Alamitos. *CA: IEEE Computer Society, 169*, 132.

International Organization for Standards. (1986). ISO 8402.

International Standards Organization/International Electrotechnical Commission (2002). *Systems Engineering – System Life Cycle Processes*. Geneva: ISO/IEC, *15288*.36-39.

James, P. T. (1996). *Total quality management: An introductory text.* New Jersey: Prentice Hall.

Jansson, G., Schade, J., & Olofsson, T. (2013). Requirements management for the design of energy-efficient buildings. *Journal of Information Technology in Construction (ITcon), 18*, 321–337.

Jansson, J. E. (2008). The importance of change management in reforming customs. *World Customs Journal, 2*(2), 41–52.

Jiao, J. , & Tseng, M. M. (1999). A requirement management database system for product definition, *Integrated Manufacturing Systems, 10*(3), 146–154.

Juran, J. M. (1992). *Juran on quality by design: The new steps for planning quality into goods and services.* New York: Simon and Schuster.

Kanji, G., & Moura e Sá, P. (2003). Sustaining healthcare excellence through performance measurement. *Total Quality Management & Business Excellence, 14*(3), 269–289.

Kanter, R. M., Jick, T. D., & Stein, B. A. (1992). The challenge of organization change: How companies experience it and leaders guide it (No. 658.406 M8553c Ej. 1 000004). *Free Press.*

Kaplan, R.S., & Norton, D.P. (2007). Balanced Scorecard. In Boersch, C., & Elschen, R. (eds) *Das Summa Summarum des Management.* Gabler.

Kaplan, R. S., & Norton, D. P. (1996). *The balanced score card.* Massachusetts, Boston: Harvard Business School Press.

Karia, N., Asaari, M. H. A. H., & Saleh, H. (2014). Exploring maintenance management in service sector: A case study. In *Proceedings of International Conference on Industrial Engineering and Operation Management*, Bali, 7–9 January 2014, pp. 3119–3128.

Kartha, C. P. (2004). A comparison of ISO 9000:2000 quality system standards, QS9000, ISO/TS 16949 and Baldrige criteria. *The TQM Magazine, 16*(5), 331–340.

Kast, F. E. (1985). *James E. Rosenzweig, Organization and Managemen.* Singapore: McGraw-Hill Books.

Kaye, M., & Anderson, R. (1999). Continuous improvement: The ten essential criteria. *International Journal of Quality & Reliability Management, 16*(5), 485–509.

Kermally, S. (1996). *Total management thinking.* Oxford, United Kingdom: Butterworth-Heinemann.

Kolawole, A. R. (2002). Developing maintenance culture in Nigeria: The role of facility management. In *National Conference of the School of Environmental Studies at the Federal Polytechnic, Ede Osun State*, 29–31 October 2002.

Kotonya, G., & Sommerville, I. (1998). *Requirements engineering: Processes and techniques.* Hoboken, NJ: John Wiley & Sons, Inc.

Kroslid, D. (1999). *In search of quality management – Rethinking and reinterpreting.* Linko"ping: Institute of Technology, Linko"ping University.

Lamprecht, J. L. (1991). ISO 9000 implementation strategies. *Quality, 30*(11), 14–17.

Lateef, O. A. (2009). Building maintenance management in Malaysia. *Journal of Building Appraisal, 4*(3), 207–214.

Lawrence, T. B., Dyck, B., Maitlis, S., & Mauws, M. K. (2006). The underlying structure of continuous change. *MIT Sloan Management Review, 47*(4), 59.

Leffingwell, D., & Widrig, D. (2000). *Managing software requirements: A unified approach.* Boston: Addison-Wesley Professional.

Leong, K. C. (2004). The essence of good facility management – A guide for maximization of facility assets' economic life and asset optimisation for reliable services and user satisfaction. *Buletin Ingenieur, 24*, 7–19.

Levy, G. N. (2006). Total quality management for rapid manufacturing. In *Proceeds of Euro-u Rapid.* Frankfurt: Germany. A5-1.

Lewis, D. (1996). The organizational culture saga – from OD to TQM: a critical review of the literature. Part 2 – applications. *Leadership & Organization Development Journal, 17*(2), 9–16.

Lewis, L. K. (2000). Communicating change: Four cases of quality programs. *The Journal of Business Communication (1973), 37*(2), 128–155.

Mabey, C., Mayon-White, B., & Mayon-White, W. M. (Eds.). (1993). *Managing change.* London: Sage.

Macdonald, J., & Piggott, J. R. (1993). *Global quality: The new management culture.* New Jersey: Pfeiffer.

Márquez, A. C. (2007). *The maintenance management framework: Models and methods for complex systems maintenance.* London: Springer-Verlag.

Maschek, T., Khazrei, K., Hempen, S., & Deuse, J. (2011). Managing continuous improvement process as an organizational task in course of a century. In *10th International Symposium on Human Factors in Organisational Design and Management*, Rhodes University, South Africa, 4–6 April 2011, pp. 23–28.

McAdams, D. P., Bauer, J. J., Sakaeda, A. R., Anyidoho, N. A., Machado, M. A., Magrino-Failla, K., … & Pals, J. L. (2006). Continuity and change in the life story: A longitudinal study of autobiographical memories in emerging adulthood. *Journal of Personality, 74*(5), 1371–1400.

McCabe, D., & Wilkinson, A. (1998). 'The rise and fall of TQM': The vision, meaning, and operation of change. *Industrial Relations Journal, 29*(1), 18–29.

Mead, L. B. (2005). The historic change in continuing education of church professionals. *The Clergy Journal, 82*(1), 3–5.

Mekasha, E. (2018). *Maintenance management framework development for competiveness of food and beverage industry: A case study on Asku Plc.* (Thesis Draft, Addis Ababa Institute of Technology, Addis Ababa University).

Mohd-Noor, N., Hamid, M. Y., Abdul-Ghani, A. A., & Haron, S. N. (2011). Building maintenance budget determination: An exploration study in the Malaysia government practice. In *2nd International Building Control Conference 2011*, Penang, Malaysia, 11–12 July 2011. *Procedia Engineering, 20*, pp. 435–444.

Mukelasi, M. F. M., Zawawi, E. A., Kamaruzzaman, S. N., Ithnin, Z., & Zulkarnain, S. H. (2012). A review of critical success factors in building maintenance management of local authority in Malaysia. In *2012 IEEE Symposium on*

Business, Engineering, and Industrial Applications, Bandung, 23–26 September 2012, pp. 653–657. IEEE.

Mylopoulos, J., Chung, L., & Yu, E. (1999). Requirements analysis: From object-oriented to goal oriented. *Communications of the ACM, 42*(1), 31–37.

Newig, J., Gaube, V., Berkhoff, K., Kaldrack, K., Kastens, B., Lutz, J., & Haberl, H. (2008). The role of formalisation, participation, and context in the success of public involvement mechanisms in resource management. *Systemic Practice and Action Research, 21*(6), 423–441.

Oakland, J. S. (1993). *Total quality management: The route to improving performance.* Oxford: Butterworth-Heinemann.

Ogunbayo, B. F., & Aigbavboa, O. C. (2019). Maintenance requirements of students' residential facility in higher educational institution (HEI) in Nigeria. In *IOP Conference Series: Materials Science and Engineering* (Vol. 640, No. 1, p. 012014). IOP Publishing.

Ogunbayo, B. F., Aigbavboa, C. O., Thwala, W., Akinradewo, O., Ikuabe, M., & Adekunle, S. A. (2022). Review of culture in maintenance management of public buildings in developing countries. *Buildings, 12*(5), 677.

Ogunbayo, B. F., Ohis Aigbavboa, C., Thwala, W. D., & Akinradewo, O. I. (2022). Assessing maintenance budget elements for building maintenance management in Nigerian built environment: A Delphi study. *Built Environment Project and Asset Management, 12*(4), 649–666.

Okolie, K. C. (2011). *Performance evaluation of buildings in educational institutions: A case of Universities in South-East Nigeria* (Published doctoral dissertation, Nelson Mandela Metropolitan University, Port Elizabeth, South Africa).

Omar, M. F., Ibrahim, F. A., & Omar, W. M. S. W. (2017). Key performance indicators for maintenance management effectiveness of public hospital building. In *MATEC Web of Conferences,* Ho Chi Minh City, Vietnam, 5–6 August 2016, pp. 01056. (Vol. 7). EDP Sciences.

Phogat, S., & Gupta, A. K. (2017). Identification of problems in maintenance operations and comparison with manufacturing operations: A review. *Journal of Quality in Maintenance Engineering, 23*(2), 226–238.

Pryor, M. G., Taneja, S., Humphreys, J., Anderson, D., & Singleton, L. (2008). Challenges facing change management theories and research. *Delhi Business Review, 9*(1), 1–20.

Robbins, S. P. (1990). *Organization theory: Structures, designs, and applications, 3/e.* Pearson Education India.

Robertson, P. J., Roberts, D. R., & Porras, J. I. (1992, August). A meta-analytic review of the impact of planned organizational change interventions. In *Academy of management proceedings* (Vol. 1992, No. 1, pp. 201–205). Briarcliff Manor, NY: Academy of Management.

Sætren, G. B., & Laumann, K. (2015). Effects of trust in high-risk organizations during technological changes. *Cognition, Technology & Work, 17*(1), 131–144.

Sanchez, L., & Blanco, B. (2014). Three decades of continuous improvement. *Total Quality Management & Business Excellence, 25*(9-10), 986–1001.

Senge, K., & Roberts, R. S. (1994). *The fifth discipline field book.* Hachette UK: Century Publishing.

Senge, P., Kleiner, A., Roberts, C., Ross, R., Roth, G., Smith, B., & Guman, E. C. (1999). The dance of change: The challenges to sustaining momentum in learning organizations. *Performance Improvement, 38*(5), 55–58.

Sharma, S. K. (2013). Maintenance reengineering framework: A case study. *Journal of Quality in Maintenance Engineering, 19*(2), 96–113.

Shiba, S., Graham, A., & Walden, D. (1993). *New American TQM*. Portland: Productivity Press.

Söderholm, P. (2005). *Maintenance and continuous improvement of complex systems: Linking stakeholder requirements to the use of built-in test systems* (Published doctoral dissertation, Luleå tekniska universitet).

Straub, A. (2010). Competences of maintenance service suppliers servicing end customers. *Construction Management and Economics, 28*(11), 1187–1195.

Szamosi, L. T., & Duxbury, L. (2002). Development of a measure to assess organizational change. *Journal of Organizational Change Management, 15*(2), 184–201.

Taguchi, G. (1995). Quality engineering (Taguchi methods) for the development of electronic circuit technology. *IEEE Transactions on Reliability, 44*(2), 225–229.

Waeyenbergh, G., & Pintelon, L. (2002). A framework for maintenance concept development. *International Journal of Production Economics, 77*(3), 299–313.

WCED, S. W. S. (1987). World commission on environment and development "Our common future. *Published Report: WCED, 17*, 1–91.

Wilkinson, A. (1998). *Managing with total quality management: Theory and practice.* London: Macmillan International Higher Education.

Wilson, D. C. (1992). *A strategy of change: Concepts and controversies in the management of change.* London: Routledge.

Wireman, T. (2005). *Developing performance indicators for managing maintenance.* New York: Industrial Press Inc.

Young, R. R. (2001). *Effective requirements practices*, Boston: Addison-Wesley. 107.

Young, E., Green, H. A., Roehrich-Patrick, L., Joseph, L., & Gibson, T. (2003). *Do K-12 school facilities affect education outcomes.* The Tennessee advisory commission on intergovernmental Relations: Staff Information Report.

Zand, D. E. (1997). *The leadership triad: Knowledge, trust, and power.* New York: Oxford University Press on Demand.

Zhu, G., Gelders, L., & Pintelon, L. (2002). Object/objective-oriented maintenance management. *Journal of Quality in Maintenance Engineering, 8*(4), 306–318.

Zimmerman, J. H. (1995). The principles of managing change. *Human Resources Focus, 72*(2), 15–16.

Zollo, M., & Winter, S. G. (2002). Deliberate learning and the evolution of dynamic capabilities. *Organization Science, 13*(3), 339–351.

6 Gaps in Maintenance Management Research

Introduction

This chapter addresses the perceived gaps in maintenance management (MM) research, which have not been appraised as an all-inclusive construct in the previous models. Although they have been mentioned in the discussion of the earlier reviewed models, this was only as a distinct variable. These observed gaps constitute the new additional constructs of the conceptualised framework for this current study. The identified gaps are maintenance culture and communication among stakeholders.

Gaps in Maintenance Management Conceptual Framework

Using the framework developed by Campbell (1998) and Wireman (2005), it is clear that most MM studies-related research findings are carried out in developed economies. The finding shows that studies on MM of buildings in developing economies are mostly developed based on their system, culture, laid down rules, processes, and procedures. Although such studies might be carried out in the developing economies, little is known of such studies that focus on the MM of buildings in the developing economies. Conversely, the few studies that focused on the MM of buildings in developing economies have not adequately provided an overview of the concept of MM compared to those done in developed economies. Hence, the result of the studies conducted in the developed economies' application and relevance will not be consistent with those of the developing countries.

This section of the study identifies the gaps in the MM conceptual framework. It should be noted that the conceptual framework, as postulated by Aigbavboa (2014), provides the perspectives from which problems are identified in an existing framework. However, it is most likely that there are some gaps in the developed conceptual framework targeting the developed countries that have failed to capture the factors affecting or influencing the MM of municipal buildings in South Africa and other developing economies at large. This section of the study attempts to address the gaps that have been identified. They include maintenance culture and communication

DOI: 10.1201/9781003344681-6

among stakeholders. This occurs in the maintenance process and operations towards meeting set maintenance objectives and user requirements.

Nonetheless, in this study, the buildings assessed and referred to are municipal (educational) buildings. These buildings are built by the municipal government and allocated to academic, administrative, and non-essential staff. However, considering these identified gaps is based on the notion that effective MM of municipal buildings cannot be achieved without a viable maintenance culture and effective communication among stakeholders in the maintenance process. This is because numerous variables account for the effective MM of buildings and not a single-track assessment.

Gap One: Understanding Maintenance Culture in Maintenance Management

Culture, as defined by the Merriam-Webster dictionary (2002), is a way of thinking, behaving, or working in a place or organisation. Olutayo and Akanle (2007) define culture as a total way of life of people. As such, culture, as defined by Gilmore (2000: 29), is:

> ... ... *that complex whole which includes knowledge, belief, art, morals, law, custom, and any other capabilities and habits acquired by man as a member of society.*

Gilmore (2000) observed that anything that differentiates human beings from nature (biology and evolution) is relevant to the culture. Gilmore (2000) further postulated that culture was what produced a distinctive identity for society, socialising members for greater internal homogeneity. Indeed, it is what is taught in the institutions that make up the social structure that is reflected in culture. Ekeh (1989) advanced the view that culture is a certain behavioural pattern that is supposed to be exhibited. Wilkins (1994), in his study, states that culture is an integrated pattern of human behaviour and interactions. Wilkins (1994) further indicates that behaviour includes thought, speech, actions, and those objects created as a result of our actions.

The basis for different cultural behaviour is society. Different societies exhibit different characteristics which are based on their culture. Thus, Olutayo and Akanle (2007) define society as a group with a culture that is organised to satisfy all human needs and interests. Olutayo and Bankole (2002) study, influenced by Marx, the founding father of sociology, shows that the culture of any society is influenced by the environment in which the people have found themselves. Also, they have been able to understand and exploit the environment (Ogunbayo, 2021). Eti, Ogaji, and Probert (2006) observed that culture endures and evolves through our learning capacity and consequently through sharing current knowledge with succeeding generations.

There are various perceptions of the term "culture"; however, Eti et al. (2006) argued that culture is perceived as the key that influences the behaviour of getting things done the right way, without which there is a hindrance to the attainment of goals. They further concluded that culture is the total of the hereditary beliefs, ideas, knowledge, and values that back the shared bases of social actions. Eti et al. (2006) state further that only through learning and validation the required behaviour becomes part of the culture as being acceptable and desirable by organisations. Wilkins (1994), Brendan (2006), and Sani, Muhammed, Shukor, and Awang (2011) observed that culture in the perspective of a work establishment is put in place when the social relationship among members influences their pattern of thinking, behaviour, and beliefs.

However, the use of maintenance culture in some existing literature for effective MM of municipal buildings has been marginalised. According to Eti et al. (2006), this has allowed huge and very expensive projects to fall into disuse in a short while. Furthermore, it has also contributed to the increase in maintenance costs, unsafe environments for users of public facilities, and negatively impacted infrastructural development (Enemuo, Ajala & Offor, 2015; Tijani, Adeyemi & Omotehinshe, 2016). The study of Eti et al. (2006) emphasised that a maintenance culture ensures regular servicing, repairs, and maintenance of working assets or established systems to guarantee their continuous usefulness. Eti et al. (2006) also observed that the inculcation of culture in the maintenance process has the advantage of increasing the quality of maintenance activities. Sani et al. (2011) also posited that the maintenance culture is unique for each organisation; therefore, it is important to develop a maintenance culture in each maintenance organisation to meet the changes in the market demands and trends. Alani (2012) opines that a culture of maintaining and sustaining infrastructures is relevant to the national development of a nation. Enofe (2009) and Iruobe (2011) posited that the dearth of a maintenance culture, especially in the public sector of most developing countries, has been the bane of an infrastructure-driven national development.

Thus, in addressing the culture gap in maintenance operations for the effective MM of buildings, the Z theory, motivation theory, and need theory were mainly engaged to address the gap in knowledge more holistically. These theories cover a wide scope of thought. It appears close to containing all the essential vital skills required for effective management in an organisation through increasing employee loyalty to the organisation by providing a job for life with a strong focus on the employee's well-being, both on and off the job. The Z theory, as described by Aithal and Kumar (2016), citing Ouchi (1981), is built on the premise that it is not technology that is important in evaluating the efficiency of the management process of an organisation. Management's effectiveness towards good performance depends on a special way of managing people (Ouchi & Price, 1978). This statement was further stressed by Holt (1990), namely that the Z theory recognised the adoption of cultural differences within an organisation without modification. Accordingly, Aithal and Kumar (2016) state that Z theory tends to show that building a happy and close working relationship between co-workers and their employer through employee support is an important aspect of a culture system. As Aithal and Kumar (2016) observed, employees, value a working environment that will improve their performance, especially when tradition, family culture, and social institution are regarded as equally important as an assignment or work.

Nevertheless, the statement of Ouchi (1981), as bolstered by the views of Aithal and Kumar (2016), posited that the Z theory is a management philosophy that deals with an organisational culture that shows how employees view their management and their set of beliefs, values, and principles. This means that the Z theory visualises an individual employee in association with the social surroundings, such as culture, family, and friends, and outlines the need to integrate it into the managerial perspective. Jerald and Baron (2008) posited that a maintenance organisation needs to (i) identify systematically how employees/users behave under a variety of conditions; (ii) know why employees/users behave as they do; (iii) project future employees/users' behaviour; and (iv) understand how the employees can be motivated and directed regarding their responsibility to enhance both individual and group performance to boost the productivity of the organisation.

Similarly, X, Y, and Z theories were prominently used as winning strategies for people management in organisational behaviour. Indabawa and Uba (2014) state that this is based on certain assumptions about human beings (employees and users) attitudes towards work or a product (building). As observed further by Indabawa and Uba (2014), X and Y theories are in sharp contrast in their view of human nature. While the Z theory ascribes reasons for organisational efficiency to effective management style and the same time, invests faith in individuals' (employees) capacity for attachment and decision making. Nevertheless, using the Z theory as one of the lead theories for the study gap show variables such as good managerial style, organisational support for employees, organisational confidence in employee capacity, better working environment, and recognising the on-job decision-making capacity of employees are what change the attitude of maintenance personnel in the maintenance operations towards effective MM activities.

Additionally, Nurun and Dip (2017), citing Herzberg's (1968) theory of motivation based on job enrichment and hygiene factors, state that there are two distinct sets of factors that can improve the employee's attitude towards achieving the set goals of an organisation. One major aim of the theory is to determine the effect of motivation on employees' attitudes towards organisational effectiveness. According to Nurun and Dip (2017), motivation inspires someone to take a desired course of action. In an organisation, it is worth noting that motivation can have a negative or positive impact.

However, Nurun and Dip (2017) posited that positive motivation always brings new ideas and conditions for better organisational performance. Nurun and Dip (2017) further postulated that effective motivation in the operation process could change the employee's attitude in a good direction, leading to the perseverance of effort towards attaining a set goal of the organisation. Nonetheless, as Rajput, Bakar, and Ahmad (2011) noted, employee attitude on organisational effectiveness stems from a need that must be satisfied. This leads to a specific change in the behaviour of the employees.

As stated by Nurun and Dip (2017), citing Herzberg (2008), motivation towards improving the attitude of the employee can be divided into two factors which include job enrichment (extrinsic) and the hygiene factor (intrinsic) motivation. Nurun and Dip (2017) postulated that the extrinsic factor could have a powerful and immediate effect, but it might not necessarily last long. Rajput et al. (2011) state that the extrinsic factors include but are not limited to recognition, achievement, the work itself, responsibility, and advancement. However, Nurun and Dip (2017) maintained that intrinsic motivation refers to internal factors such as company policy and administration, supervision, interpersonal relationships, working conditions, remuneration, status and security, job satisfaction, responsibility, freedom to act, developing skills and abilities and challenging work and opportunities for development. As Rothberg (2004) and Tella, Ayeni, and Popoola (2007) noted, extrinsic factors motivate the employee and improve their attitude when they are present in the organisation's working environment.

However, Rothberg (2004) and Tella et al. (2007) assert that intrinsic factors motivate people and leave them dissatisfied when their needs are unmet. Nurun and Dip (2017) observed that intrinsic factors are likely to have a stronger and longer-term effect because they are more concerned with the quality of working life within an organisation. Rothberg (2004) argued that their desired performance level might not be achieved unless individual employees' attitudes are motivated to use the potential around them during the employment process. Similarly, Nurun and Dip (2017) further note that in achieving organisational goals and objectives, a satisfied employee will contribute more and positively through a better attitude, whereas a dissatisfied employee could develop an attitude that can destroy the set goals and objectives. Based on this, Kian, Yusoff and Rajah (2014) postulated that satisfaction is of the essence in an organisation for the employee to develop an attitude that can boost the production output of their organisation. Tella et al. (2007) concluded that extrinsic factors motivate employees towards good behaviour and positive attitudes whenever they are given organisational tasks. Rajput et al. (2011) hypothesised that the intrinsic factors discourage employees and leave them dissatisfied when their expectations are unmet. However, they submitted that employees are challenged, and their productivity increases if the organisation's working environment has enough extrinsic factors.

Equally, Uysal, Aydemir, and Genc (2017), citing Maslow (1948) states that individual (employee/user) needs can create internal pressures that can influence behaviour change towards meeting set organisational objectives.

These theories have been used extensively in many fields, but mainly in the business and management domain to help human management, organisational development, and transformation (Abdollahian, Coan, Oh, & Yesilada, 2012; Rosenbaum, More & Steane, 2018). Nevertheless, Uysal et al. (2017) opined that Maslow's hierarchy of needs theory is one of the early theories considered relevant to individuals as a motivational tool for individual performance within an organisation. As postulated by Gignac and Palmer (2011), in any organisation, an individual cooperates with advancing the organisational culture with a passion for attaining excellent results through efficient input. Moreover, Hunter and Schmidt (1996) state that existing conceptual models and imperative empirical evidence show that motivation through need is one of the main determinants of performance, success, and efficiency. Sekhar, Patwardhan, and Singh (2013) state that motivational flow needs to be understood by managers within an organisation with the creation of a culture for the organisation and a level of motivation in which employees are always more productive.

Consequently, Uysal et al. (2017) state that motivation through needs is a general concept involving wishes, desires, needs, interests, and drives, while physiological motivations such as sexuality, thirst, and hunger are called drives. Cüceloğlu (2016) postulated that high-level drives such as the human-specific desire to achieve are called needs. Uysal et al. (2017) assert that consistent desires within an organisation are dependent on how motivated the employees are. As indicated by Campbell (1990), meeting the needs of employees through motivation in particular guides employees' inclinations to extend their efforts, the level that these efforts reach, and how long employees will stay at this level of effort. Meeting individuals' (employees and users) needs is considered a driving force in creating and designing tools that will be used to carry out meaningful organisational activities (Hancock, 2009). However, there are two dimensions to motivation by needs that can drive individuals into action towards meeting organisational set goals (Uysal et al., 2017). Hossain and Hossain (2012) postulated that these dimensions are motivation through internal and external resources. Kasser and Ryan (1996) and Ryan and Deci (2000) hypothesised that inner motivation of needs directly satisfies the innate psychological needs such as belonging, personal existence, relationship, autonomy, and competence. Hossain and Hossain (2012) posit that motivation through internal resources (employees) towards effective management comes from within the individual, such as feeling good and satisfied at the end of a good job within an organisation.

On the other hand, the external motivation (users) is the enhancement that comes from other people through material or values that increase or decrease the possibility of recurrence of behaviour in both positive and negative ways (Soyer et al., 2010). It can be further related to support awards and encouragement from another individual for sustaining an activity (Buchbinder & Shanks, 2007). However, Edrak, Yin-Fah, Gharleghi, and Seng (2013) suggested that both internal (employees) and external motivators (users) increase the job satisfaction of employees and their essential contributions to organisational activities (maintenance activities). Fundamentally, motivation by the needs theory, as postulated by Maslow (1943), has a strong and long-term influence on employees' and users' behaviour and attitude towards organisational culture (Edrak et al., 2013).

In the needs theory postulated by Uysal et al. (2017), citing Maslow's (1943) theory of needs, it is contended that human needs are unlimited. After a need is satisfied, another need will arise. Needs have a certain hierarchical order; no need or drive can be considered independently; every need is related to the satisfaction or dissatisfaction of other needs; unsatisfied needs are a great source of human motivation. Uysal et al. (2017) state that motivation results from a person's attempt at fulfilling five basic needs: physiological, safety, social, esteem, and self-actualisation. The theory, as postulated by Uysal et al. (2017), citing Maslow (1948), is based on a hierarchical model with basic needs at the bottom and higher needs at the top (see Figure 6.1).

Conversely, meeting the needs of the individual (employees and users) within maintenance organisations will increase the quality of occupational performance and the production of the mental effort that directs the knowledge and skills of both employees and the users of the maintained buildings or assets (Uysal et al., 2017). Likewise, Uysal et al. (2017) assert that provision for adequate motivation in meeting the needs of employees and users will increase their performance with a multiplying effect on maintenance activities, further boosting users' productivity. This is because the most highly skilled employees of maintenance organisations can reduce their performance if their needs are not met through adequate motivation based on their needs (Table 6.1).

Uysal et al. (2017) opined that the need theory shows that employees must be motivated based on their needs if attitudinal change towards continuous performance and productivity of developing a maintenance culture is required within an organisation. Nonetheless, using the needs theory as one of the theory gaps shows variables such as drives towards the attainment of a

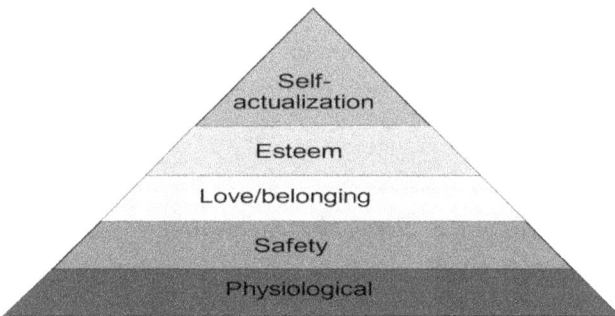

Figure 6.1 Maslow's hierarchy of needs.
Source: Amaratunga (2001); Uysal et al. (2017).

Table 6.1 Needs theory with relation to culture as a variable for effective maintenance management of buildings

Maslow's hierarchy of needs	Impact on individual	Impact on maintenance culture for buildings
Physiological needs	Meeting basic needs such as building, food.	Drives towards attainment of maintenance objectives and set goals. Efficient input.
Safety needs	Needs for a stable environment.	Creating the required tools/ environment for effective maintenance activities.
Love/social needs	Needs related to affectionate relations with others and status within a group.	Developing a maintenance culture among personnel within the organisation towards set maintenance goals.
Esteem needs	Need for self-respect, self-esteem, reputation, prestige, and the esteem of others.	Meeting maintenance activities set goals based on the organisation's objectives.
Self-actualisation	Need for self-fulfilment.	Continuous improvement to achieve good performance, success, and efficiency.

Source: Researcher's review (2022).

result, creating required tools for maintenance activity, the security of job, good working environment, good working relationship with colleagues and others, meeting maintenance set objectives and goals, good rewards for excellence performance, individual success, productivity, and efficiency were used to measure a maintenance culture within a maintenance organisation.

Thus, maintenance culture has been included in this present study to ascertain its relative contribution to the effective MM of the South African municipal buildings amid other factors that have already been researched. Maintenance culture is the change in the behaviour, mindsets, and attitude of the maintenance team towards carrying out maintenance activities and the operation of an organisation done the right way, without which there is a hindrance to the attainment of the maintenance objectives of the organisation (Abiodun, Olayemi & Joseph, 2016; Ogunbayo et al., 2021).

Challenges of Maintenance Culture in Developing Economies

In this study, before a conclusion can be reached on how the MM framework of municipal buildings is developed. It is necessary to understand why there is a divergence in MM research findings and whether the existing theoretical framework as proposed by western researchers has some gaps that have not accommodated the context of the developing nation fully in the area of maintenance culture within the maintenance process among stakeholders.

As observed by Adedokun (2011), the development of maintenance culture is one of the major forces catalyses the growth of any nation's economic, social, and technological advancement. Some existing literature for the developing countries on MM reveals the deplorable condition of municipal buildings in most developing African countries, the level of deterioration of which poses a great concern to both users and other stakeholders. Tijani et al. (2016) observed that the appearance of facilities in Nigeria's municipal buildings such as hospitals, airports, roads, and most especially educational buildings would indicate that the society lacks a cultural behaviour that ensures effective and efficient functioning of the facilities as well as fostering national development. Conversely, provision for adequate care of the hard-earned infrastructure has not gained ground in the consciousness of maintenance managers in the country over the years because of the absence of a maintenance culture.

In Nigeria, a poor maintenance culture has become a widely acknowledged problem (Mbamali,2003). This has given the country low priority to property management and maintenance activities (Wahab, 1995). The findings of Mbamali (2003), citing Usman, Gambo, and Chen (2012), affirmed that Nigeria has no functional maintenance policy and, therefore, no such culture exists. Usman, Gambo, and Chen (2012) state that an inadequate maintenance culture in almost every building in Nigeria is a notable feature. Olufunke (2011) posited that this is partly due to the problem of a maintenance culture based on societal behaviour regarding maintenance. Conversely, Usman et al. (2012) postulated that Nigeria's leading factor in the ineffective MM of municipal buildings is the declining maintenance culture.

In Ghana, the study of Nkrumah, Stephen, Takyi, and Anaba (2017) shows that warranted attention has not been given to the maintenance of municipal buildings as this has restricted Ghana's development through a gaping infrastructural deficit and poor maintenance culture of existing municipal buildings. This was further buttressed by Eghan (2014), namely that Ghana continues to invest heavily in new municipal buildings while the sustainability of the existing ones suffers from a poor maintenance culture. Kportufe (2015) opines that Ghanaians are growing aware of the lack of a maintenance culture regarding municipal buildings in the country. Kportufe (2015) stated that Ghanaians' attitude to municipal buildings is generally negative, with the common understanding that they are "nobody's property." The study of Twumasi-Ampofo et al. (2017) clearly states that in Ghana, most municipal buildings accommodating public servants have not seen any significant maintenance since construction, largely owing to a lack of attitude to a maintenance culture. Nkrumah et al. (2017) postulated that the maintenance of municipal buildings is virtually non-existent in Ghana owing to a lack of a maintenance culture.

In Kenya, Coetzee (1999), as cited by Magutu and Kamweru (2015), observed that municipal building maintenance had not received much attention because the emphasis is more on the development of new buildings.

Magutu and Kamweru (2015) further posited that a crisis looms regarding the building stocks due to the non-existence of a maintenance culture. This is because existing private buildings and other municipal infrastructure are becoming run down and losing their utility value owing to a lack of a maintenance culture. Additionally, Magutu and Kamweru (2015) opine that the maintenance policy in Kenya based on borrowed cultures from the developed countries' strongly grounded maintenance culture has been proven environmentally unfit for MM activities in Kenya.

Similar problems of maintenance culture are evident in other parts of Africa, as acknowledged in South Africa. The study of Bothma and Cloete (2000) shows that the maintenance problem concerning municipal buildings in South Africa is not unique. Bothma and Cloete (2000) observed that the maintenance of municipal buildings could not be adequately addressed owing to several structural and other factors such as lack of a MM culture. The findings of Xaba (2011) on the MM of municipal buildings in South Africa show that literature on facilities maintenance is mostly based on developed countries' maintenance systems. Bearak and Dugger (2008) postulated that the inability to meet industry and consumer demands on basic services is due to aging facilities caused by a lack of maintenance culture of the existing facilities. This assertion was buttressed by SAICE (2006), namely that since 2016 major maintenance operations for municipal buildings have been affected due largely to a lack of a proper maintenance plan and culture. Furthermore, Thurlby (2013) observed that there is little funding assigned to the maintenance of municipal buildings while new projects are funded, and existing municipal buildings are deteriorating owing to many years of maintenance neglect and the dearth of a maintenance culture. Ntjatsane and Kodongo (2017), in their study on financing infrastructure maintenance in South Africa, concluded that because of years of sub-standard maintenance, mainly due to a lack of a maintenance culture, the quality and reliability of most municipal buildings remain poor.

Literature Findings in Relation to Maintenance Culture in Maintenance Management of Buildings

On a general note, facilities, especially buildings, were developed to fulfil the needs of society and organisations. However, Suwaibatul Islamiah et al. (2012) observed that public assets, especially municipal buildings, are not maintained properly because of the nonexistence of a maintenance culture. Hence, it is essential to develop a maintenance culture based on the buildings' users and the maintenance personnel's behaviour, environmental needs, values, and cultural beliefs. Thus, Suwaibatul, Abdul Hakim, Syazwina, and Eizzatul (2012) postulated that facilities such as buildings are critical to an organisation's resources, consequently improving the working environment and well-being. Therefore, they should be given serious attention through a good maintenance attitude.

Annies (2007) observed that the maintenance problem due to the dearth of a cultural attitude towards maintenance has become an important agenda for developing countries and put pressure on their government to manage their assets and facilities (municipal buildings). Ogunbayo et al. (2022) state that the absence of maintenance culture and behaviour has led to increasing management and maintenance costs for the repair of damaged buildings. Suwaibatul Islamiah et al. (2012) observed further that maintenance problems could be solved based on the behaviour of each individual in the maintenance process of assets and facilities. Cultural attitude and behaviour change are essential to improve tenancy, skills, and diligence in maintenance work (Mbamali, 2003; Suwaibatul Islamiah et al., 2012). Individuals or groups with a maintenance culture would have a good attitude towards maintaining, preserving, and protecting private and municipal buildings (Suwaibatul & Hakim, 2011). Florence (2011) opines that a maintenance culture is not universal; it is learned or derived through individuals or groups, making maintenance a natural daily practice. Suwaibatul Islamiah et al. (2012) observed that a maintenance culture is not easy to develop. This is because maintaining a good maintenance culture takes time and occurs in response to changes in the individual (Carroll & Hannan, 2000). For society to exist, its needs culture (Abdollahian, Coan, Oh, and Yesilade, 2012). The fluid operation of society is supported by cultural norms and cultural values that guide people in making choices (Abdollahian et al., 2012; Ogunbayo et al., 2022). Just as members of a society work together to fulfil a society's needs, culture exists to meet its members' basic needs (Florence, 2011).

However, there is a dearth of studies that investigate the relationship between the maintenance cultures of stakeholders in the maintenance of municipal (educational) buildings in the South African education sector. Researchers have differed in their views about the specific variables that constitute the determining factors in developing a maintenance culture in the management of buildings. Suwaibatul Islamiah et al. (2012) identified communication, reward systems and recognition, empowerment, motivation, involvement, strategy and work planning, teamwork, a policy system, training and education, and organisational culture as determining factors in the development of a maintenance model. Nonetheless, Tijani et al. (2016) identified variables such as corruption, leadership challenges, attitudinal problems, and a lack of policy as some of the factors that cause the non-existence of a maintenance culture within an organisation and/or a country. Nevertheless, little is known of the specific variables that constitute the maintenance culture MM systems require for municipal buildings in the South African educational system. Furthermore, the researcher who studied the correlation between the MM of buildings and maintenance culture have largely analysed data using regression and/or correlation analysis and generally have the developed countries as their study area (Suwaibatul & Hakim, 2011; Tijani et al., 2016; Wilkinson, Johnstone & Townsend, 2012).

However, there is a concern about the degree the MM of buildings is determined by maintenance culture and the relative influence of maintenance culture on other variables that have been studied in previous models. This study seeks to provide the answers through the MM framework it seeks to develop for municipal buildings in the developing economies' education sector using the Delphi technique. Nonetheless, informed by the literature reviewed, the specific variables representing maintenance culture in the MM of municipal buildings in the South African education sector that this study examined include proper maintenance system, use of right skills maintenance personnel, benchmarking for maintenance performance, diligence in maintenance operations, building reliability, enhanced performance level of buildings, sustainability of building infrastructure, improved economic value of a country, attitude of ensuring regular servicing, attitude of ensuring regular repairs, attitudinal change to maintenance operations, user change of attitude to maintained buildings, commitment to change initiatives, effective MM style, prioritisation of maintenance activities, effective decision-making capacity, motivation of maintenance personnel and maintenance decision based on societal value.

Gap Two: Understanding Communication Among Stakeholders

Communication, as defined by Gizir and Simsek (2005: 198). ... is the transfer of information from one person to another that involves the exchange of facts, ideas, suggestions, and emotions between two or more people by speaking, writing, or using some other medium. It also involves the interaction of stimulus meanings through giving and receiving messages. Communication is an essential part of human life when something needs to be done at work (Gizir and Simsek, 2005: 197–221). In maintenance, communication plays an important role which involves delivering information on maintenance work practices for all members of the organisation so that every individual understands the importance of the facility and assets maintained in the organisation. According to Femi (2014), communication means different things to different people: it is a many-sided phenomenon. Femi (2014) further posited that communication is a process or a means of access to the mind of others. According to Wilson (1992), communication reduces uncertainty; in other words, it is an exchange of meanings. In maintenance activities within an organisation, individual building users hold the power to accept, reject, or be dissatisfied with maintenance operations around their facilities due to a lack of communication with the maintenance operators. As Altınöz (2009) observed, communication has a vital significance between a maintenance organisation and the users of a building. Communication is important in all maintenance operations to achieve organisational objectives (Choon Hua, Sher, & Pheng, 2005).

Holtzhausen (2002) study shows that effective communication improves job satisfaction and user trust. Kramer (1999) opines that employee commitment to an organisation could vary owing to poor communication. However, as posited by Goris (2007), effective communication improves employee job performance. Duncan and Moriarty (1998) state that communication is the human activity that links people and creates relationships, especially towards achieving a set goal. Equally, Femi (2014) states that communication is the essence of management for effective and efficient organisational work performance. Management's basic functions, including planning, organising, directing, and controlling, cannot be performed well without effective communication to attain organisational objectives and goals (Arnold & Silva, 2011). In today's organisational management, communication systems are evolving rapidly, which is crucial for the growth and success of an organisation (Femi, 2014). Human resources and management constitute the biggest challenge, unlike other inputs from other factors of production. The management of employees through organisational communication could demand the skilful handling of thoughts, feelings, and emotions to secure the highest level of productivity (Wilson, 1992). Nevertheless, Clowes (2000) postulated that the failure of organisational management to coordinate the perfect flow of communication among employees and other stakeholders might likely lead to an inefficient management process with many workers being demotivated.

However, in MM of the building and its attached facilities, communication is not restricted to its management only. It covers all activities that maintenance managers carry out to transform the users' minds (Banerji and Dayal, 2005). According to Banerji and Dayal (2005), communication in MM is a process that includes expressing, listening, and understanding. Equally, Goris, Vaught, and Pettit (2000) emphasised that communication within the process of MM is based on social life and forms the content of the organisational structure, which aims at conducting good relationships between users of facilities and their management. Pearce (1994) defines communication in relation to MM as the degree to which information about maintenance operations is transmitted among stakeholders. However, Ayatse (2005) noted that communication is important in any management process because it helps establish and disseminate an organisation's goals. Adequate and sustainable communication levels among stakeholders within the maintenance activities management will show the personnel's competencies and skills (Clowes, 2000). Furthermore, Femi (2014) posits that communication will exhibit the work behaviour of maintenance personnel which is appropriate and relevant to their job performance towards meeting the maintenance needs of the users.

Nevertheless, the value perception theory and expectancy theory were drawn upon in addressing the gap in the importance of communication among stakeholders to maintain municipal buildings effectively. These theories have been used extensively in many fields, but mainly in the business, marketing, and management domains to provide help on how to meet employees' values and job satisfaction towards meeting consumers' (users) needs and expectations (Locke, 1976; Oliver, 1980; Westbrook & Reilly, 1983; Adeniji, 2011). They have the potential to help in addressing the gap in MM of buildings research.

The value perception theory, as described by Yuksel and Yuksel (2008), citing Westbrook and Reilly (1983), indicates that expectations of a product may not correspond to what is desired or valued in a product. Yuksel and Yuksel (2008) opined that in value-perception theory, customer satisfaction is an emotional response activated by a cognitive evaluation process in which the offered perceptions are compared to one's (users) values, needs, and wants, or desires. Thus, from the value percept theory, communication among stakeholders is seen as an important drive and a morally right step towards the management of both human needs and facilities within an organisation. Schroeder (2012), the perception of people in the organisational process must be considered regarding what they think or feel within the organisational environment. This shows that for the effective MM of buildings, there is a need for interaction among stakeholders. Based on value perception theory, Brunetto and Farr-Wharton (2004) state that to improve work ethics, there is a need for a strong relationship among stakeholders towards job commitment and satisfaction. Femi (2014) states that feedback exchange within an organisation is a process of interpersonal negotiation whereby personnel is committed to working and the users' values, needs, wants, or desires are met through effective communication.

The expectancy theory, as described by Soyoung and Sungchan (2017), citing Vroom (1960), refers to expectancy as a product of motivational-based organisational relationships. The theory has been tested within both industries and management to understand better employee motivation towards meeting organisational objectives (Amaratunga, 2001; Candela, Gutierrez & Keating, 2015; Baciu, 2018). The expectancy theory has been contextualised regarding its effects on employee relationships, commitment, and organisational citizenship (Porter & Sleers, 1973; Mobley, 1997). Lunenburg (2011) states that the expectancy theory focuses on the cognitive antecedents that contribute to or detract from personal motivation towards organisational objectives. Lloyd and Mertens (2018) posited that the theory could reasonably predict workplace motivation and behaviour. The theory is based on the insinuation that individuals have a choice, and their decision could be based on which choice they perceive will lead to the best personal outcome for organisational goals (Lloyd & Mertens, 2018). However, Soyoung and Sungchan (2017), citing Vroom (1960), indicate that the expectancy theory is developed on three premises: expectancy, instrumentality, and valence.

This could be represented by the equation below:

Motivation (contribution through communication)

= Expectancy * Instrumentality * Valence

Expectancy	Instrumentality	Valence
Effort	**Performance**	**Reward**
Effective Communication	Effective MM	Proactive / Value base Maintenance

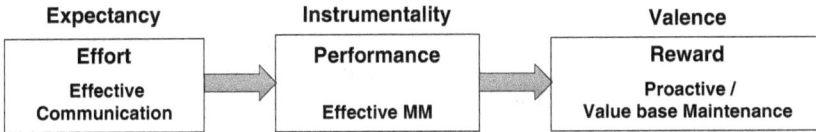

Figure 6.2 Expectancy theory with relation to communication as a variable for effective maintenance management.

Source: Adapted from Vroom (1960).

According to the theory, expectations are the stakeholders' belief that their efforts within the organisational operation will lead to specific performance or operational improvement. It is also the degree to which individuals within an organisation believe their abilities will lead to setting goals or improving a system. Lloyd and Mertens (2018) posited that expectancy theory describes a motivational force as an individual who believes that a certain effort or relationship (communication) will lead to a given performance (expectancy). The performance will lead to the attainment (instrumentality) of a desirable (valence) reward (see Figure 6.2).

Baciu (2018) explained that the expectancy theory accounts for the role of personal abilities and skills, including the existing and past relationships in an organisation. It further creates workplace momentum due to social cognition (Gyurko, 2011). The contention of Pearce (1995), based on contextual theory, affirms that a proper understanding of the passed information between the parties involved is necessary for effective communication in the management process. Thus, based on expectation theory, communication among stakeholders in the maintenance process for effective MM of buildings is regarded by Amaratunga (2001) as the key to improving service quality and services through oral and written communication, property services bulletins, databases and programmes, property services brochures and action groups.

For instance, Talib, Ahmad, Zakaria, and Sulieman (2014) posited that good verbal communication helps the maintenance organisation to understand the maintenance needs of the users of maintained buildings. However, a recent study by Choon Hua et al. (2005) found that most MM teams used electronic mail for complaining rather than immediate reports and progress reports. Nonetheless, most building users preferred to communicate in face-to-face meetings or via telephone calls (Yahya and Ibrahim, 2010). Moreover, Wordsworth and Lee (2001) observed that in an unquestioning relationship, communication regarding the MM of the building process might be through telephone, especially through face-to-face meetings rather than electronic mail. Most of the existing studies on communication on MM have generally concentrated on the briefing and maintenance process. However, Yahya and Ibrahim (2010) state that effective communication in building management depends on the system's safety, quality, and service between maintenance managers and building users, particularly during maintenance activities. Moreover, Choon Hua et al. (2005) opined that a viable communication system among stakeholders is crucial in building management. Communication for an effective maintenance process requires cooperation between maintenance managers, personnel responsible for maintenance activities and operations, and the building users, who also comprise the stakeholders (Yahya and Ibrahim, 2010).

Conversely, Clowes (2000) postulated that a lack of effective communication in the MM of buildings could affect information distribution towards sustainable maintenance activities. Accordingly, Nita Ali et al. (2002) posit that poor communication and the dearth of knowledge sharing between the maintenance team and building users are the main causes of problems, specifically in technical and documentation aspects. Also, Wordsworth and Lee (2001) state that poor communication between MM groups and building users (for example, by failing to keep customers fully informed of progress) is one of the factors that affect working efficiency and is also a reason for the relatively low productivity of the maintained building. According to Langston and Lauge-Kristensen (2013), an effective communication system in MM is critical to cope with the considerable knowledge needed to solve many complex problems and challenges. Nevertheless, Yahya and Ibrahim (2010) opined that the communication system requires the participation of the MM teams responsible for operations and maintenance activities and the building users, who are also the stakeholders of the maintained building.

Therefore, the study on communication among stakeholders in maintenance operations is essential to understand and improve the fusion between maintenance managers and the users of buildings, which is crucial to developing effective MM of buildings. However, there is a dearth of studies on this important aspect of MM of buildings. Previous studies and developed frameworks have already established some correlation between effective communication systems among stakeholders – users and MM (Clowes, 2000; Nita Ali et al., 2002; Choon Hua et al., 2005; Yahya and Ibrahim, 2010). Moreover, most studies focused on maintenance organisations and largely used regression, correlation, or both for data analysis. Nonetheless, owing to the weakness of regression and correlation analysis, this study employed a more robust tool, namely the Delphi technique, to overcome the weakness of the regression and correlation analysis.

Likewise, there is a dearth of studies that determines the relationship between the communication among stakeholders and MM of municipal buildings in the education sector of developing economies. Thus, this current study seeks to address these gaps in knowledge as it investigates the relationship between communication among stakeholders and the MM of municipal buildings in the education sector of developing economies, using the South African education sector as a case study. Additionally, the study develops a MM framework for municipal buildings in the South African education sector using the Delphi technique. Also, the study determines the relative influence of maintenance culture and an effective communication system (the new constructs this study has added) amid other known constructs that previous studies have already established.

Literature Findings in Relation to Communication in Maintenance Management of Buildings

According to Femi (2014), communication covers an individual's activities when they want to transform someone else's mind. It is also the bridge between an individual or individuals and an organisation. Additionally, Banerji and Dayal (2005) state that communication is a process that comprises expressing, listening, and understanding.

Therefore, the importance of communication during operations within an organisation, including maintenance organisations, cannot be overemphasised. Most recent empirical studies about organisational communication show a positive correlation between communication and organisational outputs such as performance, organisational commitment, organisational citizenship (users) behaviour, and job satisfaction (Husain, 2013). Conversely, Bastien (1987) and Malmelin (2007) observed that a dearth of communication within an organisation's operation might cause functionless results such as stress, a decrease in organisational commitment, organisational citizenship (occupants/users), dissatisfaction, low trust, severance intention, job dissatisfaction, and absence. In other words, Zhang and Agarwal (2009) observed that the efficiency of the organisation's set objectives can be affected based on this.

Campbell and Finch (2004: 182) posited that users' satisfaction with facilities is not determined by technical performance only but by "an elaborate set of exchange processes," including effective communication and management of users' expectations. Campbell and Finch (2004) observed that further absence of communication between personnel and users of the facility within maintenance organisation operations has led to unforeseeable changes, which result in an alteration to maintenance operation deadlines and resources management. Tyler and Bies (1990) postulated that factors such as sufficient consideration of the viewpoints of others, being bias-free, the provision of timely feedback, consistency in the criteria on which decisions are based, and effective communication on the basis for decisions were factors that influence both users and employees' perceptions of procedural fairness in organisations' operations. Additionally, Campbell and Finch (2004) hypothesised that communication allows individuals to control the submission of evidence to support their request, which is then incorporated into the decision to work according to the needs specified. Campbell and Finch (2004) further posit that facility users often cannot control outcomes of organisation operations; however, they can influence the procedure involved through communication with relevant personnel.

Accordingly, Campbell and Finch (2004) affirm that communication provides an individual with an opportunity for self-expression in the management process. Also, Gibson and Boreham (1981) posited that communication is important to organisations for two vital reasons. The first, as noted by Gibson and Boreham (1981), is fundamental in meeting economic needs (further progress), while the second is social needs (customers' concerns in the business process). Albrecht and Adelman (1987) state that supportive communication within an organisation reduces perceptions of uncertainty and stress. Campbell and Finch (2004) opined that communication in the management process signifies effective ways of upholding the views of fairness necessary to maintain satisfaction. Even if individuals (users) are not satisfied with the outcome of a process, open communication among parties involved will lead to less dissatisfaction and dysfunctional outcomes. Amaratunga (2001) postulated that there is a requirement for rapid, timely, and accurate feedback on the relations between the partners on the operational side of the supply chain in MM operations. For this purpose, an excellent communication system is required (Figure 6.3).

However, satisfaction is not the only benefit of effective communication, especially in maintenance organisations. The study of Phattanacheewapul and Ussahawanitchakit (2008) on the impact of effective communication in the management process shows that communication increases productivity, improves staff morale and organisational loyalty, and help reduces industrial disputes, especially between organisation personnel and customers.

Nevertheless, Kreps (1990) delineates that personnel gathers information relevant to their organisation's progress and changes in organisational communication processes. Phattanacheewapul and Ussahawanitchakit (2008) posited that organisational communication is transmitting the news about the work process from the organisation to employees and through employees

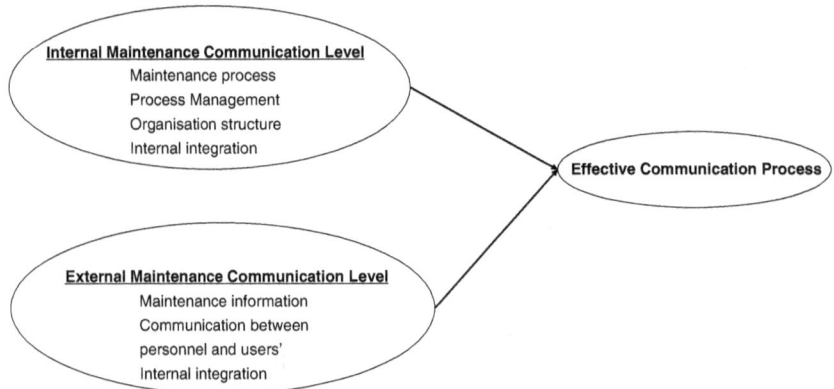

Internal Maintenance Communication Level
Maintenance process
Process Management
Organisation structure
Internal integration

External Maintenance Communication Level
Maintenance information
Communication between
personnel and users'
Internal integration

Effective Communication Process

Figure 6.3 Communication process in the maintenance organisation.

Source: Adapted from Ball (1997).

towards achieving organisational objectives. On the other hand, De Ridder (2004) postulated that the main objective of communication within an organisation is to inform the personnel about their tasks and policy issues towards meeting customer satisfaction. A similar objective as postulated by Postmes, Spears, Sakhel, and De Groot (2001) is that effective communication within an organisation will help construct a community within the organisation, especially between the workforce and a significant proportion of their customers. Barrett (2000) states that meaningful communication within an organisation's operational process informs and educates employees at all strategic units and motivates them to support operational strategies.

According to Tsai, Chuang, and Hsieh (2009), communication in organisational management is essential in generating and interpreting operational messages between employees through directional (one-way) and bidirectional (two-way) mediums. The finding of the study by Ogunbayo and Aigbavboa (2019), however, showed that a significant proportion of maintenance personnel (technical staff) lack communication skills in relating to the users of the buildings being maintained. According to Malek and Yazdanifard (2012), communication is more than saying or telling someone something; it tends to answer both customers' and employees' questions, and it further reduces employees' anxiety to keep them motivated towards achieving an organisational desired outcome. A recent study shows that many organisations fail to meet their objectives due to inadequate internal communication (Goris, 2007). Equally, Fakharyan, Jalilv, Elyasi, and Mohammadi (2012) opine that insufficient communication within the management process of an organisation creates a lack of awareness, uncertainty, rumours, and confusion.

Nevertheless, communication in MM can be viewed as a form of process control (i.e., voice effect) or as a means of allowing both users and maintenance personnel to express their views on decisions affecting them towards meeting the expectations and maintenance objectives of the organisation. Ogundipe et al. (2018) asserted that one factor that causes a delay in maintenance operation is the problem of communication between users and maintenance personnel within the axis of their duties, especially in the area of responding to the maintenance requests of users of the maintained facilities. Ogunbayo and Aigbavboa (2019) state that communication among maintenance parties will help identify factors that lead to constant maintenance requests within the maintained facilities.

The findings of Yahya and Ibrahim (2010) showed that the most effective communication took place when the reaction among stakeholders was proactive. However, Yahya and Ibrahim (2010) asserted that factors that affect efficient communication based on the impact of the proactive action in MM were related to the maintenance approach, the nature of the work, the mode of communication, adequate information, technical knowledge, personality, resources, contact with building owners and users, and working experience. The result of the study of Yahya and Ibrahim (2010) also showed that where the responses among stakeholders were not proactive, factors that influenced efficient communication were related to technical knowledge and adequate information. Additionally, Yahya and Ibrahim (2010) concluded that where the responses from the MM teams were proactive and the responses from building users were not proactive, the factors that affect efficient communication in MM were related to resources and personality.

Thontteh and Olanrele (2015) postulated that cutting costs should not always be the focus of maintenance managers but improving communication arrangements with building users to meet their expectations at all times and developing a user-driven service delivery. Buttressing this point, Thontteh and Olanrele (2015) posited that the maintenance manager should focus on users' relative satisfaction and relationship rather than cost. Furthermore, Thontteh and Olanrele (2015) suggested that the maintenance managers should interact with users, be ready to ask about their expectations and be willing to address them in time. Thontteh and Olanrele (2015) opine that understanding building users' expectations through better communication is the key to maintenance success because it will help identify gaps in service quality. Fatoye (2005) posits that users' satisfaction in building management services and performance can be evaluated through efficient communication.

Fakharyan et al. (2012) maintain that communication is a tool to build an effective management system and establish awareness of the need for change, to generate a desire to support and participate in the change. This change is impossible and might fail without effective communication. Choon Hua et al. (2005) posited that maintenance managers should ensure that the communication process is effectively carried out through serious attempts to communicate the nature and the impact of the proposed changes on operations. On the other hand, Goris et al. (2000) and Choon Hua et al. (2005) stated that most maintenance managers assume that communication within the maintenance organisation is a function of human resources or public relationships. However, based on the users' needs and expectations, communication must be a priority for every manager at every level. Nevertheless, one important objective of communication as a variable in effective MM of buildings is to improve the users' perspective of in-use facilities.

Conversely, concerning the extent to which the effective MM of buildings is improved by communication among stakeholders in the maintenance operations and the relative influence of communication in relation to other variables that have been studied in previous models, this study seeks to provide the answers through the integrated MM of the building framework it seeks to develop for municipal buildings in the education sector of the developing economies using the Delphi technique.

Informed by the literature reviewed, the specific variables representing communication among stakeholders in the MM of municipal buildings of the South African education sector that this study examined include collaboration among stakeholders, stakeholders' concerns, understanding of users' maintenance needs, good working relationships among stakeholders, quality-driven service delivery, users feedback, increase in users' participation, identify gaps in the maintenance process, users satisfaction, timely and accurate feedback, eradication of uncertainties, early defects detection, timely correction of defects, the establishment of the need for change, E-mails among stakeholders, help desk, telephone support, verbal discussion, checking maintenance information with users and web-based systems meetings and report.

Summary

The gaps observed in previous MM research frameworks, which were not assessed as all-inclusive constructs in previous models, were addressed in this chapter. The identified gaps formed the new constructs in the conceptual framework of the present study. Like any other organisation, maintenance organisations possess personalities, values, attitudes, and beliefs that guide how their operations should be carried out and how users of maintained building facilities are perceived and respected. However, the discussion in the chapter shows that the majority of existing MM frameworks were developed based on western culture, ideals, and values, with little or no contributions from the researchers from the developing economies. Similarly, the chapter identified a lack of maintenance culture and communication among stakeholders in the maintenance activities as the additional independent variables identified. The review of these two new independent variables shows that they can improve the MM of buildings. This is because a maintenance culture is capable of attitudinal change to maintenance, influenced by strong leadership, procedures, policies, and formal hierarchies. On the other hand, communication, especially between the MM team and users of the buildings, will improve maintenance operations and activities with increased productivity of the building maintained and its users. As shown in the review, the multiplying effects of identified independent variables could improve the MM system, moving away from the corrective maintenance approach towards the Value Base Maintenance (VBN) approach (see Figure 6.4).

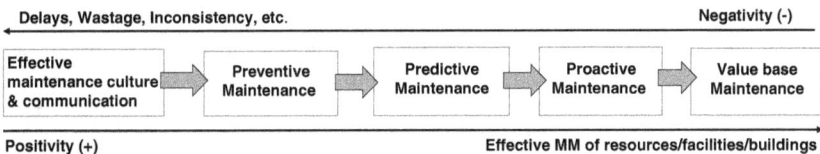

Figure 6.4 Communication and maintenance culture-based maintenance concept.

Source: Author's review (2022).

References

Abdollahian, M. A., Coan, T. G., Oh, H., & Yesilada, B. A. (2012). Dynamics of cultural change: The human development perspective. *International Studies Quarterly, 56*(4), 827–842.

Abiodun, T. S., Olayemi, A. A., & Joseph, O. O. (2016). Lack of maintenance culture in Nigeria: The bane of national development. *Civil and Environmental Research, 8*(8), 23–30.

Adeniji, A. A. (2011). *Organizational climate and job satisfaction among academic staff in some selected private universities in southwest Nigeria.* Unpublished doctoral dissertation. Ota: Covenant University.

Adedokun, M. O. (2011). Education for maintenance culture in Nigeria: Implications for community development. *International Journal of Sociology and Anthropology, 1*(8), 290–294.

Aigbavboa, C. O. (2014). *An integrated beneficiary centred satisfaction model for publicly funded housing schemes in South Africa.* Published doctoral dissertation. Johannesburg: University of Johannesburg.

Aithal, P. S., & Kumar, P. M. (2016). Comparative analysis of theory X, theory Y, theory Z, and theory A for managing people and performance. *International Journal of Scientific Research and Modern Education (IJSRME), 1*(1), 803–812.

Alani, A. (2012). Maintenance culture as threat to educational accessibility in Nigeria: Implications for sustainable open distance learning in Nigeria. *OIDA International Journal of Sustainable Development, 5*(11), 63–70.

Albrecht, T. L., & Adelman, M. B. (1987). *Communicating social support.* Washington, DC: Sage Publications, Inc.

Altınöz, M. (2009). An overall approach to the communication of organizations in conventional and virtual offices. *International Journal of Social Sciences, 4*(3), 217–223.

Amaratunga, R. G. (2001). *Theory building in facilities management performance measurement: Application of some core performance measurement and management principles.* Published doctoral dissertation. University of Salford.

Annies, A. (2007). Current issues and challenges in managing government's assets and facilities. In *Proceeding of the National Asset and Facilities Management (NAFAM) Convention,* Kuala Lumpur, 13 August 2007.

Arnold, E., & Silva, N. (2011). Perceptions of organizational communication processes in quality management. *Revista de psicologia, 29*(1), 153–174.

Ayatse, F. A. (2005). *Management information system: A global perspective.* Makurdi: Oracle.

Baciu, L. E. (2018). Expectancy theory explaining civil servants 'work motivation. Evidence from a Romanian city hall. *The USV Annals of Economics and Public Administration, 17*(2 (26)), 146–160.

Ball, S. (1997). Facilities management and supply chain management. In K. Alexander (Ed.), *CFM Best Practice 97.* Strathclyde: Strathclyde Graduate Business School.

Banerji, A., & Dayal, A. (2005). A study of communication in emergency situations in hospitals. *Journal of Organizational Culture, Communications, and Conflict, 9*(2), 35.

Barrett, P. (2000). Achieving strategic facilities management through strong relationships. *Facilities, 18*(10/11/12), 421–426.

Bastien, D. T. (1987). Common patterns of behavior and communication in corporate mergers and acquisitions. *Human Resource Management, 26*(1), 17–33.

Bearak, B., & Dugger, C. W. (2008). South Africans take out rage on immigrants. *New York Times*: 20 May.

Bothma, B., & Cloete, C. (2000). A facilities management system for the maintenance of government hospitals in South Africa. *Acta Structilia, 7*(1&2), 1–21.

Brendan, J. S. (2006). Optimising the maintenance function - It's just as much about the people as the technical solution. In J. Mathew, J. Kennedy, L. Ma, A. Tan, & D. Anderson (Eds.), *Engineering Asset Management* (pp. 568–575). London: Springer.

Brunetto, Y., & Farr-Wharton, R. (2004). Does the talk affect your decision to walk: A comparative pilot study examining the effect of communication practices on employee commitment post-managerialism. *Management Decision, 42*(3/4), 579–600.

Buchbinder, S. B., & Shanks, N. H. (2007). Management and motivation. *Introduction to health care management*. Burlington, Massachusetts: Jones & Bartlett Learning.

Campbell, J. D. (1990). Self-esteem and clarity of the self-concept. *Journal of Personality and Social Psychology, 59*(3), 538.

Campbell, J. D. (1998). *Uptime, strategies for excellence in maintenance management*. Portland, OR: Productivity Press.

Campbell, L., & Finch, E. (2004). Customer satisfaction and organisational justice. *Facilities, 22*(7/8), 178–189.

Candela, L., Gutierrez, A. P., & Keating, S. (2015). What predicts nurse faculty members' intent to stay in the academic organization? A structural equation model of a national survey of nursing faculty. *Nurse Education Today, 35*(4), 580–589.

Carroll, G. R., & Hannan, M. T. (2000). Why corporate demography matters: Policy implications of organizational diversity. *California Management Review, 42*(3), 148–163.

Choon Hua, G., Sher, W., & Sui Pheng, L. (2005). Factors affecting effective communication between building clients and maintenance contractors. *Corporate Communications: An International Journal, 10*(3), 240–251.

Clowes, A. (2000). *Barriers to information sharing in the construction supply chain*. Brisbane: Queensland University of Technology.

Coetzee, J. L. (1999). A holistic approach to the maintenance "problem". *Journal of Quality in Maintenance Engineering, 5*(3), 276–281.

Cüceloğlu, D. (2016). Savaşçı. (48. Baskı). *İstanbul: Remzi Kitapevi*.

De Ridder, J. A. (2004). Organisational communication and supportive employees. *Human Resource Management Journal, 14*(3), 20–30.

Dictionary, M. W. (2002). Merriam-webster. On-line at http://www.mw.com/homehtm

Duncan, T., & Moriarty, S. E. (1998). A communication-based marketing model for managing relationships. *Journal of Marketing, 62*(2), 1–13.

Edrak, B. B., Yin-Fah, B. C., Gharleghi, B., & Seng, T. K. (2013). The effectiveness of intrinsic and extrinsic motivations: A study of Malaysian Amway company's direct sale forces. *International Journal of Business and Social Science, 4*(9), 96–103.

Eghan, G. E. (2014). *Maintenance management of educational infrastructure in Ghana: Development of a framework for senior high schools*. Unpublished MPhil thesis. Ghana: Kwame Nkrumah University of Science and Technology.

Ekeh, P. P. (1989). The scope of culture in Nigeria. In P. P. Ekeh & G. Ashiwaju (Eds.), *Nigeria since independence—the first 25 years* (pp. 1–16). Ibadan: Heinemann.

Enemuo, O. B., Ajala, J., & Offor, R. (2015). The influence of maintenance culture in the sustainability of tourism attractions in Obudu Mountain Resort. *Journal of Tourism and Hospitality, 4*(180), 2.

Enofe, O. M. (2009). *Improving Maintenance Perception in Developing Countries*–A Case Study.

Eti, M. C., Ogaji, S. O. T., & Probert, S. D. (2006). Development and implementation of preventive-maintenance practices in Nigerian industries. *Applied Energy, 83*(10), 1163–1179.

Fakharyan, M., Jalilv, M. R., Elyasi, M., & Mohammadi, M. (2012). The influence of online word of mouth communications on tourists' attitudes toward Islamic destinations and travel intention: Evidence from Iran. *African Journal of Business Management, 6*(38), 10381–10388.

Fatoye, E. O. (2005). *An assessment of occupiers' satisfaction on quality performance of public residential estates.* Unpublished MSc. Dissertation, Department of Urban and Regional Planning. Ibadan, Nigeria: University of Ibadan.

Femi, A. F. (2014). The impact of communication on workers' performance in selected organisations in Lagos State, Nigeria. *IOSR Journal of Humanities and Social Science, 19*(8), 75–82.

Florence, T. (2011). An empirical analysis of asset replacement decisions and maintenance culture in some government organizations located in Ogbomoso and Ilorin metropolis as case study. *Journal of Management and Society, 1*(3), 01–09.

Gibson, D., & Boreham, P. (1981). Managing communication: Appropriate behaviour in medical consultations. *Australian Journal of Social Issues, 16*(1), 52–66.

Gignac, G. E., & Palmer, B. R. (2011), The Genos employee motivation assessment. *Industria and Commercial Training, 43*(2), 79–87.

Gilmore, S. (2000). Doing culture work: Negotiating tradition and authenticity in Filipino folk dance. *Sociological Perspectives, 43*(4), S21–S41.

Gizir, S., & Simsek, H. (2005). Communication in an academic context. *Higher Education, 50*(2), 197–221.

Goris, J. R., Vaught, B. C., & Pettit Jr, J. D. (2000). Effects of communication direction on job performance and satisfaction: A moderated regression analysis. *The Journal of Business Communication (1973). 37*(4), 348–368.

Goris, J. R. (2007). Effects of satisfaction with communication on the relationship between individual-job congruence and job performance/satisfaction. *Journal of Management Development, 26*(8), 737–752.

Gyurko, C. C. (2011). A synthesis of Vroom's model with other social theories: An application to nursing education. *Nurse Education Today, 31*(5), 506–510.

Hancock, D. (2009). *Oceans of Wine: Madeira and the Organization of the Atlantic World, 1640-1815.* Connecticut: Yale University Press.

Herzberg, F. (2008). *One more time: How do youmotivate employees?* Brighton: Harvard Business Review Press.

Holt, R. (1990). *Sport and the British: A modern history.* Oxford University Press.

Holtzhausen, D. (2002). The effects of a divisionalised and decentralised organisational structure on a formal internal communication function in a South African organisation. *Journal of Communication Management, 6*(4), 323–339.

Hossain, M. K., & Hossain, A. (2012). Factors affecting employee's motivation in the fast-food industry: The case of KFC UK Ltd. *Research Journal of Economics, Business and ICT, 5*, 1–30.

Hunter, J. E., & Schmidt, F. L. (1996). Intelligence and job performance: Economic and social implications. *Psychology, Public Policy, and Law, 2*(3–4), 447.

Husain, Z. (2013). Effective communication brings successful organizational change. *The Business & Management Review*, *3*(2), 43.

Indabawa, S. L., & Uba, Z. (2014). Human relations and behavioral science approach to motivation in selected business organizations in Kano metropolis Nigeria. *Human Relations*, *6*(25), 168–173.

Iruobe, O. J. (2011). Effective maintenance of engineering infrastructure for national development: A case study of building. *IRCAB Journal of Science and Technology*, *1*(1), 127–133.

Jerald, G., & Baron Robert, A. (2008). *Behavior in organizations*. New Jersey: Prentice Hall.

Kasser, T., & Ryan, R. M. (1996). Further examining the American dream: Differential correlates of intrinsic and extrinsic goals. *Personality and Social Psychology Bulletin*, *22*(3), 280–287.

Kian, T. S., Yusoff, W. F. W., & Rajah, S. (2014). Job satisfaction and motivation: What are the difference among these two. *European Journal of Business and Social Sciences*, *3*(2), 94–102.

Kportufe, G. S. (2015). Lack of maintenance culture of public buildings in the capital city of Ghana-Accra. *Journal of Culture, Society and Development*, *12*, 94–103.

Kramer, M. W. (1999). Motivation to reduce uncertainty: A reconceptualization of uncertainty reduction theory. *Management Communication Quarterly*, *13*(2), 305–316.

Kreps, G. L. (1990). Applied health communication research. *Applied Communication Theory and Research*, London: Routledge (Taylor and Francis). 271–296.

Kian, T. S., Yusoff, W. F. W., & Rajah, S. (2014). Job satisfaction and motivation: What are the difference among these two. *European Journal of Business and Social Sciences*, *3*(2), 94–102.

Langston, C., & Lauge-Kristensen, R. (2013). *Strategic Management of Built Facilities*. London: Routledge.

Lloyd, R., & Mertens, D. (2018). Expecting more out of Expectancy Theory: History urges inclusion of the social context. *International Management Review*, *14*(1), 28–43.

Locke, E. A. (1976). The nature and causes of job satisfaction. *Handbook of industrial and organizational psychology*. Chicago: Rand McNally College Publishing Company.

Lunenburg, F. C. (2011). Expectancy theory of motivation: Motivating by altering expectations Sam Houston State University. *International Journal of Management, Business, And Administration*, *15*, 1–6.

Magutu, J., & Kamweru, K. (2015). The phenomenon of building maintenance culture: Need for enabling systems. The crisis of building maintenance in Kenya. *Global Journal of Engineering, Design & Technology*, *4*(5), 8–12.

Malek, R., & Yazdanifard, R. (2012). Communication as a crucial lever in change management. *International Journal of Research in Management & Technology*, *2*(1), 52–57.

Malmelin, N. (2007). Communication capital: Modelling corporate communications as an organizational asset. *Corporate Communications: An International Journal*, *12*(3), 298–310.

Maslow, A. H. (1948). Some theoretical consequences of basic need-gratification. *Journal of Personality*, *16*, 402–416.

Maslow, A. (1943). Preface to Motivation Theory. *Psychosomatic Medicine*, *5*(1), 85–92.

Mbamali, I. (2003). The impact of accumulation deferred maintenance on selected buildings of two federal universities in the Northwest zone of Nigeria. *Journal of Environmental Science*, 5(1), 77–83.

Mobley, H. (1997). Intermediate linkages in a relationship between job satisfaction and employee turnover. *Journal of Applied Psychology*, 62(2), 237.

Nita Ali, K., Sun, M., Petley, G., & Barrett, P. (2002). Improving the business process of reactive maintenance projects. *Facilities*, 20(7/8), 251–261.

Nkrumah, E. N. K., Stephen, T., Takyi, L., & Anaba, O. A. (2017). Public infrastructure maintenance practices in Ghana. *Review of Public Administration and Management*, 5(234), 2.

Ntjatsane, M. C., & Kodongo, O. (2017). *Financing of infrastructure maintenance in South Africa*. Published doctoral dissertation. Johannesburg: Wits Business School, University of Witwatersrand.

Nurun Nabi, I. M., & Dip TM, H. A. (2017). Impact of motivation on employee performances: A case study of Karmasangsthan bank Limited, Bangladesh. *Arabian Journal of Business Management Review*, 7(293), 2.

Ogunbayo B. F., Aigbavboa C. O., Thwala W., Akinradewo O., Ikuabe M., & Adekunle S. A. (2022). Review of culture in maintenance management of public buildings in developing countries. *Buildings*, 12(5), 677.

Ogunbayo, B. F., & Aigbavboa, O. C. (2019). Maintenance requirements of students' residential facility in higher educational institution (HEI) in Nigeria. In *Proceedings of 1st International Conference on Sustainable Infrastructural Development*, Ota, 24–28 June 2019, IOP Conference Series: Materials Science and Engineering, 640, (1) 012014. IOP Publishing.

Ogunbayo, B. F., Aigbavboa, C. O., Amusan, L. M., Ogundipe, K. E., & Akinradewo, O. I. (2021). Appraisal of facility provisions in public-private partnership housing delivery in southwest Nigeria. *African Journal of Reproductive Health*, 25(5s), 46–54.

Ogunbayo, B. F., Ohis Aigbavboa, C., Thwala, W. D., & Akinradewo, O. I. (2022). Assessing maintenance budget elements for building maintenance management in Nigerian built environment: A Delphi study. *Built Environment Project and Asset Management*, 12(4), 649–666.

Ogunbayo, S. B. (2021). The effects of molestation and violence against women during the Covid-19 pandemic. *Alternate Horizons*. pp. 1–5. Available online: file: file:///C:/Users/hp/Downloads/ALTERNATE+HORIZONS+ARTICLE+Aug+Ogunbayo.pdf

Ogundipe, K. E., Ogunde, A., Olaniran, H. F., Ajao, A. M., Ogunbayo, B. F., & Ogundipe, J. A. (2018). Missing gaps in safety education and practices: academia perspectives. *International Journal of Civil Engineering and Technology (IJCIET)*, 9(1), 273–289.

Oliver, R. L. (1980). A cognitive model of the antecedents and consequences of satisfaction decisions. *Journal of Marketing Research*, 17(4), 460–469.

Olufunke, A. M. (2011). Education for maintenance culture in Nigeria: Implications for community development. *International Journal of Sociology and Anthropology*, 3(8), 290–294.

Olutayo, A. O., & Akanle, O. (2007). Modernity, mcdonaldisation, and family values in Nigeria. *The Nigerian Journal of Sociology and Anthropology*, 5(46), 53–72.

Olutayo, A. O., & Bankole, A. O. (2002). The concept of development in historical perspective: The third world experience. *Currents and perspectives in sociology* (pp. 105–122). Lagos: Malthouse Press Ltd.

Ouchi, W. G. (1981). Organizational paradigms: A commentary on Japanese management and Theory Z organizations. *Organizational Dynamics, 9*(4), 36–43.

Ouchi, W. G., & Price, R. L. (1978). Hierarchies, clans, and theory Z: A new perspective on organization development. Organizational Dynamics, 7(2), 25–44.

Pearce, W. B. (1994). *Interpersonal communication: Making social worlds.* New York: Harper Collins College Division.

Pearce, W. B. (1995). A sailing guide for social constructionists. In W. Leeds-Hurwitz (Ed.), *Social approaches to communication* (pp. 88–113). New York: Guilford.

Phattanacheewapul, A., & Ussahawanitchakit, P. (2008). Organizational justice versus organizational support: The driven factors of employee satisfaction and employee commitment on job performance. *Journal of Academy of Business and Economics, 8*(2), 114–123.

Porter, L. W., & Steers, R. M. (1973). Organizational, work, and personal factors in employee turnover and absenteeism. *Psychological Bulletin, 80*(2), 151–315.

Postmes, T., Spears, R., Sakhel, K., & De Groot, D. (2001). Social influence in computer mediated communication: The effects of anonymity on group behavior. *Personality and Social Psychology Bulletin, 27*(10), 1243–1254.

Rajput, A., Bakar, A. H. A., & Ahmad, M. S. (2011). Motivators used by foreign and local banks in Pakistan: A comparative analysis. *International Journal of Academic Research, 3*(2).

Rosenbaum, D., More, E., & Steane, P. (2018). Planned organisational change management: Forward to the past? An exploratory literature review. *Journal of Organizational Change Management, 31*(2), 286–303.

Rothberg, G. (2004). The role of ideas in the manager's workplace: Theory and practice. *Management Decision, 42*(9), 1060–1081.

Ryan, R. M., & Deci, E. L. (2000). Intrinsic and extrinsic motivations: Classic definitions and new directions. *Contemporary Educational Psychology,* 25(1), 54–67.

SAICE (South African Institution of Civil Engineering) (2006). The SAICE *Infrastructure Report Card for South Africa.* Valley, Midrand: SAICE Report.

Sani, S. I. A., Muhammed, A. H., Shukor, F. S., & Awang, M. (2011). Development of maintenance culture: A conceptual framework. In *International Conference of Management Proceeding,* Penang, Malaysia, 13–14 July 2011, pp. 1007–1013.

Schroeder, C. (2012). *Coming in from the margins: Faculty development's emerging organizational development role in institutional change.* Sterling, VA: Stylus Publishing, LLC.

Sekhar, C., Patwardhan, M., & Singh, R. K. (2013). A literature review on motivation. *Global Business Perspectives, 1*(4), 471–487.

Soyer, F., Can, Y., Güven, H., Hergüner, H., Bayansalduz, M., & Tetik, B. (2010). An analysis of the relationship between success motivation and team alignment in sports. *International Journal of Human Sciences, 7*(1), 225–238.

Soyoung, P. A. R. K., & Sungchan, K. I. M. (2017). The linkage between work unit performance perceptions of US federal employees and their job satisfaction: An expectancy theory. *Transylvanian Review of Administrative Sciences, 13*(52), 77–93.

Suwaibatul Islamiah, A. S., & Hakim, A. M. (2011). Key factors in developing maintenance culture of public asset management. In Proceedings of the

International Building and Infrastructure Technology Conference. Penang, Malaysia 7–8, pp. 281–287.

Suwaibatul Islamiah, A. S., Abdul Hakim, M., Syazwina, F. A. S., & Eizzatul, A. S. (2012). An overview development of maintenance culture. In Proceedings of *3rd International Conference on Business and Economic Research*. Bandung Indonesia. 12–13 March 2012, pp. 2206–2217.

Talib, R., Ahmad, A. G., Zakaria, N., & Sulieman, M. Z. (2014). Assessment of factors affecting building maintenance and defects of public buildings in Penang, Malaysia. *Architecture Research*, *4*(2), 48–53.

Tella, A., Ayeni, C. O., & Popoola, S. O. (2007). Work motivation, job satisfaction, and organisational commitment of library personnel in academic and research libraries in Oyo State, Nigeria. *Library Philosophy and Practice*, *9*(2), 113–118.

Thontteh, E. O., & Olanrele, O. O. (2015). Occupants' perception of quality design and standard in building services for effective property management. *International Journal of Property Sciences (E-ISSN: 2229-8568)*, *5*(1).

Thurlby, R. (2013). Managing the asset time bomb: A system dynamics approach. *Proceedings of the Institution of Civil Engineers-Forensic Engineering*, *166*(3), 134–142.

Tijani, S. A., Adeyemi, A. O., & Omotehinshe, O. J. (2016). Lack of maintenance culture in Nigeria: The bane of national development. *Civil and Environmental Research*, *8*(8), 23–30.

Tsai, M. T., Chuang, S. S., & Hsieh, W. P. (2009). An integrated process model of communication satisfaction and organizational outcomes. *Social Behavior and Personality: An International Journal*, *37*(6), 825–834.

Twumasi-Ampofo, K., Ofori, P. A., Tutu, E. O., Cobinah, R., Twumasi, E. A., & Kusi, S. (2017). Maintenance of government buildings in Ghana: The case of selected public residential buildings in Ejisu-Ashanti. *Journal of Emerging Trends in Economics and Management Sciences*, *8*(3):146–154.

Tyler, T. R., & Bies, R. J. 1990. Beyond formal procedures: The interpersonal context of procedural justice. In J. S. Carroll (Ed.), *Applied social psychology and organizational settings* (pp. 77–98). Hillsdale, NJ: Lawrence Erlbaum.

Usman, N. D., Gambo, M. J., & Chen J. A. (2012). Maintenance culture and its impact on the construction of residential buildings in Nigeria. *Journal of Environmental Sciences and Resources Management*, *4*, 69–81.

Uysal, H., Aydemir, S., & Genc, E. (2017). Maslow's hierarchy of needs in 21st century: The examination of vocational differences. *Research on Science and Art in 21st Century Turkey*, *211*(1), 211–227.

Vroom, V. H. (1960). The effects of attitudes on perception of organizational goals. *Human Relations*, *13*(3), 229–240.

Wahab, K. A. (1995). Adequate and affordable housing Nigeria in the 21st century, housing today. *The Journal of the Association of Housing Science and its Application*, *4*(3).

Westbrook, R. A., & Reilly, M. D. (1983). Value-percept disparity: An alternative to the disconfirmation of expectations theory of consumer satisfaction. In R. P. Bagozzi & A. M. Tybout (Eds.), *NA – Advances in consumer research volume 10* (pp. 256–26). Ann Abor, MI: Association for Consumer Research.

Wilkins, R. H. (1994). *The quality-empowered business*. Des Moines, IA: Prentice Hall Direct.

Wilkinson, A., Johnstone, S., & Townsend, K. (2012). Changing patterns of human resource management in construction. *Construction Management and Economics*, *30*(7), 507–512.

Willson, D. (1992). Diagonal communication links within organization. *The Journal of Business Communications*, *29*(2), 129–143.

Wilson, D. C. (1992). *A strategy of change: Concepts and controversies in the management of change*. London: Routledge.

Wireman, T. (2005). *Developing performance indicators for managing maintenance*. New York: Industrial Press Inc.

Wordsworth, P., & Lee, R. (2001). *Lee's building maintenance management*. London: Blackwell Science.

Xaba, M. I. (2011). The possible cause of school governance challenges in South Africa. *South African Journal of Education*, *31*(2), 201–211.

Yahya, M. R., & Ibrahim, M. N. (2010). Strategic and operational factors influence on building maintenance management operation process in office high rise buildings in Malaysia. In *1st International Conference on Sustainable Building and Infrastructure*, Kuala Lumpur, Malaysia, 5–17, June 2010.

Yuksel, A., & Yuksel, F. (2008). *Tourist satisfaction and complaining behavior: Measurement and management issues in the tourism and hospitality industry*. New York: Nova Science Publishers.

Zhang, H., & Agarwal, N. C. (2009). The mediating roles of organizational justice on the relationships between HR practices and workplace outcomes: An investigation in China. *The International Journal of Human Resource Management*, *20*(3), 676–693

7 Maintenance Management of Municipal Buildings in Developing Economies: An African Experience

Introduction

Maintaining buildings and other auxiliary facilities and services is one of the main challenges faced by most African countries, especially the maintenance management (MM) of municipal buildings in the educational sector. This chapter discusses maintenance policies and other maintenance issues, including the maintenance of municipal buildings in the education sector of Ghana and South Africa. Also, the role played by the different bodies such as the government and other maintenance organisations and departments involved in building maintenance was looked into. Finally, a background review on maintenance activities in developing economies and a summary of the lesson learned are presented.

Municipal Buildings

As defined by Fulmer (2009), municipal buildings are any type of buildings accessible to the public and funded by public sources. Numerous recent studies have justified municipal building management's significance in the past (Klumbyte, Bliudzius, & Fokaides, 2020; Ogunbayo & Mhlanga, 2022). Edmundas, Zenonas, Jurate, and Tatjana (2021) sustain that municipality authorities manage public properties with various buildings, including social and economic infrastructure services. Municipalities usually control infrastructure, land, schools, health facilities, and social housing (Edmundas et al., 2021). The maintenance of municipal buildings is a phenomenon that transcends all over the world, and its significance cannot be overemphasised (Ampofo, Amoah, & Peprah, 2020).

Municipality buildings represent significant taxpayers' money; thus, in developing economies, especially in Africa, many municipal buildings have not seen any substantial maintenance since they were constructed (Wood, 2005; Adesoji, 2011). This has resulted in the deterioration and abandonment of some educational buildings provided by the municipalities (Ogunbayo, Aigbavboa, Thwala, Akinradewo, 2022). Several municipal buildings in developing economies, including educational buildings, were springing up as a

DOI: 10.1201/9781003344681-7

mark of a developing nation (Ampofo et al., 2020). These educational buildings make up the municipal buildings that support teaching and learning (Aghimien, Adegbembo, Aghimien, & Awodele, 2018). According to Kyeremateng (2008), besides providing a conducive environment for teaching and learning, educational buildings' expenditure constitutes a greater proportion of the developing economies' investment in education.

However, these municipal educational buildings provided by the municipal council cannot remain new throughout their entire life (Niemelä, Kosonen & Jokisalo, 2016). Zavadskas, Turskis, Šliogerienė, and Vilutienė (2021) observed that maintenance problems creep in before a municipal building is completed. Iyagba (2005) contended that municipal buildings remain a key concern to all stakeholders, including their users. However, Rahman and Vanier (2004) maintain that the effective maintenance of the municipal buildings ensures sustainable management, improving the utilisation of municipal properties. Thus, the safety and performance of users who regularly use municipal buildings in the education sector can always be certain. Löfsten (1998) noted that when municipal buildings in the education sector are neglected, defects occur, leading to extensive and avoidable damage to the building fabric. In this study, municipal buildings are referred to as educational buildings, which were further discussed extensively in the follow-up sections.

Educational Buildings

According to Anugwo, Shakantu, Saidu, and Adamu (2018), educational buildings are essential facilities that help provide suitable academic administration. Michael, Omole, Peter, Babalola Olufemi, and Olusoji (2018) posit that in seeking admission into educational institutions, the state of the educational buildings was among the key facilities being considered by parents and students. Additionally, Aghimien et al., (2018) postulated that educational buildings are the physical structures that provide residential comfort for the users. Becker (1990) opines those educational buildings are facilities through which knowledge can be produced or managed. Emetarom (2004) states that educational buildings aid teaching. Learning and research in educational buildings, as noted by Young, Green, Roehrich-Patrick, Joseph, and Gibson (2003), help establish a well-disciplined, learning, and secured learning environment. Wahab, Basari, and Samad (2013) affirm that educational buildings must create a conducive and suitable environment that adequately supports and encourages learning, innovation, and teaching. Ab Wahab et al. (2013) further opined that failing to supply educational institutions' essential facilities would affect their value, students, staff, community, and other stakeholders. Understandably, Okolie (2011) noted that educational building provision would help to provide a better platform for quality education.

The Commission for Architecture and the Built Environment, as cited by Heritage (2009), provides assessment criteria for the design of educational buildings, and these include the following:

- Provision of identity and context for the educational institution buildings;
- Making the best use of the site for the educational institution buildings through the provision of the site plan;
- Provision of school grounds (outdoor spaces);
- A clear drawing of buildings (organisation);
- Buildings (massing, form, and appearance working together);
- Interiors (excellent spaces for teaching and learning);
- Resources (organising convincing environmental strategies);
- Safety (creating a secure and welcoming place);
- Long life, loose fit (adapting and evolving in the future); and
- Successful whole (a design that works all round).

The studies of Sanoff and Powell (2003: 12) show that the performance level of educational buildings is critical to educational effectiveness because it represents a substantial percentage of its "operational costs, assets, and user requirement." Also, Okolie (2011) opines that educational buildings are built and designed for specific needs already determined to a large extent before being put into use. The educational buildings create important vital features that enable teachers to teach efficiently (Price & Pitt, 2012). This shows that it is an important resource that can influence the cause of learning. However, Mayaki (2005) posted that to measure educational institutions' maintenance success, the educational buildings should achieve the main purpose. Based on this, educational buildings as an educational tool are planned and designed as a base for transmitting knowledge and raising learning capacity (Sanoff & Powell, 2003). The design of modern educational institution buildings strongly emphasises stimulation and adapted learning together with suitable learning environments and support spaces with various learning and teaching styles (Sanoff & Powell, 2003).

Furthermore, Michael et al. (2018) observed that adequate provision of buildings within the educational environment has potential merits. Oluwunmi et al. (2015) postulated that some of these merits include punctuality, class peace, and tranquility derivable from a campus environment. Additionally, Okolie (2011) asserts that the provision of educational buildings within institutions with other support facilities such as Internet connectivity, constant power supply, and potable water, among others, will attract high-quality faculty from far and wide. Nevertheless, Asiyai (2012) opines that the standard and quality expected of educational buildings largely depend on establishing a standard management team. This is because the MM of educational buildings is far above traditionally mere managing of buildings to the holistic reality of being woven into an institution's core and support services (Price & Pitt, 2012). Consequently, Oluwunmi et al.

(2015) posited that effective MM of educational buildings falls with the management of the educational institutions. Michael et al. (2018) noted that the inefficient MM of these buildings constitutes a challenge to the management of educational institutions.

One of the greatest challenges facing the sustainable development of infrastructure in developing countries is infrastructure management (Obamwonyi, 2009). In sustaining major infrastructure within developing countries, viable maintenance policies and legislation are necessary to maintain the original intent and purpose of the infrastructure, especially buildings, in terms of aesthetics, functions, health, safety, and others (Obamwonyi, 2009). As Tan, Shen, and Langston (2012) noted, in most developing countries, the need for maintenance functions increases as their economy grows. Based on this, Tan et al. (2012) postulated every developing country must strengthen its maintenance to meet the changing business environment. According to Tan et al. (2012), factors such as social responsibility, sustainability, legislation, obsolescence, adaptive reuse, and building aging are some factors that affect the maintenance market of developing economies. Adenuga, Olufowobi, and Raheem (2010) noted that the attitude to the maintenance approach has been more reactive than proactive. This, as postulated by Zubairu (2000) and Adenuga (2012), has led to the appalling conditions of buildings and auxiliary facilities in most developing countries.

The study by Estache and Fay (2007) shows that maintenance needs for developing countries have been estimated to be between 1.5% and 3.3% of the gross domestic product (GDP). However, Obamwonyi (2009) states that in developing countries, most administrators and maintenance managers still see maintenance from a narrow-minded perspective because they believe maintenance is expensive and has no impact on business. Additionally, most maintenance managers managing the administration of infrastructure of developing economies lack the required knowledge, experience, and foresight needed for long-term planning.

In Malaysia, for instance, Abdul Lateef Olanrewaju (2012) observed that buildings have not been effectively maintained despite the government's commitment to public and privately owned building maintenance. Buttressing this point, Ahmad (2006) posited that in Malaysia, regardless of the size and owners of the buildings, maintenance problems in buildings are common. They further note that despite the vast resources committed to maintenance activities, many buildings in Malaysia still lack adequate maintenance. The results of the study conducted by Kayan (2006) to assess the present conditions of buildings in Kuala Lumpur show that high numbers of buildings face problems of maintenance that need urgent attention to avoid further decay and deterioration. In Petaling Jaya, also in Malaysia, Syamilah (2006) examined the buildings of more than 50 schools. The results show that there was a lack of adequate maintenance of school facilities. According to Gasskov (1992), Malaysia is losing billions of ringgits owing to the poor maintenance of public buildings and amenities. Nonetheless, Lateef (2009) hypothesised that although expenditure on maintenance is increasing, there is still a huge backlog of building maintenance in Malaysia.

Equally, in Zambia, the study of Zulu and Chileshe (2010) shows a need for maintenance personnel to improve their service quality. This notion was based on the fact that the Zambian building industry is characterised by poor maintenance performance. As Zulu and Chileshe (2010) observed, many building projects are completed late, over budget, and of poor quality, leading to increased maintenance activities immediately after completion.

Oladapo (2005) suggested that many developing countries have poor building project completion rates due to corruption, inadequate managerial skills, conflict of interest, and the lack of a maintenance plan. The maintenance approach among developing countries has been more reactive rather than proactive (Mukelasi, Zawawi, Kamaruzzaman, Ithnin, & Zulkarnain, 2012). This has resulted in appalling conditions of the buildings and auxiliary facilities in the majority of the developing countries (Okolie, 2011). Recently, Obamwonyi (2009) noted that major developments in developing countries are associated with modern technology from developed countries. Consequently, based on this, the maintenance process should be given top priority in developing countries because the proper functioning of the technologies depends on developing the proficiency and viability of the maintenance policies to benefit from the developed countries' modern technology.

Benefit of Maintenance Management of Educational Buildings

The need to improve how knowledge can be produced by providing educational buildings that will encompass the immediate environment that makes life worthwhile for its users cannot be overemphasised. Educational buildings, like any other physical facility, deteriorate with age and at various rates. Educational facilities are one of the factors necessary to produce a quality student and an academic base (Okolie, 2011).

However, maintaining the existing educational buildings is of utmost importance to meet acceptable quality standards that can facilitate knowledge transfer and carry out other academic activities effectively and efficiently (Oluwunmi et al., 2015). According to Lateef, Khamidi, and Idrus (2010), educational buildings provide residential needs for students and staff and support and stimulate teaching, research, learning, and other academic-related activities. In designing these, Mutlaq (2002) posited that minimum standards for size, lighting and heating, acoustics, and a productive environment must be met so that the teaching and learning process can proceed normally. Research conducted by CABE (2005) shows that the overall service regarding the performance and productivity of both students and academic staff of the educational institutions is linked to the effective MM of the educational buildings. Moreover, Mutlaq (2002) and Robinson, Rudisill, and

Goodway (2009) opine that improper management of educational buildings and the physical environment may cause stress to their users.

As an important strategy, educational institutions these days use the nature, design, and performance of their buildings to entice students. Baharun (2002) observed that educational institutions in the past relied on the applications sent to them through the educational ministries and others. Educational agencies now use the design and performance of their buildings as one of the variables to attract students (Baharun, 2002). An empirical study by Heitor (2005) shows that students' and teacher attitude and performance are impacted by the level of MM of the educational buildings and their auxiliary services. Okolie (2011), in his study, postulated that creating MM for educational buildings is a complex issue that deals with the maintenance needs of the user of facilities within the educational environment. Thus, Okolie (2011) states further that a well-designed MM will provide a support platform for users of educational buildings. It will also address a broad spectrum of user-related issues such as creating personal user-access needs, technology and equipment, temperature and noise control, and a physically comfortable environment with adequate lighting (Okolie, 2011; Ogunbayo, 2021).

Additionally, Robinson and Robinson (2009) observed that maintaining educational buildings through stable MM entails a close collaboration among the stakeholders through a viable maintenance objective, policy, and strategies. Robinson and Robinson (2009) further state that developing comprehensive MM for educational buildings should be based on functional/technical requirements and aspirational goals with commitment from all stakeholders, including the maintenance unit, institution management, and users. Ogunbayo et al. (2018) state that the maintenance of educational buildings should be based on the MM system that will respond to the maintenance needs of the users. This can be achieved by establishing stable management with an integrated resources infrastructure approach that will significantly support teaching, research, and the quality of campus life (Yacob, 2005; Ogunbayo et al., 2018). However, Yacob (2005) posits that for effectiveness and efficiency, MM of educational buildings requires four major factors that need to be considered. They include the following:

- Effective maintenance organisation structure and general responsibilities of MM;
- Provision of policies and standards for maintenance;
- Provision of maintenance planning; and
- Provision of maintenance budgeting and cost control.

However, Buys and Nkado (2006) affirms that apart from the provision of standard education by the educational institutions attracting learners, their facilities, including buildings, also attracted them. Olanrewaju, Fang, and Tan (2018) opined that the benefit of MM in educational buildings is to provide the following:

- To meet its functional performance as expected by the users.
- To provide technical support to the functional requirement of the available space.
- To guarantee a conducive space suitable for the activity to be carried out within the educational buildings.
- To delay and avoid deterioration and decay within the educational buildings.
- To maximise the service life of educational buildings.
- To ensure optimal performance of educational institution buildings' life cycle.
- To meet the users' expectations (effectively) with fewer resources (efficiently).

Additionally, Then and Tan (2002) postulated that the general benefit of MM in educational buildings is about fitness for users' purposes. The Task Force on Education and Society (2000) maintained that this could be achieved by maintaining clean and orderly educational institution buildings by MM units. Also, Okolie (2011) states that the management of educational institutions has a responsibility to provide educational opportunities through well-developed and maintainable buildings that facilitate students to obtain academic and professional competencies. Oladapo (2006) affirms that the maintenance of educational buildings through preventive measures ensures academic productivity for the users.

Also, Karia, Asaari, and Saleh (2014) postulated that the benefit of providing a MM team for the maintenance of educational buildings is essentially a preventive management philosophy that provides opportunities to retain the life, quality, and value of educational buildings. Nevertheless, the Task Force on Higher Education and Society (2000) posited that the management of educational buildings depends upon a good management context of planning, coordinating, and controlling operations. Akhlaghi (1996) and Ogunbayo and Mhlanga (2021) contended that the MM of educational buildings would ensure that they meet the users' maintenance performance and expectation (effectiveness) with fewer resources (efficiently). The main goal for the maintenance managers of educational buildings is to relieve their users of the burden of ensuring the buildings and support components run effectively and concentrate on their core activities, thereby boosting their performance and productivity (Buys & Nkado, 2006).

Barriers to Maintenance Management of Educational Buildings

The building constitutes a basic human need, an important component of man's welfare, survival, life sustenance, and investment value. According to Onibokun (1982), housing (buildings) is ranked second to food in the hierarchy of man's needs because it profoundly impacts people's efficiency, health, and social behaviour satisfaction. Ogunbayo, Aigbavboa, Amusan, Ogundipe, and Akinradewo state that for buildings to exert these important influences, they need constant management for effectiveness. However, in the educational sectors of most developing countries, the demand for educational buildings has far outweighed the growth in the construction of new buildings and expansion of the existing ones (Michael et al., 2018). Lateef et al. (2010) posit that the achievement of the prime objectives of the educational buildings will seriously be affected if there is any inadequacy in the MM processes.

It is noteworthy to state that the performance and functions of the buildings where we learn, teach, and carry out our research are reflections of our well-being and the quality of our education. Howard (2006) states that many barriers affect the effective MM of the educational buildings and attached components. As postulated by Howard (2006), pressure on finance is one-factor affecting educational buildings, which has reduced overhead costs, hence support services, including the MM of educational buildings. Owoeye (2000) and Owuamanam (2005) observed that the lack of effective MM in the educational buildings would spell doom for their users in achieving their academic pursuits and residency.

The study by Kotzé and Nkado (2003) shows that overlooking maintenance activities for educational buildings have led to an increase in the cost of maintenance operations and the waste of both natural and financial resources, including human resources. Banful (2004) postulated that overlooking maintenance in educational buildings will increase the cost of operation, premature replacement of building components, and reduced asset life. Kotzé and Nkado (2003) opined that deferred maintenance and lack of proper phasing of the maintenance workload are barriers that give rise to uneconomical MM practice within educational buildings and other auxiliary facilities. Ogunbayo et al. (2018) opined that corruption in contracting out major key services of educational facilities to external service providers is one of the major barriers facing effective MM of the educational buildings. Spedding (1994) argues that contracting out some of the support services functions of educational buildings can create an element of risk. Nevertheless, Alexander (1996) states that enabling educational institutions' maintenance managers to focus on buying standard material required for repairs within an approved budget will provide a platform for accurately tailoring the contracting structure.

The study of Zubairu (2010) showed that one of the barriers to the maintenance of educational buildings is the non-visible preventive maintenance culture for the maintained buildings and other facilities within the educational environment. Lavy (2008) postulated that planning challenges are barriers that affect the effective running of maintenance activities within educational institutions. Moreover, Oluwunmi et al. (2015) concluded that the lack of training of maintenance personnel on the maintenance process, procedure, execution, and material selection is a barrier to the MM of educational buildings. Asiyai (2012) states that excess pressure on available buildings through delayed maintenance is a barrier to the MM of educational buildings. Adewunmi, Omirin, Famuyiwa, and Farinloye (2011) postulated that focusing on securing investment returns on educational buildings has created an inefficient maintenance system for their buildings. Accordingly, Clark (2008) opined that against satisfying the end-users need, most management of educational buildings uses their buildings to achieve economic goals.

Nevertheless, Ogunbayo (2021) conjectured that the productivity of the users of this building academically had been affected by many negative consequences due largely to the bad operational planning and lack of maintenance objectives for the MM of the educational buildings. Ab Wahab et al. (2013) posited that once the maintenance requests of buildings are not well managed, it constitutes health challenges and affects the users' performance level. Moreover, Ab Wahab et al. (2013) maintained that neglecting maintenance requests from users and a lack of maintenance policy are barriers to educational building maintenance activities.

Lateef (2009) also states that the absence of quality monitoring of maintenance operations and proper maintenance policies for educational buildings and their other components are fundamentally serious issues. Ikediashi, Ogunlana, Boateng, and Okwuashi (2012) posited that the leading basic challenges to MM in the educational building include the absence of a good MM system and a usage guide, dearth of maintenance policies and regulations, and the use of substandard replacement maintenance material. Adegbite (2007) study results show that the MM of educational institutions is greatly affected by overcrowding because it puts pressure on available facilities within educational buildings. Also, Ogunbayo and Aigbavboa (2019) postulated that factors such as an error in construction method and design, vandalism, deterioration due to age, and poor maintenance culture are barriers to the MM of the educational buildings. Asiabaka (2008) observed that the absence of a facility inventory base on MIS for the components of new and existing facilities is a barrier to the MM of educational buildings. Additionally, Odediran, Gbadegesin, and Babalola (2015) submitted that there is a lack of proactive, result-oriented maintenance strategy, whereas most of the approaches to the MM in educational institutions are always based on a reactive maintenance system. Also, the findings of the study of Swanson (2001) buttressed by those of Cholasuka, Bhardwa, and Antony (2004) confirm that a lack of an aggressive and proactive maintenance model is a significant factor that affects the performance and effective maintenance of the educational buildings.

Many scholars have further identified different barriers to effective MM of the educational institution buildings affecting their efficiency and effectiveness. Amusan et al. (2014) found that corruption in the contract process, absence of replacing materials, management bottleneck, poor record-keeping, policy somersaulting, and wrong maintenance strategy were barriers to the effective MM of educational buildings. Moreover, Odediran et al. (2015) state that educational buildings are deteriorating owing to non-continuous improvement of the maintenance process. Also, Asiyai (2012) noted that ineffective human resources management in the maintenance units of educational institutions is a barrier to the maintenance of educational buildings. Ab Wahab et al. (2013) state that some of the barriers to the MM of educational buildings include untimely replacement of damaged building components, lack of a maintenance schedule, lack of maintenance strategies, and the problem of finance. Lateef et al. (2010) observed that some barriers affecting the success of the MM of educational buildings include lack of long-term plans for maintenance, lack of dedicated financial provision for maintenance services, lack of MISs, and lack of adequate record keeping. Additionally, Asiyai (2012) postulated that barriers such as lack of personnel training, poor management support, maintenance planning and strategy, and lack of MM models for educational institutions' structures have made it difficult to maintain their buildings effectively. Adenuga (2012) opines that the deficiency in the MM of educational buildings originated from the institutions' management structure which is based on an inherent government structure that promotes inefficiencies and inflexibility through the imposition of bureaucratic impediments to operational effectiveness.

Nonetheless, Ekundayo and Ajayi (2009) posit that the key barriers to the MM educational buildings are a lack of proper maintenance regulation and accountability, faulty tendering processes, poor funding, and unstable maintenance arrangements. Adegbite (2007) categorically stated that overcrowding is a major threat to available facilities' usage within educational institutions. Also, Odediran et al. (2015) research findings show that in most developing economies, there are not adequate managerial officers with the required knowledge, skill, and experience in the management and administration of educational buildings. Odediran et al. (2015) concluded that most developing economies' educational buildings are deplorable and obsolete. While the majority of the MM units faced challenges of poor funding, lack of standard maintenance policies and models, the wrong maintenance approach/strategy, and the problem of implementation of maintenance policies.

Ikediashi et al. (2012) concluded in their study that for decades now, there have been difficulties with funding and poor national maintenance template that will guide developing economies' educational buildings towards meeting maintenance requirements for both academic and residential purposes. Okolie (2011) posited that these barriers have seriously undermined the objectives of national policy on education and academic productivity within the developing economies' education sector.

Maintenance Management and Policies in Developing Countries

Ghana

This section of the study reviews the maintenance system in Ghana. It provides a historical overview of the maintenance framework in Ghana. It also attempts to focus on the challenges of maintenance activities of buildings by reviewing past and present maintenance policies in Ghana. The chapter also explores policies and agencies supporting the maintenance programme and its enactment in Ghana, such as the government, private sector, and others. Furthermore, the maintenance of educational buildings in Ghana is also reviewed. Lastly, a summary of the lesson learned to date from the literature is presented.

Educational System in Ghana

The history of education in Ghana dates back to 1592. Over the centuries, education has had different goals, from spreading the Gospel to creating an elite group to run the colony. After Ghana gained independence in 1957, the education system underwent a series of reforms modelled on the British system. Especially the reforms in the 1980s geared the education system away from being purely academic to being more in tune with the nation's manpower needs. According to Apusigah (2017), the present structure of education, which starts at the age of six years, consists of six (6) years of primary education, three (3) years of Junior Secondary School, three (3) years of Senior Secondary School, and four (4) years of university or courses at other tertiary institutions. The first nine (9) years form the basic education and are free and compulsory (Apusigah, 2017) (Map 7.1).

At independence, the nation's leaders appeared clear on the kind of human resources required for leading and facilitating the rapid socio-economic development of the newly "liberated" country (Aryeetey & Kanbur, 2017). The close connection that has been established between educational delivery and national development, dating back to the pre-independence era, continues to drive the nationwide search for socio-economic improvement and social progress. In Ghana, like elsewhere, education has been viewed generally as the panacea to national development (Manuh, Gariba, & Budu, 2007). As a result, efforts to accelerate Ghanaian national development have been characterised by not just infrastructural, industrial, and agricultural development projects but also by educational reforms that have been meant to produce the much-needed human resources for driving the development process (Aryeetey & Kanbur 2017).

Even before independence, the early nationalists, with the support of the Ghanaian public, had compelled the colonial government to start the first university, the University College of the Gold Coast (UCGC), now the University of Ghana (UG) (Bening, 2005). Antwi (1992) explained the

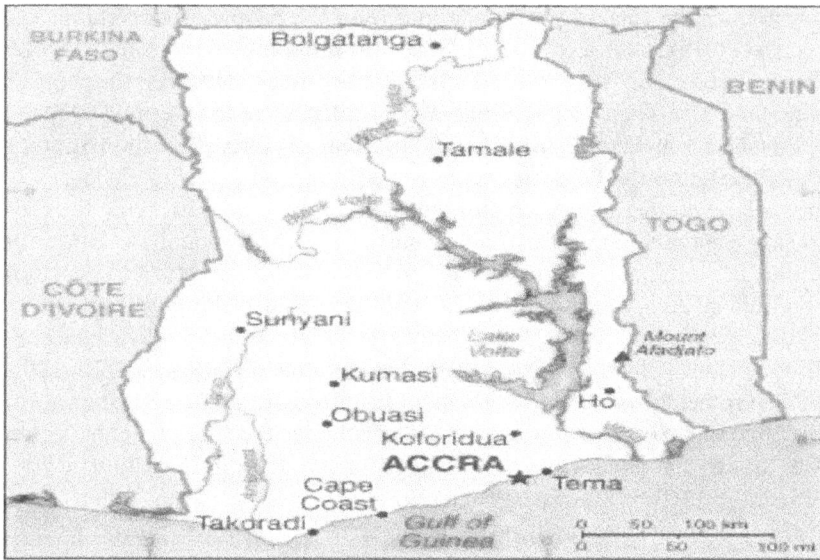

Map 7.1 Map of the Republic of Ghana.

Source: CIA Factbook (2018).

Image Link https://www.cia.gov/the-world-factbook/static/dce1eb4dd2dce442eb5ecaa1b019e11c/ec46e/GH-map.jpg

motivation for the demand for education in the then British colonies, including the Gold Coast: The social and economic development program in the colonies required an increasing number of men and women with professional qualifications in all spheres of human endeavour. Many professionals and specialists needed to be locally trained and recruited (Antwi, 1992).

Moreover, Ghana's private education institutions have experienced rapid growth since the turn of the 21st century. The rapid expansion in Ghana's public and private education reflects the huge demand for education with effective buildings and other facilities within the country (Yusif, Yussof & Osman, 2013). Also, four of the six leading public educational institutions have instituted a continuing education programme to provide training for qualified student applicants who could otherwise not gain access to university education (Osei-Owusu & Awunyo-Vitor, 2012). This is because of limited academic facilities, inadequate programmes, and inadequate financial resources to acquire the knowledge and skills for professional development (Osei-Owusu & Awunyo-Vitor, 2012).

Evolution of Maintenance Policy in Ghana

Since Ghana attained independence in 1957, all successive ruling governments have pursued, with varying degrees of success, several policies and programmes to accelerate the economy's growth and raise the people's living standards. Gyampo (2015) states that these policies and programmes were aimed to accelerate the economy's growth and infrastructure that will increase the people's living standards. These include Ghana Vision 2020: The First Step (1996–2000), the First Medium-Term Plan (1997–2000), Ghana Poverty Reduction Strategy (2003–2005), and the Growth and Poverty Reduction Strategy (2006–2009). Under these strategic programmes, substantial progress was made towards realising economic and infrastructure sustainability (National Development Planning Commission [NDPC], 2010). However, with the return to constitutional rule in 1993, successive governments have provided policy frameworks and development plans to guide Ghana's overall economic and social development in line with Article 36 (5) of the Constitution (NDPC, 2010). In 1995, the Government presented to Parliament the first Coordinated Programme of Economic and Social Development Policies under the title "Ghana: Vision 2020," aimed at making Ghana a middle-income country in 25 years. The First Medium-Term Development Plan (1997–2000) was based on Vision 2020, which focused on infrastructure development, urban development, and an enabling environment (NDPC, 2010). One of the key goals of this policy is rehabilitating and expanding infrastructural facilities within the educational buildings.

The history of development planning in Ghana goes back to the early 1920s with the preparation of the first Ten-Year Development Plan (1920–1930) by the then colonial Governor, Sir Gordon Guggisberg. The tradition of planning continued in the post-colonial era with the preparation of several plans, including the Seven-Year Development Plan (1963–1970) in the Nkrumah era (Gyampo, 2015). More than ten plans were prepared between 1940 and 1986 (NDPC, 2010). All those plans were centrally designed by Government bureaucrats with little or no consultation or participation of the stakeholders, beneficiaries, or the public at large (Government of Ghana, 2003). Nevertheless, the maintenance policies and laws mainly draw their sources from the provision of the 1992 Constitution of the Republic of Ghana. It is enshrined in the constitution that under chapter 11(1b), a long-term national development imperative should be provided by the parliament of Ghana through the directive principles of state policy (Constitution of Ghana, 1992). This requires that every government pursue policies that ultimately lead to the "establishment of a just and free society," where every Ghanaian would have the opportunity to live a long, productive, and meaningful life (Constitution of Ghana, 1992).

In 1994 a new decentralised development planning system was established by Acts of Parliament, namely the National Development Planning (System) Act, 1994 (Act 480) and the accompanying National Development Planning Commission Act, 1994 (Act 479) (Ghana Parliament, 2016). These Acts sought to democratise the development process by creating space for stakeholders to participate in the decision-making process towards public infrastructure, including educational institutions' infrastructure. The primary aim of establishing the new system was to put in place a mechanism that would allow for broad participation in the development planning process at all levels of society, including education, businesses, and others (Ghana Parliament, 2016). Currently, no standalone national maintenance policy regulates infrastructure maintenance, particularly educational buildings (Nkrumah et al., 2017). Recently, draft policies have been rolled out to enhance greater access to decent buildings and bridge other infrastructure deficits confronting the country (International Labour Organisation (ILO), 2009). The ILO (2009) report on Ghana noted that most of the intervention in maintaining infrastructure and other utility national facilities has been in direct government approach or providing subsidies to state agencies. Examples include roof and wall protection loan schemes initiated by the government. To harmonise the development of infrastructure projects in the country to realise the long-term goal of national development in all facets, the government of Ghana initiated a National Infrastructure Plan (NIP) (Bosompen, Montes Belot, Joya, & Mejía de Pereira, 2012). The NDPC is coordinating the NIP. It has the vision of building world-class infrastructure assets that are dependable, efficient, resilient, accessible, functional, and inclusive with the capacity to support the Ghana economy, which includes the educational sector, and as well improving the quality of life of all Ghanaians (Bosompen et al., 2012). The NIP is proposed to span 40 years (2018–2057) which will be implemented via a public-private partnership (PPP). As the ILO (2009) reported, appropriate legislative policies and government support would need to be provided to ensure that the private sector charts how to execute the infrastructure plan. This would be achieved through enhanced equipment capacity, skilled development, and the local production of a wide range of construction materials, among others.

As documented by the NDPC (2015), some of the infrastructures to be provided include institutional development (construction industry development, infrastructure maintenance policy and strategy, human resources development framework, national infrastructure database, procurement administration reforms, land administration) and logistics infrastructure (institutional infrastructure, technology infrastructure, facilities infrastructure) among others. The purpose of the NIP policy is to ensure a very harmonised and sequential provision of critical infrastructure within the country on a sustainable basis, away from what currently pertains. However, to effectively address the infrastructure deficit based on the NIP policy in the country, the infrastructure investment fund was set up by the government of Ghana (NDPC, 2015). The government has proceeded to pass the Ghana Infrastructure Investment Fund (GIIF) Act, 2014 (Act 877), to make this fund operational. The main objective of the GIIF is to manage, mobilise, coordinate, and provide financial resources for investment in a diversified portfolio of infrastructure projects, including educational infrastructure for national development. The seed amount for the fund was US $250 million (Government of Ghana, 2013). However, towards stable infrastructure for economic development, education facilities, including student and instructor accommodation, are one of the social infrastructures that qualify to be funded by the GIIF (NDPC, 2015).

Maintenance Policy and Legal Framework in the Maintenance of Educational Buildings in Ghana

As captured in one of the addresses delivered by Ghana's first President, Dr. Kwame Nkrumah, to the legislative assembly to identify the critical role of quality education in national development in its pre-independence days. He stated that "only with a population so educated can we hope to face the tremendous problem which confronts any country attempting to raise the standard of living in a tropical zone" (McWilliams & Kwamena-Poh, 1975: 9). Owusu, Aigbavboa, and Thwala (2016) posit that since this period, the critical role of relevant and quality education in national development has never been lost in Ghana. According to Addae-Mensah (2000), education in Ghana has been given a transforming and liberatory impetus with a special mandate to serve the immediate needs of national development.

Since the country's independence in 1957, education has been taken very seriously until now: the country has established over twenty public and private higher educational institutions (Addae-Mensah, 2000). In Ghana, higher educational institutions comprised universities, polytechnics, colleges of education, nursing colleges, and professional institutions (Bingab, Forson, Mmbali, & Baah-Ennumh, 2016). Owusu et al. (2016) observed that these educational institutions found it difficult to admit all qualified students over the years, owing largely to a lack of infrastructure development on various university campuses in Ghana. Nevertheless, to meet the growing needs of its population, especially their education needs, there is a mounting demand for stepping up infrastructure development in the country's public educational institutions (Owusu et al., 2016). Kadingdi (2006) postulated that although the early years of Ghana's independence experienced massive educational infrastructural development.

According to the World Bank (2004), the 1961 Education Act and the Accelerated Development Plans for education expedited educational institutions' steady and rapid growth qualitatively and numerically shortly after independence. This makes Ghana's education system gain the enviable reputation of being one of the best and most respected in Africa. However, the decline in human capacity and infrastructural development within the country's educational institutions started in the 1970s and 1980s due to economic decline, which saw a sharp decline in Ghana's education system. It was evident in the myriads of problems experienced by the sector, such as poor maintenance of educational institution buildings, inadequate and poor education infrastructure, non-availability of replacement material for maintenance activities, and others (Akyeampong, 2010). The economic decline makes it difficult for the management of the educational institution to utilise the already insufficient funds to maintain their facilities effectively (Simpeh, 2018).

Additionally, Adu-Agyem and Osei-Poku (2012)) postulated that this leads to low prioritisation of maintaining educational institutions' infrastructure within the country. As noted by Akyeampong (2010), to improve the educational infrastructure within the country, an aspect of community participation in the provision of school infrastructure was introduced in the 1987 educational reform. Owusu et al. (2016) noted that the root problem of the MM of educational institution buildings and other auxiliary facilities in Ghana is funding, lack of policy, and resource constraints. Huaisheng, Manu, Mensah, Mingyue, and Oduro (2019) posited that addressing the issue of MM in educational buildings in Ghana requires doing things out of the ordinary, thinking outside the box, and being innovative.

In solving this mirage of problems associated with MM of educational buildings, Owusu et al. (2016) observed that the Ghana government, through the education ministry in recent years, introduced another way of solving maintenance of educational institutions infrastructure in the country called the Ghana Education Trust Fund (GET Fund). The GET is a public trust established by an act of parliament (Ghana Parliament, 2016). As noted further by Owusu et al. (2016), the core mandate of the GET is to provide funding to supplement government efforts in the provisions of required educational infrastructure and facilities from the pre-tertiary to the tertiary level within the country. As stated in the Ghana Education Trust Fund Act 2000 Act 581, the main objectives of the GET Fund are to provide financial support to MDAs under the Ministry of Education for the development and maintenance of essential educational facilities and infrastructure (Ghana government, 2003). The GET fund is helping the Ghana Education Ministry and Ghana as a country towards providing adequate and sound educational infrastructure at every level of education (Government of Ghana, 2013).

Challenges Associated with Implementation of Maintenance Policy in Ghana Education Sector

According to Obeng-Odoom and Amedzro (2011) and Quayson and Akomah (2016), a poor maintenance policy towards public infrastructure, especially buildings, contributes to inadequate housing provision and a reduction in property value in Ghana. As noted further by Adu-Agyem and Osei-Poku (2012), one of the critical challenges confronting the maintenance of buildings in Ghana is the non-availability of a standard maintenance policy. However, public institutions' input in national development cannot be over-emphasised. As observed by Cobbinah (2010), despite heavy investment in public buildings, public institutions allow their structure to care for themselves without any sustainable maintenance plan or policy to preserve the quality of the buildings. Nkrumah et al. (2017) postulated that the lack of national maintenance guidelines by the relevant authorities in Ghana often leads to the reduced lifespan of these buildings. As buttressed by Owusu et al. (2016), this invariably defeats the purpose for which a building is set up as a factor of production and accommodation. Yinghua, Sylvester, Bonsu, Minkah, and Zhenjiang (2018) postulated that although Ghana is doing well economically, there are serious shortfalls with the provision of infrastructure and the maintenance of the present ones. They identified challenges associated with the successful implementation of maintenance laws and policies in Ghana, including the following:

- The negative attitude of stakeholders in the direction of rehabilitation and maintenance of Ghana's constructions and facilities;
- Corruption in the process and development of public buildings in Ghana;
- Lack of political will from the government to implement policies and laws on maintenance which is evidenced in various developmental policies and trust funds on infrastructure promulgated;
- Negative attitude amongst those exercising authorities or political power towards establishing maintenance laws or policy; and
- Over-centralisation of maintenance decisions for public buildings, including educational buildings in Ghana.

Additionally, in his study, Simpeh (2018) affirms that inadequate finance is a major challenge that affects implementing policies towards maintenance practices in Ghana. Moreover, Wuni, Agyeman-Yeboah, and Boafo (2017) postulated that inadequate human resources (skilled) are another factor that affects the implementation of maintenance policies and laws in Ghana. However, in their study, Nkrumah et al. (2017) concluded that the leading challenge towards implementing a maintenance policy and laws in Ghana was the lack of a national maintenance programme.

Lessons Learned From Ghana Maintenance Studies

Lessons learned from the relevant literature survey relating to Ghana include the following:

- Maintenance is an important integral part of governance because it helps protect national assets and monuments. However, the attributes that determine the MM of buildings are unknown.
- There is a dearth of studies that develop models to aid the MM of buildings in the Ghanaian educational sector.
- From the review, it can be deduced that in Ghana, the provision of MM units is not part of the criteria for establishing educational institutions in Ghana, as is the case in developed economies.

- There is a lack of political will on the part of government and political players to establish maintenance policies and laws.
- Ghana has no standalone national maintenance policies or laws that holistically address the MM of the educational institutions' facilities and other public buildings in the country. However, the Ghana Education Trust Fund Act, 581 Act 2000 provides funding for developing and maintaining essential educational institutions' facilities and infrastructure.
- Again, no independent authority in Ghana is responsible for implementing maintenance laws and policies.
- There is a lack of standalone maintenance laws or policies that can guide the development of maintenance practices towards protecting public buildings, including educational buildings, from deterioration.

Nigeria

This section reviews the literature on MM studies in Nigeria. It gives an account of MM studies in the Nigerian education sector. Additionally, it highlights the evolution of maintenance policies and laws in Nigeria, the maintenance policy and legal framework in maintaining educational buildings in Nigeria's education sector, and the challenges associated with implementing a maintenance policy in Nigeria.

The Nigerian Education Sector

Education, as described by Olujuwon and Perumal (2015), is a fundamental tool for constructing the knowledge economy and society. Fafunwa (2018) posited that education is the most valuable of all capital that is invested in human beings. However, Salami (2010) observed that the potential of education to fulfil its responsibility in developing countries has frequently been slowed down by the challenges of finance, efficiency, equity, equality, and governance.

In Nigeria, the government in the early 1980s and 1990s implemented a series of far-reaching education reforms (Okebukola, 2006). Through these reforms, the structure of secondary education was significantly altered. Before these new reforms were formulated, secondary school education closely resembled the British system consisting of GCE '0' levels followed by two years of GCE 'A'-level courses (Okebukola, 2006). However, Nigeria's education system since the late 1990s has followed a 6-3-3-4 structure. The new structure means the duration for primary education is six years, while the period for secondary education is also six years, made up of two cycles of three years for both junior and senior secondary school. The duration of tertiary education is an average of four years (Eleweke, 2002) (Map 7.2).

Education policy in Nigeria is centrally determined by structure, curriculum, and school year (Eleweke, 2002). However, Nigeria's education management is dictated by Nigeria's political structure grounded on

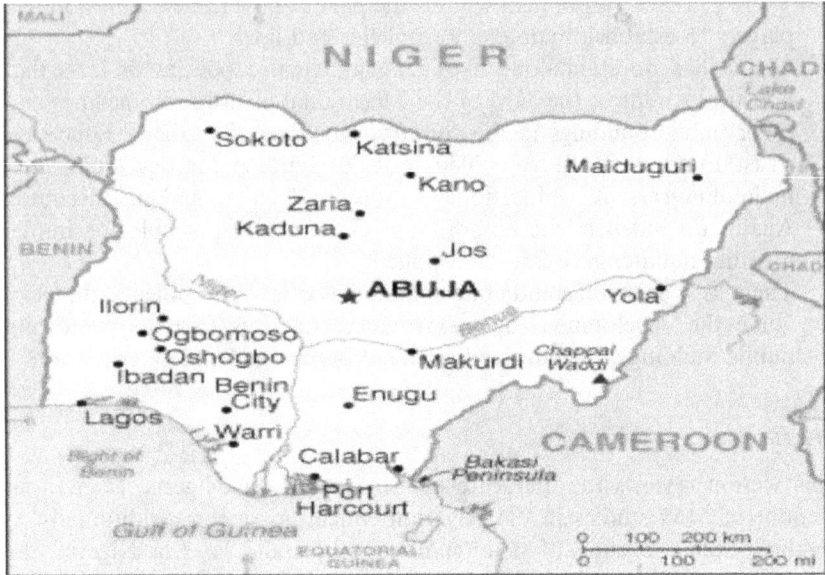

Map 7.2 Map of Federal Republic of Nigeria.

Source: CIA Factbook (2018).

Image link https://www.cia.gov/the-world-factbook/static/40093e4806b520e46381eafb78c8383b/a2c49/NI-map.jpg

federalism (Okolie, 2011). Subsequently, the administrative mechanism for education in the country devolves some authority to the state and local governments. The responsibility for managing primary education was shared among the three tiers of government of the country, namely the federal government, state government, and local government (Okolie, 2011).

The aims of education in Nigeria, as outlined by the Federal Ministry of Education (2004) section 8 (59), are as follows:

- Development and inculcation of proper values for the survival of the individual and the society;
- Contribution to national development through high-level manpower training;
- Development of the intellectual capabilities of the individual to understand and appreciate their local and external environment;
- Promotion and encouragement of scholarship and community services;
- Acquisition of both physical and intellectual skills enabling the individual to be a self-reliant and useful member of society; and
- For national unity and the promotion of national and international understanding and interaction.

Conversely, the Federal Ministry of Education harmonises educational policies and procedures of all states of the federations through the National Council of Education(NCE) (Odia & Omofonmwan, 2007). In educational matters for the country, the NCE is the highest policy-making body, consisting of the Federal Ministry of Education and all state commissioners of education. The NCE is further supported by the Joint Consultative Committee (JCC) on education. The Joint Consultative Committee comprises all the federal and state directors of education, chief executives of education, parastatals, and directors of universities and institutes of education (Odia & Omofonmwan, 2007). The Joint Consultative Committee advises the NCE on educational issues and is headed by a Federal Ministry of Education director. Another parastatal under the Ministry of Education is the National Universities Commission (NUC). The National Universities Commission is responsible for developing the universities in the country (Odia & Omofonmwan, 2007).

EDUCATION PHILOSOPHY IN NIGERIA

The Federal Government of Nigeria (2004) declared in its National Policy on Education that education is an instrument par excellence for achieving national development. However, Okolie (2011) and Olujuwon and Perumal (2015) in their study noted that the Nigerian national policy on education highlights the following as the philosophy of Nigerian education:

* Education is a tool for the promotion of a progressive united Nigeria;
* Education is a mechanism for national development;
* Education fosters the worth and development of the individual, for each individual's sake and the general development of the society; and
* Every Nigerian child shall have the right to equal educational opportunities irrespective of any real or imagined disabilities, according to their abilities.

Accordingly, Okoroma (2006) stated that the National Policy on Education (NPE) is embedded in Nigeria's educational philosophy as articulated through the nation's objectives. The main national objectives are provided through the Second National Development Plan, recognised as the necessary foundation for the national policy on education (Federal Republic of Nigeria, 1999). They include the building of the following:

* A great and dynamic economy;
* A free and democratic society;
* A united, strong, and self-reliant nation;
* A just and egalitarian society;
* A land of bright and full opportunities for all citizens.

Nonetheless, the philosophical base of the national policy on education is to develop an individual into an effective and sound citizen (Federal Government of Nigeria, 2004). Furthermore, it is also themed towards integrating the individual into the community by providing equal access to educational prospects for all citizens of the country at the primary, secondary, and tertiary levels (Okoroma, 2006). Abiogu (2014) posited that for any meaningful growth and development, there needs to have an educational policy based on a philosophy developed on sound educational planning. Abiogu (2014) affirms that this is necessary because education constitutes an essential aspect of the social realities of a nation and any society.

In Nigeria, educational institutions stand out as the knowledge storehouse (Omomia et al., 2014). As noted by Omomia et al. (2014), the consequence of this is that it is a place where the right knowledge is disseminated to students who will invariably contribute to the country's sustainable development. As stated by Olaniyan (2001), the education institutions in the context of Nigeria are expected to create scarce needed human capital with enhanced skills that can lead to good productivity, technological innovation, and growth of the economy. The World Bank (1999) posited that the Nigerian educational institutions are fundamental to the constitution of a knowledge economy and society. Ibukun (1997) postulated that the major relevance of educational institutions in Nigeria is the development of the economy and the provision of the much-needed manpower to accelerate national growth. Nevertheless, education institutions stand out as the focal point of general development (Omomia et al., 2014).

For these educational institutions to meet the set goals of government towards national development. The national policy on education (Federal Government of Nigeria, 2004) highlighted specifically that the major goals of educational institutions in Nigeria should include the following:

- To contribute to national development through high-level relevant manpower training;
- To promote national and international understanding and interaction;
- To forge and cement national unity;
- To develop and inculcate proper values for the survival of the individual and society;
- To acquire both physical and intellectual skills which will enable individuals to be self-reliant and useful members of society;
- To develop the intellectual capacity of individuals to understand and appreciate local and external environments; and
- To promote and encourage scholarship and community service.

However, Okolie (2011) observed that the extent of student enrolment is not represented by the numerical strength of the Nigerian educational institutions. The overall enrolment rates of Nigerian educational institutions far exceeded government policy guidelines (NUC, 2020). This shows that the annual number of candidates seeking enrolment into the Nigerian educational institution far outnumbers the available space (Okolie, 2011). Due to the overpopulation of educational institutions, there is a great deal of pressure on the available educational facilities, including buildings.

CHALLENGES OF THE EDUCATION SYSTEM IN NIGERIA

According to Ukpia and Ereh (2016), educational institutions in developing countries, particularly in Africa, continue to face serious issues that have profound implications for educational management. Okoli (2015) states that from inception, Nigerian educational institutions faced many challenges, including education policies, accommodation, student unionism funding, and struggling to grapple with admission issues. Other scholars such as Ojogwu and Alutu (2009) and Ajayi and Ekundayo (2008) postulated that the challenges of improving the educational quality in Nigeria are severely exacerbated by declining resources, lack of funding, poor work environment, overpopulation, poor management, and inadequate physical infrastructure management

The World Bank (1994) stated that the challenges facing the educational institutions in Nigeria could be traced to the following:

- Deteriorated infrastructure/facilities/equipment for teaching, research, and learning;
- Brain drain, student unrest, and constant strikes by both students and academic staff;
- The increasing rate of graduate unemployment;
- The erosion of university autonomy and academic freedom; and
- The decline in public expenditure.

Nevertheless, the World Bank (1994) postulated that some of the core challenges faced by the educational institutions in Nigeria could be listed as follows:

- Lack of funding;
- Deterioration of existing infrastructure;
- Overpopulation;
- Poor working environment,
- Policy somersaulting;
- Corruption;
- Inadequate physical planning;
- Cultism;
- Political instability;
- Poor management/leadership;
- The increasing shortage of quality academics; and
- Lack of proper planning towards the establishment.

Ajayi and Ekundayo (2008) noted that challenges facing educational institutions in Nigeria demand proactive and strategic institutional planning and require a responsive and reflective approach to systems management. The deterioration of existing educational institutions' infrastructure is one of the major challenges facing the education system in Nigeria. To eliminate these challenges, constant adaptation to the rapid change in MM is necessary. With a radical transformation on all fronts for the educational institutions to meet the government target. The application of a MM framework that will provide a maintenance process for existing and proposed educational buildings must be given priority for an effective education system.

OVERVIEW OF MAINTENANCE OF EDUCATIONAL BUILDINGS IN NIGERIA

One of the cardinal aims of the educational institutions is to provide skilled, higher-level manpower, which is vital to the economic and national development of Nigeria. Educational buildings are one of the essential resources needed within the educational institutions to bring about more skilled graduates to achieve the educational philosophy of Nigeria. According to Ojara (2013), the success of every educational institution in Nigeria will depend on the availability of the right type of physical facilities, especially buildings. Ojara (2013) states further that students' overall academic achievement depends on the environmental nature of where students reside and learn. Educational buildings with quality materials and effective maintenance make learning interesting and motivating. Iyamu, Imasuen, and Osakue (2018) postulated that the quality of educational institutions' structure depends on several factors of which building facilities are supreme. Iyamu et al. (2018) further state that educational buildings are crucial because they foster effective teaching and thus boost good results. This shows that the quality and standard of educational buildings should preferably be rooted in provision efficiency. However, Abdulrahman (2013) observed that there is a need for both government and private-owned educational institutions to improve the maintenance approach of their facilities.

The educational system should boast about the numbers established within the country, the quality of facilities, and a better management system (Abdulrahman, 2013). As noted by Ojara (2013), poor funding of the educational institutions by the government has affected the maintenance unit of the public educational institutions in Nigeria from effectively maintaining the educational buildings. However, Okolie (2011) states that in Nigeria, the state of physical infrastructure in the public institutions is not encouraging. Nonetheless, the study of Cobbinah (2010) shows that the effect of the dwelling resources in the Nigerian educational system has a huge impact on the maintenance budgets for the educational buildings.

As noted by Moja (2000), a shortage of space currently exists in all the sub-sectors of the educational system in Nigeria. Moja (2000) further posited that there is pressure on education facilities that did not expand at the same rate as the school population within the education system in Nigeria. Many scholars have been worried about the state of the educational buildings due to their undesirable condition (Odetunde, 2004). The findings of the study of Asiyai (2012) show that in some of the institutions of learning in Nigeria, students are learning in dilapidated buildings that are poorly illuminated, furnished, ventilated, and environmentally depressed disabling. Moja (2000) observed that in the early days of the Third Republic, Nigeria's government acknowledged that existing physical facilities must be upgraded to improve educational quality.

Nevertheless, the problem with the existing educational buildings started due to a lack of maintenance and repairs on reported maintenance needs within the educational system (Ogunbayo & Aigbavboa, 2019). It is noteworthy that the quality of education offered within these educational institutions is negatively impacted by the present condition of the buildings. Moreover, the funds needed to build new educational buildings and maintain existing buildings are even higher (Moja, 2000).

The defects in educational buildings such as sagging beams, foundation failure, leaking roof, door and windows' defects, floor slab failure, defacing of wall surfaces, and rising dampness in structure have led to a reduction in students' attendance in class and difficulties in school administration (Kunya, 2012). Additionally, Ipingbemi (2010) contended that educational buildings' flaws include the use of inferior building materials and poor technical skills, improper design of fire suppression systems, lack of regular check-ups of the facilities, and improper management of the facilities. Ajayi and Ekundayo (2008) postulated that the proliferation of education policies is one of the challenges of effective MM of educational buildings in the Nigerian education sector. Similarly, Abdulrahman (2013) affirms that problems of MM of educational buildings in Nigeria are rooted in many inadequacies, especially in educational policies. Based on this, Odia and Omofonmwan (2007) suggested that the problem of policy somersaulting has led to the underfunding of the educational buildings of the educational sector in the country and the abandonment of other physical facilities. In many of these educational institutions, classrooms, libraries, and laboratories are sub-standard while both instructional and living conditions have deteriorated.

Another major problem with the maintenance of educational buildings is overpopulation (Abdulrasheed and Bello, 2015). This is caused by the over-admitting of students beyond the institution's capacity. Ultimately, it puts pressure on the available facilities, including buildings. Mgbekem (2004) contended that an unmanaged social demand for admission to educational institutions is putting pressure on the educational facilities within Nigeria, whereas most of the infrastructure has deteriorated. Furthermore, the dearth of effective MM of educational buildings in Nigeria is attributed to the lack of a standalone maintenance policy for the education sector (Abdulrasheed and Bello, 2015). The report of the World Bank (2008) shows that the lack of a standalone maintenance Act to regulate maintenance operations in Nigeria has also affected the maintenance of educational infrastructures in the country.

Also, Kunya (2012) opined that knowledge gaps in MM studies due to lack of research had been the bedrock for the lack of a viable MM system for the educational buildings in Nigeria's education sector. Thus, this current study addresses this gap in knowledge by developing a MM framework for municipal buildings in the education sector of developing economies using the South African education sector.

Evolution of Maintenance Policy in Nigeria

After the independence of Nigeria on the 1st of October 1960, several economic and development plans were undertaken by the civilian government of the First Republic and the successive military government and later the civilian government of the Fourth Republic, leading to the development of the most required facilities for national development. As Iheanacho (2014) noted, in Nigeria, the instrument of both diagnosis and remedy to better infrastructure is developing a better developmental plan. This shows that a nation can hardly achieve reasonable development without an infrastructural development plan backed by an effective maintenance policy.

In Nigeria, to meet the constitutional requirements of its citizens, infrastructural development planning has been a consistent occurrence. The First National Development Plan (1962–1968), the Second National Development Plan (1970–1974), the Third National Development Plan (1975–1980), and the Fourth National Development Plan (1981–1985) were all established to meet the infrastructure needs of the general public such as good housing, roads, and hospitals, among others (Lawal & Oluwatoyin, 2011). However, from 1960–1985, no maintenance policy was part of the national development plan for the sustainability of the developed infrastructure (Abdulrasheed and Bello, 2015). This led to the early deterioration and abandonment of most of the public infrastructure developed, such as the National Stadium Lagos.

Maintenance policies and laws in Nigeria draw their source from the provision of the 1999 Constitution of the Federal Republic of Nigeria. The Constitution enjoins the parliament under sections 16(2b) and 16(2d) to enact relevant laws and, in consultation with the executive arm of government. They also ratify laws to control the national economy to secure the maximum welfare, freedom, and happiness of every citizen based on social justice and equality of status and opportunity (Federal Republic of Nigeria, 1999). The Constitution of the Federal Republic of Nigeria state that adequate and suitable infrastructure should be provided for all the citizen of the country (Federal Republic of Nigeria, 1999). Nevertheless, there is no standalone national policy or law on maintenance that regulates infrastructural management in the educational sector and other national assets in Nigeria (Ugwu, Okafor & Nwoji, 2018). Nonetheless, there are authorities and agencies to monitor and carry out maintenance operations on the national assets of the government. The authorities and agencies include the Federal Road Maintenance Agency (FERMA), the National Emergency Management Agency (NEMA), and the maintenance unit of the Federal Ministry of Works, and Housing, among others (Odediran, Adeyinka, Opatunji, & Morakinyo, 2012).

In the early 1990s, the Federal Government of Nigeria, in the quest for another national development plan backed with a viable maintenance system, proposed another long-term development plan with little or nothing to show for it (Akpobasah, 2004). However, between 2003 and 2019, the Federal Government introduced a different ambitious plan for upgrading the existing public infrastructure, including educational buildings in the country (Akpobasah, 2004). Such policies include the National Economic Empowerment and Development Strategy (2003–2007), Vision 20: 2020 (2007), the Transformation Agenda (2011–2015), and, more recently, the Economic Recovery and Growth Plan (ERGP) 2017 (Ministry of Budget and National Planning, 2017). These policies and programmes attempt to implement the rudiments of a developmental state and design instruments that can accelerate growth towards effective and sustainable infrastructure.

However, a notable step forward in developing the maintenance framework for public buildings and other public assets was in 2018, when the present democratically elected government introduced a policy known as the maintenance framework of public buildings. However, the notion that Nigeria does not have a maintenance culture provoked the decision to prescribe this new maintenance framework by the present government (Aidoghie, 2018). The new maintenance policy and framework seek to institutionalise a maintenance culture in Nigeria (Forrest, 2019). Conversely, there are institutions in Nigeria, for example, the Nigeria Institute of Building (NIOB), the Nigeria Institute of Engineering (NIE), the Council of Registered Builders of Nigeria (CORBON), and the Real Estate Developers Association of Nigeria (REDAN) among others, which mount pressure on the Federal government to develop a maintenance framework for public buildings in Nigeria.

Maintenance Policy and Legal Framework in the Nigeria Education Sector

The Education Tax Act No. 7 of 1993 was promulgated because of a widely known decline in educational standards, especially the high level of degradation in infrastructure and other facilities at all levels of the Nigerian educational system (Education Trust Fund, 1993). However, the Education Tax Act established the Education Tax Fund (ETF) No. 7 of 1993 in 1998, amended by Act 40 (22nd Dec) 1998, which imposed a 2% tax on all assessable profits of all companies registered in Nigeria. The collection of the tax fund started as early as 1994 (Ugwuanyi, 2014). Nonetheless, the details were stipulated in the Companies Income Tax Act or the Petroleum Profit

Act Tax, as the case may be. The amendment based on Act No. 40 1998 led to the establishment of the ETF, which serves as an intervention agency with a project management system for improving the education quality in Nigeria (Soetan, 2018). Additionally, the Act states that the Federal Inland Revenue Service (FIRS) is empowered to assess and collect the Education Tax (Soetan, 2018). Equally, the ETF administers the tax imposed by the Act and distributes the amount to educational institutions at federal, state, and local government levels to carry out infrastructural development of their institutions as provided in Section 5(1) (a) to (g) of the Act No. 7 of the Fund (Oraka, Ogbodo & Raymond, 2013).

Today, there is no institution in Nigeria, especially public tertiary institutions, where there are no ETF projects (Rufai, 2012). In 2008, the Tertiary Education Trust Fund (TETFUND) was established (Uzondu, 2012). The TETFUND was established by an Act of the National Assembly in June 2011 as an intervention agency under the TET fund Act, namely the Tertiary Education Trust Fund Act, 2011 (TETfund News Panaroma, 2013). The Act replaced the Education Tax Fund Act Cap. E4 Laws of the Federation of Nigeria 2004 and the Education Tax Fund (Amendment) Act No. 17, 2003 (Uzondu, 2012). The Trust Fund was responsible for imposing, managing, and disbursing the tax to public tertiary institutions in Nigeria (Oraka et al., 2013). Guidelines on assessing the TETFund, as noted by Oraka et al. (2013), show that it was set up to administer and disburse education tax collections to government educational institutions. To enable the TETFUND to achieve the above objectives, the TETFUND Act 2011 imposes a 2% Education Tax on the assessable profits of all registered companies in Nigeria. In 2008, the TETFund came into being (Uzondu, 2012).

Recently, another effort by the Nigerian government towards infrastructural sustainability in the education section is the National Economic Empowerment and Development Strategy (NEEDS) (National Planning Commission, 2004). According to Okolie (2011), the NEEDS is a reform-based plan policy themed towards economic recovery, infrastructural growth, and development. Okolie (2011) noted that the NEEDS report recognises that the Nigerian education system is dysfunctional, characterised by low standards and institutional decay. Based on this, the Federal Government's policy thrusts for educational infrastructure development were developed to ensure that 60% of schools have conducive teaching and learning environments (National Planning Commission, 2004).

Challenges Associated with Implementation of Maintenance Policy in the Nigeria Education Sector

According to Moja (2000), the limited political will to implement policies and laws is one of the major challenges associated with implementing maintenance policy in the Nigerian education sector. This is evidenced in the attitude of both legislative and executive arms of government in Nigeria

(Ugwu et al., 2018). Also, Nigeria has no independent authority responsible for implementing maintenance laws and policies (Ugwuanyi, 2014; Moja, 2000). Nevertheless, some specific bodies ensure compliance with sector-specific Acts that seek, among others, to encourage effective MM of public assets in Nigeria. Examples of such authority are the authorities and agencies, including the FERMA, NEMA, and the Federal Ministry of Works and Housing maintenance unit, among others (Ugwu et al., 2018).

Another challenge facing Nigeria's maintenance laws and policy development and implementation is the absolute alignment to Western patterns and the notion of maintenance policy development (Onah, 2006). The lack of a clear maintenance vision is also one of the biggest obstacles to the development plans for maintenance policies and laws in Nigeria. Eneh (2011) acknowledged that it is the foundational basis for Nigeria's disjointed development planning mission. Other factors that are responsible for the lack of implementation of maintenance policies and laws include, among others, corruption and corrupt practices, erratic and conflicting government policies which result in policy changes and abandonment, plan indiscipline and unnecessary partnerships, poor and inadequate feasibility studies in planning and a lack of accurate data (Eneh, 2011).

Lesson Learned From Nigeria Maintenance Studies

Lessons learned from the relevant literature survey relating to Nigeria include the following:

- Maintenance is recognised as one of the important aspects of managing educational buildings in the Nigerian educational sector. It is generally acknowledged that effective maintenance of educational buildings in Nigeria will improve student output towards achieving a government development plan for the country.
- Nigeria has no standalone national maintenance policy or law that holistically addresses maintenance-related issues in the nation. However, there was a developmental strategy towards the sustainability of existing public assets such as the FERMA, NEMA, and the maintenance unit of the Federal Ministry of Works, and Housing, among others.
- Nigeria has no independent authority responsible for implementing maintenance laws or policies.
- There is a lack of standalone maintenance policies and laws for the Nigerian educational sector towards effective MM of the educational institution buildings and other education infrastructure.

Maintenance Policy Issues: Ghana and Nigeria

Maintenance has been recognised as all works related to repairs, replacement, and other redecoration performed on any building with the main aim of increasing the useful economic life of the building (British Standard: EN 13306, 2001, Ogunbayo et al., 2022). A well-designed maintenance policy is a strategy within which countries decide on maintenance tasks to ensure that work considered necessary on public infrastructure is carried out with maximum economy to satisfy the criteria for effectiveness and efficiency. Above all, it defines the ground rules for allocating resources, which include man, material, and money, against the alternative type of maintenance activities available to management. Given these, some policy considerations are vital for Ghana and Nigeria in the maintenance of public assets, including educational buildings, towards meeting the maintenance needs and expectations of their citizens (Nkrumah et al., 2017, Ogunbayo, 2021). The policy consideration should include the following:

- There should be capacity development for the implementing institutions of the maintenance policies and laws as enshrined in the Constitutions of both Ghana and Nigeria.
- Educational institutions should provide maintenance policies for their institutions as enshrined in their countries' maintenance policies and laws before establishing or obtaining their license of operation.
- Furthermore, there should be political will from the government and other political players on maintenance policies and laws' implementation.
- Also, there is a need for training and re-training of maintenance personnel and other skilled workers within the countries towards improving the human capacity for the effective maintenance of national assets as enshrined in the constitutions of both countries.

Finally, there is a need to establish or enact standalone national maintenance policies and laws that will address the maintenance of national assets, including educational buildings.

Summary

In this chapter, maintenance policy in developing economies was discussed with an emphasis on Ghana and Nigeria. As shown in the literature, recently Ghana government has established a maintenance policy that will put in place a mechanism that would allow for broad participation in the development planning process at all levels of society, including education, business, and others. The main aim of these new policies is to enhance greater access to decent buildings and bridge other infrastructure deficits confronting the country. In order to meet the objectives of this new maintenance policy, the Ghana government went ahead and established the Ghana Infrastructure Investment Fund. This is done to provide financial resources for investment in the maintenance of old and new national assets, including the educational buildings within the country. Also, from the literature review, it is revealed that the success of every educational institution in

Nigeria will depend on the availability of the right type of physical facilities, especially buildings. The quality of education in any educational institution is also dependent on several factors of which building facilities are supreme. Thus, the attributes determining the effective MM of educational buildings are unknown. There is a dearth of studies that develop a framework to aid effective MM of educational buildings in the Nigerian education sectors. The literature also reveals that Nigeria has no standalone national maintenance policy or law that holistically addresses maintenance-related issues in the country. However, there was a developmental strategy towards the sustainability of existing public assets through institutional agencies such as FERMA and the TETfund, among others. Additionally, the literature shows that there is no local content law for the Ghanaian and Nigerian education sectors or their countries to carry out the MM of educational buildings and other public assets effectively. However, the review showed that the attributes that determine effective MM of educational buildings differ from one country to another and from one education section to another.

References

Ab Wahab, Y., Basari, A. S. H., & Samad, A. (2013). Building maintenance management preliminary finding of a case study in ICYM. *Middle East Journal of Science Research, 17*(9), 1260–1268.

Abdul Lateef Olanrewaju, A. (2012). Quantitative analysis of defects in university buildings: User perspective, *Built Environment Project and Asset Management, 2*(2), 167–181.

Abdulrahman, M. Y. (2013). Historical development of universities in Nigeria: Chronology and the journey so far. *African Journal of Higher Education Studies and Development, 1*, 54–72.

Abdulrasheed, O., & Bello, A. S. (2015). Challenges to secondary school principal' leadership in northern region of Nigeria. *British Journal of Education, 3*(3), 1–5.

Abiogu, G. (2014). Philosophy of education: A tool for national development. *Open Journal of Philosophy,* 4(3), 372–377.

Addae-Mensah, I. (2000). *Education in Ghana: A tool for social mobility or social stratification?* Accra: Ghana Academy of Arts and Sciences.

Aghimien, D. O., Adegbembo, T. F., Aghimien, E. I., & Awodele, O. A. (2018). Challenges of sustainable construction: a study of educational buildings in Nigeria. *International Journal of Built Environment and Sustainability, 5*(1), 33–46.

Akhlaghi, F. (1996). Ensuring value for moneyin FM contract services. *Facilities, 14*(1/2), 26–33.

Adegbite, J. G. O. (2007). The Education Reform Agenda: Challenges for tertiary education Administration in Nigeria. Being: *A paper presented at the Sixth Annual Seminar of the Conference of registrars of College of education in Nigeria (South-West Zone)*, at the College of Education, Ikere-Ekiti, Ekiti State, pp. 12–13 June 2007.

Adenuga, O. A. (2012). Maintenance management practices in public hospital-built environment: Nigeria case study. *Journal of Sustainable Development in Africa, 14*(1), 185–201.

Adenuga, O. A., Olufowobi, M. B., & Raheem, A. A. (2010). Effective maintenance policy as a tool for sustaining housing stock in downturn economy. *Journal of Building Performance, 1*(1), 93–109.

Adesoji, D. J. (2011). Everlasting Public Hosing Performance: Providing a bases for residential. Quality Improvement in Nigeria. *Middle East Journal of Scientific Research, 9*(2), 225–232.

Adewunmi, Y., Omirin, M., Famuyiwa, F., & Farinloye, O. (2011). Post-occupancy evaluation of postgraduate hostel facilities. *Facilities, 29*(3/4), 149–168.

Adu-Agyem, J., & Osei-Poku, P. (2012). Quality education in Ghana: The way forward. *International Journal of Innovative Research and Development, 1*(9), 164–177.

Ahmad, R. B. H. (2006). Maintenance Management and Services (Case Study: PERKESO, Buildings in Peninsular of Malaysia), Unpublished Master's Thesis, University Technology Malaysia.

Aidoghie, P. (2018). Federal government of Nigeria moves to ensure public buildings maintenance. *The Sun Newspaper, 18*. 4 December.

Ajayi, I. A., & Ekundayo, H. T. (2008). The deregulation of university education in Nigeria: Implications for quality assurance. *Nebula, 5*(4), 212–224.

Akpobasah, M. O. S. E. S. (2004). Development strategy for Nigeria. A paper presented at *Overseas Development/Nigerian Economic Summit Group Meeting on Nigeria*, London, 16–17 June, pp. 1–20.

Akyeampong, K. (2010). 50 years of educational progress and challenges in Ghana. *Create Pathways to Access Research Monograph, 33*, University of Sussex: CREATE.

Alexander, K. (1996). Facilities management: A strategic framework. *Facilities management: Theory and Practice*, London: Routledge (Taylor and Francis). 10–25.

Ampofo, J. A., Amoah, S. T., & Peprah, K. (2020). Examination of the current state of government buildings in senior high schools in Wa Municipal. *International Journal of Management & Entrepreneurship Research, 2*(3), 161–193.

Amusan, L. M., Owolabi, J. D., Tunji-Olayeni, P. F., Peter, N. J., & Omuh, I. O. (2014). Assessing the effectiveness of maintenance practices in public schools. *European International Journal of Science and Technology, 3*(3), 103–109.

Antwi, M. K. (1992). *Education, society, and development in Ghana.* Unimax

Anugwo, I. C., Shakantu, W., Saidu, I., & Adamu, A. (2018). Potentiality of the South African construction SMME contractors globalising within and beyond the SADC construction markets. *Journal of Construction Business and Management, 2*(1), 41–49.

Apusigah, A. A. (2017). Forging African unity in a globalizing world: A challenge for Postcolonial nation-building. *Journal of Social Science, 3*(1), 13–26.

Aryeetey, E., & Kanbur, S. R. (Eds.). (2017). *The economy of Ghana sixty years after independence.* London: Oxford University Press.

Asiabaka, I. P. (2008). The need for effective facility management in schools in Nigeria. *New York Science Journal, 1*(2), 10–21.

Asiyai, R. I. (2012). Assessing school facilities in public secondary schools in Delta State, Nigeria. *African Research Review, 6*(2), 192–205.

Baharun, R. (2002). *A study of market segmentation in tertiary education for local public higher learning institutes.* Universiti Teknologi Malaysia.

Banful, E. (2004). A stitch in time saves nine, Cultivating a maintenance culture in Ghana. A Paper Presented at a Seminar on Maintenance Culture in Ghana, Accra, 16 March, pp. 1–2.

Becker, F. (1990). Facility management: A cutting-edge field? *Property Management*, 8(2), 108–116.

Bening, R. B. (2005). *University for Development Studies in the history of higher education in Ghana*. Accra: Centre for Savana Art and Civilisation.

Bingab, B., Forson, J., Mmbali, O., & Baah-Ennumh, T. (2016). The evolution of university governance in Ghana: Implications for education policy and practice. *Asian Social Science*, 12(5), 147–160.

Bosompen, K., Montes Belot, R., Joya, W. B., & Mejía de Pereira, A. C. (2012). *Assessment of the Viability of PPPs and Sub-national Lending in Ghana*. World Bank.

Buys, F., & Nkado, R. (2006). A survey of maintenance management systems in South African tertiary educational institutions. *Construction Management and Economics*, 24(10), 997–1005.

CABE, B. (2005). *The impact of office design on business performance*, Commission for Architecture, and the Built Environment: British Council for Offices.

Cholasuke, C., Bhardwa, R., & Antony, J. (2004). The status of maintenance management in UK manufacturing organisations: Results from a pilot survey. *Journal of Quality in Maintenance Engineering*, 10(1), 5–15.

Clark, J. (2008). PowerPoint and pedagogy: Maintaining student interest in university lectures. *College teaching*, 56(1), 39–44.

Cobbinah, P. J. (2010). *Maintenance of Buildings of Public Institutions in Ghana. Case Study of Selected Institutions in the Ashanti Region of Ghana*. Published master thesis. Kumasi: Kwame Nkrumah University of Science and Technology.

Ekundayo, H. T., & Ajayi, I. A. (2009). Towards effective management of university education in Nigeria. *International NGO Journal*, 4(8), 342–347.

Eleweke, C. J. (2002). A review of issues in deaf education under Nigeria's 6-3-3-4 education system. *Journal of Deaf Studies and Deaf Education*, 7(1), 74–82.

Emetarom, U. G. (2004). Provision and Management of Facilities in Primary Schools in Nigeria: Implications for Policy Formulation. In Paper presented at the annual national congress of Nigerian Educational Administration and Planning (NEAP). University of Ibadan. 28th–31st October.

Eneh, O. C. (2011). Nigeria's Vision 20: 2020-issues, challenges, and implications for development management. *Asian Journal of Rural Development*, 1(1), 21–40.

Estache, A., & Fay, M. (2007). *Current debates on infrastructure policy*. The World Bank.

Fafunwa, A.B. (2018). *History of education in Nigeria*. London: Routledge (Taylor & Francis).

Federal Government of Nigeria (2004). *National Policy on Education* (4th ed.). Yaba: NERDC Press.

Federal Ministry of Education (2004). *National Policy on Education*. Lagos: Education research and development council.

Federal Republic of Nigeria. (1999). *Constitution* of the Federal Republic of Nigeria.

Forrest, T. (2019). *Politics And Economic Development in Nigeria: Updated Edition*. Routledge.

Fulmer, J. (2009). What in the world is infrastructure? *PEI Infrastructure investor*, 1(4), 30–32.

Fund, E. T. (1993). *Establishment of Education Tax Fund*: A Supplement to Nigerian Official Gazette No. 7 of 1993. Lagos: Government Printer.

Gasskov, V. (1992). *Training for Maintenance in Developing Countries. Training Discussion* (pp. 1–37). Geneva, Switzerland: International Labour Office, Paper No.97.

Ghana, Constitution (1992). *The 1992 constitution of the Republic of Ghana.* Accra: Government Printer.

Ghana, Parliament (2016). *List of Acts of the Republic of Ghana from 1993-2017.* Accra: Government Printer.

Government of Ghana (2003). *The Coordinated Programme for the Economic and Social Development of Ghana (2003-2012).* Accra, Ghana: Government printer.

Government of Ghana (2013). *Annual Progress Report of the Implementation of the Ghana Shared Growth and Development Agenda (GSGDA) for 2010 – 2013.* Accra, Ghana: Government printer.

Gyampo, R. (2015). Winner-takes-all politics in Ghana: The case for effective council of state. *Journal of Politics & Governance, 4*(1), 20–28.

Heitor, T. (2005). Potential problems and challenges in defining international design principles for school. *Evaluating Quality in Educational Facilities, 48,* 44–54.

Heritage, E. (2009). Commission for Architecture and the Built Environment. Large digital screens in public spaces: Joint guidance from English Heritage and CABE.

Howard, M. (2006). *Best practices maintenance plan for school buildings.* Idaho: State Department of Education.

Huaisheng, Z., Manu, B. D., Mensah, I. A., Mingyue, F., & Oduro, D. (2019). Exploring the Effect of School Management Functions on Student's Academic Performance: A Dilemma from Public Senior High Schools in Ghana. *Journal of Arts and Humanities, 8*(6), 33–45.

Ibukun, W. O. (1997). *Educational management: Theory and practice.* Ado-Ekiti: Green Line Publishers.

Iheanacho, E. N. (2014). National Development Planning in Nigeria: An endless search for appropriate development strategy. *International Journal of Economic Development Research and Investment, 5*(2), 49–60.

Ikediashi, D. I., Ogunlana, S. O., Boateng, P., & Okwuashi, O. (2012) Analysis of risks associated with facilities management outsourcing: A multivariate approach, *Journal of Facilities Management, 10*(4), 301–316.

International Labour Organisation (ILO) (2009). *Background studies on infrastructure sector in Ghana.* Cape Coast, South Africa: Directorate of Research, Innovation and Consultancy University of Cape Coast.

Ipingbemi, O. (2010). *Facility Management* Unpublished M.Sc. Housing Development and Management Lecture Notes. Nigeria: University of Ibadan.

Iyagba, R. A. (2005). *The menace of sick buildings: A challenge to all for its prevention and treatment.* Lagos, LA: Inaugural Lecture, University of Lagos Press.

Iyamu I. F., Imasuen K., & Osakue E. F. (2018). Analysis of maintenance culture in public primary and secondary schools in Edo State. *International Journal of Vocational and Technical Education Research, 4*(2), 23–30.

Kadingdi, S. (2006). Policy initiatives for change and innovation in basic education programmes in Ghana. *Educate~, 4*(2), 3–18.

Karia, N., Asaari, M. H. A. H., & Saleh, H. (2014). Exploring maintenance management in service sector: A case study. In Proceedings of International Conference on Industrial Engineering and Operation Management, Bali, 7–9 January 2014, pp. 3119–3128.

Kayan, B. (2006). Building maintenance in old buildings conservation approach: an overview of related problems. *Journal of Design and Built Environment, 2*(1), 41–56.

Klumbyte, E., Bliudzius, R., & Fokaides, P. (2020). Development and application of municipal residential buildings facilities management model. *Sustainable Cities and Society, 52,* 101804.

Klumbytė, E., Bliūdžius, R., Medineckienė, M., & Fokaides, P. A. (2021). An MCDM model for sustainable decision-making in municipal residential buildings facilities management. *Sustainability, 13*(5), 2820.

Kotzé, M., & Nkado, R. (2003). An investigation into the use of facilities management in institutions of higher learning in South Africa. In 1ST CIDB Postgraduate Conference 2003. Port Elizabeth, South Africa, 12–14 October 2003, 70–76.

Kunya, S. U. (2012). Maintenance Management Unpublished MTech Construction Management Lecture Notes. In *Building Programme, Faculty of Environmental Technology.* Bauchi, Nigeria: Abubakar Tafawa Balewa University of Technology.

Kyeremateng, C. (2008). An Appraisal of the Maintenance culture of some selected Second Cycle Schools in Kumasi Metropolis. *Unpublished Dissertation) Building Technology Department, KNUST, Library Ref. No. BT606.*

Lateef, O. A. (2009). Building maintenance management in Malaysia. *Journal of Building Appraisal, 4*(3), 207–214.

Lateef, O. A., Khamidi, M. F., & Idrus, A. (2010). Building maintenance management in a Malaysian university campus: a case study. *Construction Economics and Building, 10*(1-2), 76–89.

Lavy, S. (2008). Facility management practices in higher education buildings: A case study. *Journal of Facilities Management, 6*(4), 303–315

Lawal, T., & Oluwatoyin, A. (2011). National development in Nigeria: Issues, challenges, and prospects. *Journal of Public Administration and Policy Research, 3*(9), 237–241.

Löfsten, H. (1998). Maintenance of municipal infrastructure. *Journal of infrastructure systems, 4*(4), 139–145.

Manuh, T., Gariba, S., & Budu, J. (2007). *Change and transformation in Ghana's publicly funded universities.* Partnership for Higher Education in Africa. Oxford, UK: James Currey and Accra, Ghana: Woeli Publishing Services.

Mayaki, S. S. (2005). Facility performance evaluation. In International Conference Organized by the Nigerian Institute of Building. Nasarawa, 20–22 April 2005. pp. 34–47.

McWilliam, H. O. A., & Kwamena-Poh, M. A. (1975). *The development of education in Ghana: An outline.* Longman.

Mgbekem, S. J. A. (2004). *Management of university education in Nigeria.* Calabar: University of Calabar Press.

Michael, A. N., Omole David, O., Peter, A., Babalola Olufemi, D., & Olusoji, M. T. (2018). Appraisal of the state of Health of Residential Building Facilities in a Private University in Nigeria. *International Journal of Engineer Technology. and Management Research, 5*(4), 153–167.

Ministry of Budget and National Planning (2017). *Economic Recovery & Growth Plan-ERGP 2017-2020.* Accra: Government Press.

Moja, T. (2000). *Nigeria education sector analysis: An analytical synthesis of performance and main issues* (vol. 3, pp. 46–56). New York, NY: World Bank Report.

Mukelasi, M. F. M., Zawawi, E. A., Kamaruzzaman, S. N., Ithnin, Z., & Zulkarnain, S. H. (2012). A review of critical success factors in building maintenance management of local authority in Malaysia. In 2012 *IEEE Symposium* on

Business, Engineering, and Industrial Applications Bandung, 23- 26 September 2012. pp. 653–657. IEEE.

Mutlaq, M. A. (2002). *A Study of the Relationship between School Building Conditions and Academic Achievement of Twelfth Grade Students in Kuwaiti Public High Schools.* Unpublished PhD thesis. America: The Virginia Polytechnic Institute and State University.

National Development Planning Commission (2010). *Medium-term national development policy framework: Ghana shared growth and development agenda (GSGDA), 2010-2013.* Accra: Government Printer.

National Development Planning Commission (2015). *Ghana Millennium Development Goals* 2015 Report. Accra: NDPC

National Planning Commission (2004). *National economic empowerment and development strategy (NEEDS).* Abuja: National Planning Commission.

National Universities Commission (NUC) (2020). *Manual of Accreditation Procedures for Academic Programmes in Nigerian Universities* (MAP). Abuja: National Universities Commission.

Niemelä, T., Kosonen, R., & Jokisalo, J. (2016). Cost-optimal energy performance renovation measures of educational buildings in cold climate. *Applied Energy, 183,* 1005–1020.

Nkrumah, E. N. K., Stephen, T., Takyi, L., & Anaba, O. A. (2017). Public Infrastructure Maintenance Practices in Ghana. *Review Public Administration Management, 5*(234), 2.

Obamwonyi, M. (2009). *Improving maintenance perception in developing countries: a case study.* Published Bachelor Thesis. Linnaeus University: School of Engineering Department of Terotechnology.

Obeng-Odoom, F., & Amedzro, L. (2011). Inadequate housing in Ghana. *Urbani izziv, 22*(1), 127.

Odediran, S. J., Adeyinka, B. F., Opatunji, O. A., & Morakinyo, K. O. (2012). Business structure of indigenous firms in the Nigerian construction industry. *International Journal of Business Research and Management, 3*(5), 255–264.

Odediran, S. J., Gbadegesin, J. T., & Babalola, M. O. (2015). Facilities management practices in the Nigerian public universities. *Journal of Facilities Management, 13*(1), 5–26.

Odetunde, C. (2004). The state of higher education in Nigeria. Retrieved on 20th February 2016 from http://www.nigerdeltacongress.Com/sertive/state-of-higher-education

Odia, L. O., & Omofonmwan, S. I. (2007). Educational system in Nigeria problems and prospects. *Journal of Social Sciences, 14*(1), 86-85.

Ogunbayo, B. F., & Aigbavboa, O. C. (2019). Maintenance requirements of students' residential facility in higher educational institution (HEI) in Nigeria. In *Proceedings of 1st International Conference on Sustainable Infrastructural Development*, Ota, 24–28 June 2019, IOP Conference Series: Materials Science and Engineering, 640, (1) 012014. IOP Publishing.

Ogunbayo, B. F., Aigbavboa, C. O., Amusan, L. M., Ogundipe, K. E., & Akinradewo, O. I. (2021). Appraisal of facility provisions in public-private partnership housing delivery in southwest Nigeria. *African Journal of Reproductive Health, 25.*

Ogunbayo, B. F., Aigbavboa, C. O., Thwala, W., Akinradewo, O., Ikuabe, M., & Adekunle, S. A. (2022). Review of culture in maintenance management of public buildings in developing countries. *Buildings, 12*(5), 677.

Ogunbayo, B. F., Ajao, A. M., Alagbe, O. T., Ogundipe, K. E., Tunji-Olayeni, P. F., & Ogunde, A. (2018). Residents' facilities Satisfaction in Housing Project Delivered by Public Private Partnership (PPP) In Ogun State, Nigeria. *International Journal of Civil Engineering and Technology (IJCIET)*, *9*(1), 562–577.

Ogunbayo, B. F., Ohis Aigbavboa, C., Thwala, W. D., & Akinradewo, O. I. (2022). Assessing maintenance budget elements for building maintenance management in Nigerian built environment: A Delphi study. *Built Environment Project and Asset Management*, Vol. ahead-of-print No. ahead-of-print. 10.1108/BEPAM-06-2021-0080

Ogunbayo, S. B. (2021). *States of education, funding, and democracy in Nigeria.* Alternate Horizons.

Ogunbayo, S. B., & Mhlanga, N. (2021). Assessment of public primary school teachers computer literacy and usage in teaching and learning. *Journal of Computer Adaptive Testing in Africa*, *1*(1), 1–14.

Ogunbayo, S. B., & Mhlanga, N. (2022). Effects of training on teachers' job performance in Nigeria's public secondary schools. *Asian Journal of Assessment in Teaching and Learning*, *12*(1), 44–51.

Ojara, E. S. (2013). *The Challenges of Housing Maintenance in Nigeria. Unpublished HND Project.* Nigeria: Department of Building Technology, The Federal Polytechnic Ado Ekiti.

Ojogwu, C. N., & Alutu, A. N. G. (2009). Analysis of the learning environment of university students on Nigeria: A case study of University of Benin. *Journal of Social Sciences*, *19*(1), 69–73.

Okebukola, P. (2006). Principles and policies guiding current reforms in Nigerian universities. *Journal of Higher Education in Africa/Revue de l'enseignement supérieur en Afrique*, *4*, (1), 25–36.

Okolie, K. C. (2011). *Performance evaluation of buildings in Educational Institutions: A case of Universities in South-East Nigeria.* Published doctoral dissertation. Port Elizabeth, South Africa: Nelson Mandela Metropolitan University.

Okoroma, N. S. (2006). Educational policies and problems of implementation in Nigeria. *Australian Journal of Adult Learning*, *46*(2), 243–263.

Oladapo, A. A. (2005). *An evaluation of the maintenance management of the staff housing estates of selected first-generation universities in Southwestern Nigeria.* Unpublished PhD. Thesis. Nigeria: Dept. of Building, Obafemi Awolowo University, Ile-Ife.

Oladapo, A. A. (2006). A study of tenants' maintenance awareness, responsibility, and satisfaction in institutional housing in Nigeria. *International Journal of Strategic Property Management*, *10*(4), 217–231.

Olaniyan, O. (2001). Public finance and higher education in Nigeria. In *Proceedings of the 12th General Assembly of Social Science Academy of Nigeria*, Yola, 3–7 July, pp. 21–28.

Olanrewaju, A., Fang, W. W., & Tan, Y. S. (2018). Hospital building maintenance management model. *International Journal of Engineering and Technology*, *2*(29), 747–753.

Olujuwon, O., & Perumal, J. (2015). The impact of teacher leadership practices on school management in public secondary schools in Nigeria. In Edited by Y. Ono, *59thYear Book on Teacher Education* (pp. 472–481). Japan: A publication of the International Council on Education for Teaching and Naruto University of Education.

Oluwunmi, A. O., Akinjare, O. A., Ayedun, C. A., & Akinyemi, O. (2015). Management efficiency of private university facilities: A study of Covenant University's residential Estate. *European Scientific Journal, 8*(4), 109–126.

Omomia, O. A., Omomia, T. A., & Babalola, J. A. (2014). The history of private sector participation in university education in Nigeria (1989-2012). *Research on Humanities and Social Sciences, 14*(8), 425, 439.

Onah, F. O. (2006). *Managing public programmes and projects.* Nsukka: Great Ap Express Publishers Limited.

Onibokun, A. G. (1982). *In search of solutions: A comprehensive review of housing literature and research in Nigeria.* Ibadan: Nigerian Institute of Social and Economic Research (NISER).

Oraka, L. A., Ogbodo, E. A., & Raymond, U. S. (2013). Effect of tertiary education tax fund (TETFund) in management of Nigerian tertiary institutions. *International Journal of Educational Administration, Planning and Research, 6*(1), 36–43.

Osei-Owusu, B., & Awunyo-Vitor, D. (2012). *Teachers' perception on sustainability of distance education in Ghana.* Evidence from Ashanti Region.

Owoeye, J. S. (2000). *The effect of interaction of location, facilities, and class size on academic achievement of secondary school students in Ekiti State, Nigeria.* Unpublished doctoral thesis. Ibadan: University of Ibadan.

Owuamanam, D. O. (2005). Threats to academic integrity in Nigerian universities. In Lead paper presented at the conference of the National Association of Educational Researchers and Evaluators, University of Ado-Ekiti, June 13–17.

Owusu, K., Aigbavboa, C. O., & Thwala, D. W. (2016) Built-operate-transfer (Bot) model for public universities in Ghana for accelerating infrastructural development. In J. N. Mojekwu, G. Nani, L. Atepor, R. A. Oppong, M. O. Adetunji, L. Ogunsumi, U. S. Tetteh, E. Awere, S. P. Ocran, and E. Bamfo-Agyei (Eds). *Proceedings of 5th Applied Research Conference in Africa. (ARCA) Cape Coast*, Ghana, 25–27 August 2016.

Price, S., & Pitt, M. (2012). The influence of facilities and environmental values on recycling in an office environment. *Indoor and Built Environment, 21*(5), 622–632.

Quayson, J. H., & Akomah, B. B. (2016). Maintenance of residential buildings of selected public institutions in Ghana. *African Journal of Applied Research (AJAR), 2*(1).

Rahman, S., & Vanier, D. J. (2004, May). Life cycle cost analysis as a decision support tool for managing municipal infrastructure. In *CIB 2004 triennial congress,* (Vol. 2, No. 1, pp. 11–18). Ottawa: National Research Council (NRC).

Robinson, L. E., Rudisill, M. E., & Goodway, J. D. (2009). Instructional climates in preschool children who are at-risk. Part II: Perceived physical competence. *Research quarterly for exercise and sport, 80*(3), 543–551.

Robinson, L., & Robinson, T. (2009). An Australian approach to school design.

Rufai, R. A. (2012). *Federal Government Allocates N25 Billion to Tertiary Institutions.* Abuja: The National Universities Commission.

Salami, S. O. (2010). Emotional intelligence, self-efficacy, psychological well-being and students' attitudes: Implications for quality education. *European Journal of Educational Studies, 2*(3), 247–257.

Sanoff, A. P., & Powell, D. S. (2003). *Restricted access: The doors to higher education remain closed to many deserving students.* Indianapolis, IN: Lumina Foundation for Education.

Simpeh, F. (2018) Challenges faced by university hostel managers in the Greater Accra Region of Ghana In J. N. Mojekwu, G. Nani, L. Atepor, R. A. Oppong, M. O. Adetunji, L. Ogunsumi, S. P. Ocran, and E. Bamfo-Agyei (Eds). *Proceedings of 7th Applied Research Conference in Africa. (ARCA) Conference*, Nairobi, Kenya, 1–3 August 2018. pp. 39 – 49.

Soetan, T. O. (2018). Trends in higher education financing: Evidence of dwindling government supports and a case for the aggressive marketing of higher education programs using the 9 Ps of marketing. *Journal of Marketing Management, 6*(2), 34–43.

Soleimanzadeh, S., & Mydin, M. A. O. (2013). Building maintenance management preliminary finding of a case study in ICYM. *Middle-East Journal of Scientific Research, 17*, 1260–1268.

Spedding, A. H. (1994). Prioritization of building maintenance work. In CIB W7, Symposium, Tokyo, Vol. 2, October 14–15, 1994. pp 1263–1270.

Swanson, L. (2001). Linking maintenance strategies to performance. *International Journal of Production Economics, 70*(3), 237–244.

Syamilah, B. Y. (2006). *Maintenance management through strategic planning for public school in Malaysia.* Unpublished Master's Thesis, University Technology Malaysia.

Tan, Y., Shen, L., & Langston, C. (2012). Competition environment, strategy, and performance in the Hong Kong construction industry. *Journal of Construction Engineering and Management, 138*(3), 352–360.

Task Force on Higher Education & Society (2000). *Higher Education in Developing Countries: Peril and Promise.* Washington DC: The World Bank.

TETfund News Panaroma (2013). *Tertiary Education Trust Fund.* February Edition, www.tetfund.gov.ng

Then, D. S., & Tan, T. (2002). Measuring Operational Building Asset Performance: Concepts and Implementation. In *Proceedings of CIB W60/W96 Hong Kong: Joint Conference on Performance Concept in Building and Architectural Management.* Hong Kong, 6- 8 May 2002. pp. 381–395.

Ugwu, O. O., Okafor, C. C., & Nwoji, C. U. (2018). Assessment of building maintenance in Nigerian university system: A case study of University of Nigeria, Nsukka. *Nigerian Journal of Technology, 37*(1), 44–52.

Ugwuanyi, G. O. (2014). Taxation and tertiary education enhancement in Nigeria: An evaluation of the education tax fund (ETF) between 1999-2010. *Journal of Economics and Sustainable Development, 5*(6), 131–141.

Ukpia, U. E. & Ereh, C. E. (2016). Current challenges and the needed competencies in the management of university education in Nigeria. *British Journal of Education, 4*(2), 74–86.

Uzondu, J. (2012). Funding, Not Problem of Education in Nigeria. *Leadership Forum Nigerian New World.* Monday, April 23.

Wood, B. R. (2005). Towards innovative building maintenance. *Structural Survey, 23*(4), 291–297.

World Bank (1999). Knowledge Economy. *World Development Report.* New York: Oxford University Press.

World Bank (2004). *Books, buildings and learning outcomes: An impact evaluation of World Bank support to basic education in Ghana.* Washington, DC: World Bank.

World Bank (1994). Higher education: *The lessons of experience.* Washington, DC: World Bank report.

World Bank (2008). *Accelerating catch-up: Tertiary education for growth in sub-Saharan Africa*. The World Bank.

Wuni, I. Y., Agyeman-Yeboah, S., & Boafo, H. K. (2017). Poor Facility Management in the Public Schools of Ghana; Recent Empirical Discoveries. *Journal of Sustainable Development Studies*, *11*(1), 1–30.

Yacob, S. (2005). *Maintenance management system through strategic planning for public school in Malaysia*. Published doctoral dissertation. Universiti Teknologi Malaysia.

Yinghua, C., Sylvester, A. P. F., Bonsu, M. O. A., Minkah, A. Y., & Zhenjiang, P. R. (2018). Practice of Maintenance Management of Infrastructures on Sports Stadia in Ghana. *Journal of Public Administration and Governance*, *8*(4), 250–263.

Young, E., Green, H. A., Roehrich-Patrick, L., Joseph, L., & Gibson, T. (2003). Do K-12 school facilities affect education outcomes. The Tennessee advisory commission on intergovernmental Relations: Staff Information Report.

Yusif, H., Yussof, I., & Osman, Z. (2013). Public university entry in Ghana: Is it equitable? *International Review of Education*, *59*(1), 7–27.

Zavadskas, E. K., Turskis, Z., Šliogerienė, J., & Vilutienė, T. (2021). An integrated assessment of the municipal buildings' use, including sustainability criteria. *Sustainable Cities and Society*, *67*, 102708.

Zubairu, N. S. (2000). *Maintenance of Government Office Buildings in Nigeria–A Post Occupancy Evaluation Approach*. Unpublished doctoral dissertation. University of Lagos.

Zubairu, S. N. (2010). The National Building Maintenance Policy for Nigeria: The Architects' Perspective. In *Compilation of Seminar Papers presented at the 2010 Architects Colloquium-Architecture and the National Development Agenda III* (pp. 1–12). Lagos: Architects Registration Council of Nigeria.

Zulu, S., & Chileshe, N. (2010). Service quality of building maintenance contractors in Zambia: A pilot study. *International Journal of Construction Management*, *10*(3), 63–81.

8 Maintenance Management of Municipal Buildings in the South African Education Sector

Introduction

This section of the study reviews maintenance and provides a historical overview of the maintenance framework in South Africa. It attempts to focus on the challenges of maintenance activities of buildings by reviewing past and present maintenance policies in South Africa. The policies and agencies supporting the maintenance programme and its enactments in South Africa, such as the government, private sector, and others, are also explored. Furthermore, the maintenance of educational buildings in South Africa is also reviewed. Lastly, a summary of the lesson learned to date from the literature is presented.

Educational System in South Africa

The Anglo-Zulu War (1879) incorporated the Zulu kingdom's territory into the British Empire. Subsequently, the Afrikaner republics were incorporated into the British Empire after their defeat in the Second South African War (1899–1902). However, the British and the Afrikaners ruled together from 1910 under the Union of South Africa, which became a republic in 1961 after a whites-only referendum (Central Intelligence Agency World Fact Book, 2018). In 1948, the National Party was voted into power and instituted a policy of apartheid – billed as "separate development" of the races – which favoured the white minority at the expense of the black majority and other non-white groups. The African National Congress (ANC) led the opposition to apartheid and many top ANC leaders, such as Nelson Mandela, spent decades in South Africa's prisons. Internal protests and insurgency, as well as boycotts by some Western nations and institutions, led to the regime's eventual willingness to negotiate a peaceful transition to majority rule (Central Intelligence Agency World Fact Book, 2018) (Map 8.1).

The first multi-racial elections in 1994 following the end of apartheid ushered in majority rule under an ANC-led government. Nevertheless, the racial differentiation in education had taken root from the beginning of the European colonisation of South Africa, with separate schools established

DOI: 10.1201/9781003344681-8

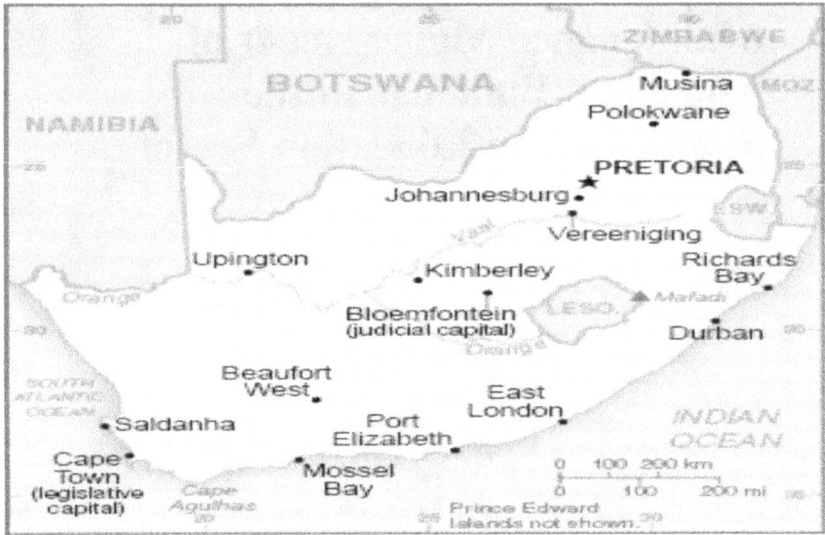

Map 8.1 Map of Republic of South Africa.

Source: CIA Factbook (2018).

Image Link https://www.cia.gov/the-world-factbook/static/f58294449f4f723e4b69e9be867d8b5e/SF-map.jpg

for the different "racial" groups (Frey, 2018). Essentially, the education system was designed to elevate Europeans above all other groups, who were programmed and socialised for subordinate roles in the European-dominated commercial and administrative systems to further the goals of South Africa's colonial overlords (Kallaway, 2002).

However, most early schools established for Africans were missionary institutions (Kallaway, 2002). By 1945 approximately 4,400 church-related schools were offering instruction in South Africa, compared with only 230 government schools. African culture, history, religion, values, botany, zoology, medicine, and so on provided no points of reference for African education, whether in the secular system of missionary schools or not (Kallaway, 2002). European-administered education for Africans was designed for alienation and underdevelopment, with all genuine educational opportunities reserved for the Europeans (Mouton, Louw & Strydom, 2012). However, South Africa has since struggled to address apartheid-era imbalances in its public buildings, including educational buildings located among the white minority settlement and the larger black settlement areas (Tebele, 2016).

Evolution of Maintenance Policy in South Africa

The South Africa maintenance laws and policies legal foundation is in the Constitution of the Republic of South Africa. Article (1) of the 1993

Constitution of the Republic provides the subsequent enactments of Acts and policies as needed (Constitution of Republic of South Africa, 1996). Nevertheless, all Acts and policies depend on the Constitution of the Republic, which is considered supreme to all laws (Constitution of the Republic of South Africa, 1996). Section 24 of the South African Constitution indicates that every citizen of South Africa has the right to an environment that is not harmful to their health (Republic of South Africa, 2007). Because of the imbalance in infrastructure that characterised the Republic, in 1994, the democratic government of South Africa embarked on an ambitious plan to put matters right by addressing the backlog of deterioration in the public infrastructure (Construction Industry Development Board [CIDB], 2007a).

In addressing the issue, in 2006, the South African government, through the Constitution of the Republic of South Africa (Act No. 108 of 1996), approved a maintenance policy, namely the National Infrastructure Maintenance Strategy (NIMS), for the maintenance of its public buildings (CIDB, 2007b). The main purpose of the NIMS was to set an overarching policy for sector-based initiatives and describes the framework for a co-ordinated program of action (Republic of South Africa, 2006). The NIMS policy was enacted to meet the government's vision of delivering infra-structure services to all (CIDB, 2007b). As noted by Construction Industry Development Board (2007b), the NIMS policy was formulated because there is strong evidence that much of the infrastructure of both pre- and post-1994 vintage is not being properly maintained. Wall (2008) further states that in South Africa, older infrastructure is often not being renovated and renewed when needed, and there is insufficient planned preventive maintenance on new infrastructure. At this point, it was evident that the Republic needs a holistic NIMS for public buildings, including the educa-tional institution buildings, without which many institutions are unlikely to improve their maintenance policies and practices (Wall, 2008).

In developing the NIMS policy, it draws extensively on maintenance fra-mework documents developed by the following institutions, the Department of Public Works (DPW), the Construction Industry Development Board (CIDB), and the Council for scientific and industrial research. According to CIDB (2007a), the four thrusts of the NIMS policy implementation that will lead to this achievement include the following:

- Strengthening the regulatory framework governing planning and bud-geting for infrastructure maintenance;
- Strengthening monitoring, evaluation, and reporting, and feeding this into a process of continuous improvement;
- Developing the maintenance industry; and
- Assisting institutions with non-financial resources.

However, to boost this maintenance policy, South Africa adopted a na-tional policy in 2012 intending to transform the economic landscape and delivery of basic services, including educational institution buildings within the country (CIDB, 2013). The Government Immovable Asset Management Act, No. 19 of 2007 (GIAMA), was enacted to strengthen maintenance laws further. The policy aims to provide a uniform framework for maintaining and managing immovable assets held or used by a national or provincial department (CIDB, 2007b; Republic of South Africa, 2006). Before then, the State Property Agency (SPA) was established by the government to manage the state's immovable assets (CIDB, 2007b). The main objective of the GIAMA is to provide a uniform framework for the maintenance and management of immovable assets held or used by a national or provincial department, including educational institution buildings (Phathela & Cloete, 2018). According to CIDB (2013), the GIAMA standard establishes a system of principles or practice specifications for the management and care of immovable assets after initial construction or acquisition. Nevertheless, other maintenance policies and laws were continuously amended to improve maintenance activities within the country's public infrastructure. The policies include the National Building Regulations and Building Standards Act (Act No. 103 of 1977), the National Environmental Management Act (Act No. 107 of 1998), as amended the Infrastructure Delivery Management Strategy (IDMS) Toolkit (2010), and many more others.

Thus, in South Africa, there is a legal framework for maintenance across all public infrastructure, both movable and non-movable, regulated by these policies. The main challenge is developing the capacity of institutions to implement the NIMS of 2006 and the Government Immovable Asset Management Act No. 19 of 2007 to the letter without fear or favour as enshrined in the Constitution of the Republic of South Africa (Phathela & Cloete, 2018).

Maintenance Policy and Legal Framework in the Maintenance of Educational Buildings in South Africa

The principal legislation that regulates the maintenance of educational institution buildings is the 1996 Constitution of the Republic of South Africa. However, the South African Schools Act No. 84 (1996) states that the Department utilises education facilities to provide educational activities, including learning, teaching, administrative work, and ancillary services such as student accommodation. Section 29 of the South Africa Constitution Act No. 108 of 1996 considers everyone's right to basic education (Constitution of the Republic of South Africa, 1996). During apartheid, the principal legislation guiding the education sector was the Bantu Education Act, Act No. 47 of 1953 (Cloete, 2014). The Act established a separate educational system for the black majority with the required skills to serve their people in the homeland (Cloete, 2014). In 1959, the Extension of University Education Act No. 45 was enacted. The Act further established separate educational institutions for Blacks, Indians, Coloureds, and Whites (Xaba, 2011). These educational institutions were separated on racial and ethnic lines (Xaba, 2011, Ogunbayo & Mhlanga, 2022).

However, in 1979 the two earlier Acts were repealed by section 45 of the Education and Training Act, Act No. 90 of 1979 (Xaba, 2011). This new Education law was passed in line with the provisions of section 10 of the South Africa Constitution, which postulates that "everyone has inherent dignity and the right to have their dignity respected and protected" (Constitution of Republic of South Africa, 1996). However, the dignity stated in the South African Constitution spreads to the quality of the facilities that need to be provided where educational activities would take place. This includes teaching and learning spaces, administrative services, learner boarding facilities, and facilities for educational support (RSA, Department of Education, 2015). These education facilities and services must be provided in conducive, safe, decent, and accessible facilities and be located in a supportive, complementary, and ideal environment for learning and teaching activities (RSA, Department of Education, 2015).

Equally, Goal 24, as specified in the Action Plan-Towards the Realization of Schooling 2025, showed that the main goal of the Department of Basic Education (DBE) is to ensure that the environment and physical infrastructure of educational institutions inspire learners to come to school and learn (DBE, Republic of South Africa, 2015). The South African Schools Act (Act No. 84 of 1996) (SASA), section 12(1), as amended, entails that the Provincial Education Department Member of the Executive Council (MEC) provides funds for the maintenance of public schools out of funds appropriated by the Provincial Legislature (DBE, Republic of South Africa, 2015). This is because education facilities must follow the Norms and Standards for Education Facilities (NSEF) provisions for acceptable minimum standards and various design aspects (DBE, Republic of South Africa, 2015). Additionally, Section 38(1) (d) of the Public Finance Management Act (PFMA) (Act No. 1 of 1999) stipulates that ownership of public schools, including educational institutions together with their residential facilities, remains the Provincial of Education Departments (PEDs) (PFMA, RSA, 1999). The PEDs make the decisions on the use of these educational facilities, with their usage and funding for their development, maintenance and upkeep, closure, and disposal if they are no longer required (PFMA, RSA, 1999).

However, as noted from the review, different maintenance policies are guiding the maintenance and management of education facilities in South Africa. Hence, it is noteworthy that there were challenges still facing the effective MM of education facilities. As stated in the KZN Department of Education Maintenance Strategy (2016), the key challenges that hamper the maintenance policies towards effective delivery of education facilities incorporating educational institutions' infrastructure in South Africa include the following:

- The dearth of proper maintenance units at a provincial level;
- The dearth of resources relating to transport, tools, and materials;
- Poor maintenance planning, budget, and implementation;
- Undefined reporting lines and lines of accountability;
- Limited capacity at both head office and district levels; The dearth of custodian commitment to maintenance;
- Incorrect and inappropriate norms and standards;
- School governing bodies not prioritising maintenance;
- Lack of close supervision of implementing agents; and
- Lack of clear responsibility lines between user and custodian.

Challenges Associated with Implementation of Maintenance Policy in South Africa

As Sebola (2014) noted, South Africa has faced many challenges regarding maintenance policies and laws implementation. Sebola (2014) observed one of these challenges: the irony that there are excellent written public policies such as maintenance policies and laws that have been identified as comprehensive and progressive under the South African Constitution. Tebele (2016) hypothesised that the problem with maintenance policies and laws appears to be that even though they describe the "what," the "how" context is not duly explained. Tebele (2016) states further that it is important for policymakers to explain the "how" because the primary focus of maintenance policies and laws is putting the maintenance policy into effect. Sebola (2014) asserted that for effecting the implementation of public policies such as maintenance policies and laws, a course of action must be put in place regarding how to implement the maintenance policy.

Looking at a national policy such as the National Development Plan (NDP) from the perspective of policymakers, the NDP is a well-designed and well-structured national policy. Nonetheless, there is no course of action towards implementing these policies (National Planning Commission, 2012). In another instance, in 2007, the CIDB published the NIMS; however, one problem with implementing the NIMS policy is the lack of documentary evidence. In another instance towards providing a uniform maintenance policy for the management of immovable assets that a national or provincial department uses, in 2008, a new maintenance policy, "the government immovable asset management Act (GIAMA)," was enacted. The main objectives of the GIAMA, as noted by the policies and laws that guide it, include the following:

- Provide a uniform immovable asset management framework to promote accountability and transparency within government;
- Ensure effective immovable asset management within government; and
- Ensure coordination of the use of immovable assets with service delivery objects of a national or provincial and the efficient utilisation of the immovable asset.

However, one of the challenges of implementing the GIAMA as a main-tenance policy or law guiding the maintenance of national assets, as observed by Boshoff and Peters (2013), is that although it came into existence in 2008, its scope excludes the local government arm of governance. From the point raised, McLaughlin (1987) posits that one major issue with the implementation of maintenance policies and the law is that policymakers have to rely on bureaucrats to implement what they have drafted because of a lack of knowledge to carry out the policies drafted by them. McLaughlin (1987) observed further that no matter how efficient and effective a drafted maintenance policy looks, Parliament must approve it before it becomes a law. Nevertheless, Broadnax (1976) posits that the leading challenges facing implementing maintenance policies and laws in South Africa are no involvement of any relevant stakeholders in drafting the maintenance policies. This has led to a lack of support for the maintenance policies. Mkhize (2017) noted that systematic problems relating to the lack of consultation and participation around policy choices in all spheres of government (national, provincial, and local government) are challenges faced in implementing its maintenance policies in South Africa.

However, Brynard and Netshikhophani (2011) state that another challenge facing the implementation of the maintenance policies and laws is the fact that it is time-consuming; that is why it is implemented in such a manner that will gratify every stakeholder involved. Because this decision was taken unilaterally, this might be why public policy implementation, including maintenance policies, is failing in South Africa (Mkhize, 2017). Similarly, McLaughlin (1987) opines that for any policy, including a maintenance policy, to be successful and free of challenges, local capacity should be available and the will of all concerned stakeholders, especially public officials. Tebele (2016) advised that for challenge-free maintenance policies, there ought to be enough human capacity as well as the readiness of relevant stakeholders to bring the intended policy to fruition.

Another known challenge raised by Human (1998) regarding public policy implementation, including maintenance policy, in South Africa, is the problem of adopting and applying the Western method and solution to "African" problems. Ferim (2013) opined that the Western countries' history is unlike that of Africa; with many years of practicing stable democracy, it is not advisable to be implementing Western solutions in South Africa, especially on maintenance policy. Additionally, Solomon (2015) states that the ghost of colonialism is still plaguing Africa. Hence Cloete (2006) observed that using western policy to guide South African policy decisions will mean having a heterogeneous nation and the skeletons left by the apartheid government.

Consequently, Human (1998) contends that the problem with adopting Western countries' policies for developing South African maintenance policies and laws is that policies framed at the time do not fit the environment in which they must be applied. Nevertheless, Ferim (2013) concluded that for maintenance policies and laws to be effective in South Africa, the commitment of all concerned is necessary, especially the state and local capacity.

Lessons Learnt From South Africa Maintenance Studies

Lessons learned from the relevant literature relating to the Republic of South Africa include the following:

- Maintenance policy and legal frameworks draw their source from the Constitution of the Republic of South Africa.
- Some legal frameworks and laws regulate the maintenance of all public infrastructure, both movable and non-movable, in South Africa.
- Apart from general maintenance policies and laws, other policies are guiding the maintenance of educational buildings and other educational assets in South Africa.
- The maintenance policies and laws in South Africa are aligned with the main objectives of maintenance policies and laws globally.
- The main challenge to maintaining policies and laws in South Africa is the capacity development of implementation institutions of different maintenance policies and laws enshrined in the Constitution of the Republic of South Africa.

Summary

The literature review showed that in South Africa, apart from the main Constitution, that guides the provision of laws and policies towards sustainability of the infrastructure and other national assets. The democratic government of South Africa embarked on an ambitious plan to put matters right by addressing the backlog of deterioration in the public infrastructure. Through the Constitution of the Republic of South Africa in 2006, the government of South Africa approved a maintenance policy for its public buildings. The policy was enacted to meet the government's vision of delivering infrastructure services to all. Nevertheless, a fundamental finding from the reviewed literature showed that no standalone maintenance law or policy guides the development of maintenance practice towards protecting educational buildings and other public assets.

References

Boshoff, L., & Peters, S. (2013). Challenges, Constraints and Best Practices in Rehabilitating Water and Electricity Distribution Infrastructure. Technical Report: Submission for the Division of Revenue 2014/2015.

Broadnax, W. D. (1976). Public policy: Its formulation, adoption, implementation, and evaluation. *Public Administration Review, 36*(6), 699–703.

Brynard, P. A., & Netshikhophani, A. F. (2011). Educator training challenges in implementing the National Curriculum Statement Policy. *African Journal of Public Affairs*, *4*(3), 61–71.

CIDB (2007b). NIMS. Available online from http://www.cidb.org.za

CIDB (2013). Standard for Developing Skills on Infrastructure Contracts. *Board Notice 180 of 2013*. Pretoria: Government Gazette 36760, 23.

CIDB, C. (2007a). *Construction Industry Master Plan 2006-2015 (CIMP)*. Pretoria: Construction Industry Development Board.

Cloete, F. (2006). Fundamentals of Evaluation research. *South African Journal of Public Administration*, *41*(3), 682–693.

Cloete, N. (2014). The South African higher education system: Performance and policy. *Studies in Higher Education*, *39*(8), 1355–1368.

Department of Basic Education, Republic of South Africa (2015). *Action plan to 2019: Towards the realisation of schooling 2030*. Pretoria: Taking forward South Africa's National Development Plan 2030.

Ferim, V. (2013). African Solutions to African Problems. *The African Union ten years after: Solving African problems with Pan-Africanism and the African Renaissance* (pp. 143–155). Pretoria: Africa Institute of South Africa.

Frey, B. B. (Ed.). (2018). *The SAGE encyclopedia of educational research, measurement, and evaluation*. Sage Publications.

Government Acts: Republic of South Africa (1996). *Constitution of the Republic of South Africa, Acts 63*. Pretoria: Government Printer.

Human, P. (1998). *Yenza: A blueprint for transformation.* Cape Town: Oxford University Press, USA.

Kallaway, P. (Ed.). (2002). *The history of education under apartheid, 1948-1994: The doors of learning and culture shall be opened.* South Africa: Pearson South Africa.

Kwazulu-Natal (2016). *KZN maintenance strategy.* Department of Education: Republic of South Africa.

McLaughlin, M. W. (1987). Learning from experience: Lessons from policy implementation. *Educational Evaluation and Policy Analysis*, *9*(2), 171–178.

Mkhize, F. (2017). *Cost management of public sector construction projects in KwaZulu-Natal.* Published doctoral dissertation. University of KwaZulu-Natal.

Mouton, N., Louw, G. P., & Strydom, G. L. (2012). A historical analysis of the post-apartheid dispensation education in South Africa (1994-2011). *International Business &Economics Research Journal*, *11*(11), 1211–1221.

National Planning Commission (2012). *Executive summary: National development plan 2030. Presidency.* Republic of South Africa.

Ogunbayo, S. B., & Mhlanga, N. (2022). Effects of Training on Teachers' Job Performance in Nigeria's Public Secondary Schools. *Asian Journal of Assessment in Teaching and Learning*, *12*(1), 44–51.

Phathela, A. V., & Cloete, C. E. (2018). The impact of the Government Immovable Asset Management Act (GIAMA) on the Department of Public Works, South Africa. RELAND*: International Journal of Real Estate & Land Planning*, *1*, 285–291.

Profile, N. D. (2018). Central Intelligence Agency (CIA) World Fact Book.

Republic of South Africa (2006). The National Infrastructure Maintenance Strategy Act No. 108 of South Africa.

Republic of South Africa (2007) Government Immovable Asset Management Act No. 19 of 2007, Parliament.

Republic of South Africa (2018). *Norms and Standards for Education Facilities.* Department of Basic Education.

Republic of South Africa (1996). Act No. 84, 1996: South African Schools Act, 1996. *Government Gazette, 377*(17579).

Republic of South Africa (1999). Public Finance Management Act, No. 1 of 1999.

Sebola, M. P. (2014). Administrative policies for good governance in Africa: Makers, implementers, liars, and no integrity. *Journal of Public Administration, 49*(4), 995–1007.

Solomon, H. (2015). African solutions to Africa's problems? African approaches to peace, security, and stability. *Scientia Militaria: South African Journal of Military Studies, 43*(1), 45–76.

Tebele, M. M. (2016). *Problems and challenges related to public policy implementation within the South African democratic dispensation: A theoretical exploration.* Unpublished doctoral dissertation. South Africa: North-West University.

Wall, K. (2008). *Focus Note from the 2008 DBSA Infrastructure Barometer: Maintenance, Saving Money.* Johannesburg: Development Bank of South Africa.

Xaba, M. I. (2011). The possible cause of school governance challenges in South Africa. *South African Journal of Education, 31*(2), 201–211.

9 Methodological Framework for Developing a Maintenance Management Conceptual Model

Introduction

This chapter gives a detailed methodological framework for the study. The chapter covers research design and methodology, qualitative and quantitative studies.

Quantitative Versus Qualitative Research

According to Remenyi, Williams, Money, and Swartz (1998), qualitative, quantitative, or mixed studies are not different in terms of their impact on research and the generalisation of findings. Aigbavboa (2014) postulated that the study seeks to achieve the research aim. Consequently, Jean (1992) hypothesised that qualitative, quantitative, or mixed studies are influenced by the research paradigm, the aim, and the major philosophical considerations or positions (ontology or epistemology) of the research. King (1994) opined that the research paradigm is a set of beliefs a researcher uses in explaining or making sense of a phenomenon in a segment of the world or the world at large (see Figure 9.1). Maguire (1987) noted that the research paradigm guides a researcher's study or conceptual framework to make sense of or seek to understand a real-life phenomenon globally. Regardless of the research method adopted, it is worth emphasising that qualitative and qualitative methods have their basis in ontological and epistemological assumptions. Saunders et al. (2006) posited that the ontological assumptions focus on the researcher's assumptions of how the world operates. Jean (1992) states that the epistemological assumptions relate to the nature of knowledge and the methods of abstracting the knowledge. Crotty (1998) and Aigbavboa (2014) contend that ontology is about what things are, whereas epistemology is concerned with how we know things.

Fellows and Liu (2015) affirm that research increases knowledge; it is a process of enquiry an investigation, and it is systematic and methodical. However, Ellram (1996) further postulated that research explains, explores, predicts, and/or describes issues. Amaratunga (2001) stated that quantitative research methodology is usually used for prediction and description

DOI: 10.1201/9781003344681-9

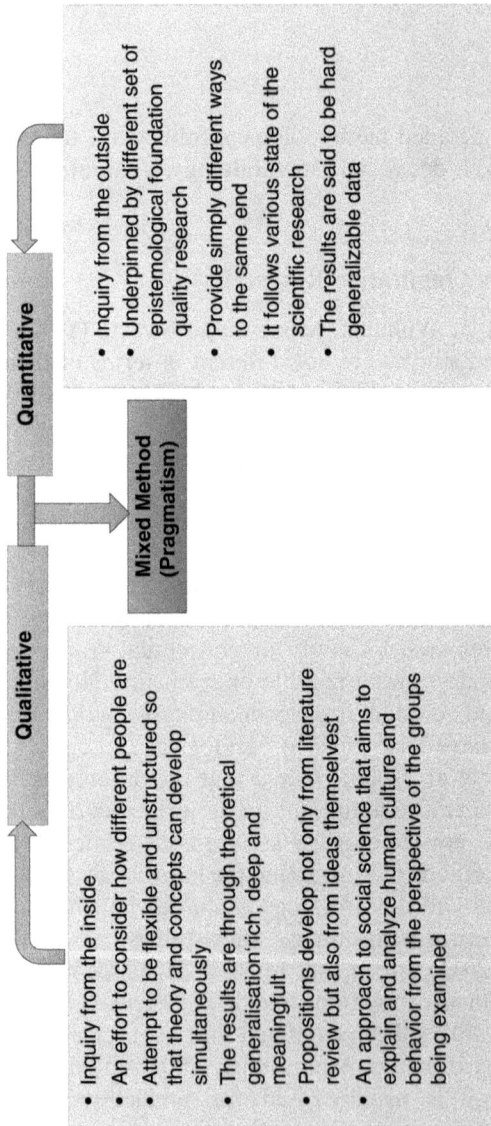

Figure 9.1 Quantitative and qualitative methods: key features.

Source: Adapted from King (1994).

research. Statistical techniques are utilised to describe or predict relationship patterns among variables that constitute reality or phenomena. Ellram (1996) opined that exploration or explanation studies often use a qualitative approach to research. Qualitative research is conducted based on intense and prolonged contact with a "field" or life situation (Amaratunga, 2001). However, the quantitative approach to research grows out of a strong academic tradition that places considerable trust in numbers representing opinions or concepts (Amaratunga, 2001).

Additionally, Jonker and Pennink (2010) affirm that at the beginning of the research, the quantitative approach is initialised using a closed question that results in a problem definition that usually emerges. The question elaborates based on an amalgamated relevant existing theory (Jonker & Pennink, 2010). Jonker and Pennink (2010: 43) posited that "after this elaboration, the problem is more or less definite and most of the time elaborated in a conceptual model."

It is worth noting that quantitative research is research based on a closed question "... can be started with an already existing model with a precise academic discipline" (Amaratunga, 2001). Jonker and Pennink (2010: 69) suggested that based on this ideal, often combined with the work of others, new questions are formulated and tested. Amaratunga (2001) and King (1994) noted that qualitative research concentrates on words and observations to express reality and attempts to describe people in natural situations. Moreover, Jonker and Pennink (2010) opine that an open question usually initialises qualitative research. In some instances, Bryman (2016) postulated that the nature of the study would require using both qualitative and quantitative approaches to research. In such instances, Bryman (2016) contended that a combination of the two research methods is used (pragmatism or mixed method). Furthermore, Amaratunga (2001) postulated that both quantitative and qualitative research is best thought of as complementary and ideal for research of many kinds (see Figure 9.1). In much of the research literature, pragmatism or mixed-method is referred to as triangulation (Jack & Raturi, 2006).

For this study, a qualitative research method was adopted, in line with the aim of the current study. The next section discusses the two philosophical positions of research – ontology and epistemology.

Philosophical Positions in Research Methodology

Like any human activity, research is grounded on philosophical perspectives, either explicitly or implicitly. Amaratunga (2001) observed that the quality of research could seriously be affected if the philosophical issue of the research is overlooked. Understanding the philosophical standing of research is particularly important in helping researchers clarify alternative methods and designs for specific research and identifying which are more suitable to work in practice (Easterby-Smith, 1991). Creswell (2010) stated that a researcher's philosophical position is embedded in many considerations.

Accordingly, Crotty (1998) postulated that the choice of a research method regarding philosophical positions is influenced and supported by two sets of assumptions, considerations, or positions. As Aigbavboa (2014) and Amaratunga (2001) noted, the two main research assumptions are ontological and epistemological. Creswell (2010) affirms that the research philosophy greatly influences the methodology used for research. However, it is plausible that a suitable philosophical position is essential in establishing an appropriate methodology for a study. Therefore, it is important to understand this study's ontological and epistemological positions.

Ontological Position or Ontology

According to Creswell (2010), the assumption underlying ontology is concerned with the nature of reality; thus, it conveys the researcher's view of the nature of reality or being. According to King (1994), an ontological position focuses on the researcher's assumptions of how the world operates and the commitments held on particular views. As noted by Creswell (2010) and Saunders et al. (2007), under the ontological consideration, two main views: exist the realist and idealist perspectives. They are both sometimes referred to as objectivism or subjectivism.

Creswell (2010) and Vanderstoep (2009) postulated that the realists examine a social phenomenon using a predetermined structure, while the idealists believe that the research reality is based on different perceptions from different observers. Saunders et al. (2007) hypothesised that objectivism assumes that social entities exist independently of social actors, while subjectivism seeks to understand the meanings individuals attach to social phenomena. The realists' (objectivism) position is proportionate to the positivist philosophy, while the idealist position portrays the interpretive view of reality (Creswell, 2010; Bryman, 2016). Burrell and Morgan (1994) explained that the opposite view of objectivism is subjectivism or constructivism. It assumes that the world consists of names, labels, and concepts that are used to create the meaning of reality. Based on the subjectivist view, reality is not discovered but constructed by human beings as they become involved with the world they live in (Crotty, 1998).

However, Amaratunga (2001) postulated that the aim of the positivist regime, which is the possibility of discovering universally applicable generalisations, is rejected in realism. Additionally, positivists' beliefs, which centre on the assumption that the objects of experience are atomic, independent events, are also rejected in a realism approach. The possibility of discovering universally applicable generalisations, which is the aim of the positivist regime, is rejected in realism (Pacitti, 1998). The realism approach also rejects the positivists' beliefs, which centre on the assumption that the objects of experience are atomic, independent events. Summarily, the main differences between the positivist and the realism viewpoints are described in Figure 9.2.

Positivist Paradigm	Subject	Realism Paradigm
• The world is objective • Science is value free • Observer is independent	Main Belief	• The world is socially constructed and subjective • Observer is part of what is observed • Science is driven by human
• Focus on facts • Look for causality and fundamental laws • Phenomena reduction to simplest element • Hypotheses formulation and testing	Focus in research	• Focus more on meaning • Focus on understanding what is happening • Look at totality of situations • Idea is developed through data induction

Figure 9.2 Positivist and realism paradigm key features.

Source: Adapted from Easterby-Smith (1991).

According to Burrell and Morgan (1994), the objectivity stance of ontology holds that to understand reality well enough to explain it, one should study the world. This is the stance of this present study, namely the causal relationship of the attributes that constitute maintenance management (MM) of municipal buildings in the education sector in developing economies. Crotty (1998) observed that the objectivity stance is very much aligned with the positivism philosophical theory. Hence, it is also referred to as positivism or the positivist stance. The positivism philosophical theory is based on empiricism. Thus, deductive and devoid of introspective and intuitive knowledge. The use of the scientific method of enquiry aids in discovering the true meaning of realities or phenomena in the world (Jean, 1992; Saunders et al., 2007). It is more tilted towards quantitative research methods – counting and measuring (Saunders et al., 2007). The scientific method of enquiry is devoid of elements of bias on the researcher's part in the research process, thus giving an outcome that can be scientifically verified (Saunders et al., 2007, Aigbavboa, 2014). The outcome of the scientific method of enquiry generates rules and theories that explain reality or aid in understanding realities in the world (Maguire, 1987). Thus, this current study takes an objective or ontological position and aims to develop a conceptual framework that explains the attributes that determine the MM of municipal buildings and aids educational institutions in effectively maintaining the educational buildings in the South African education sector.

However, the objectivity stance of ontological consideration is not without criticism. Maguire (1987) critiqued that the objectivity stance that reality exists outside an individual's conception is inconsistent since reality is socially and humanly created. Aigbavboa (2014) also argued that the human conception or the knower, the subject of the knowledge, cannot be separated from the known object, thus making ontology's objectivity stance elusive. This led to the development of the subjectivity stance of ontology. The subjectivity stance is very much aligned in thought to constructivism or constructivity theory. This informs that people's knowledge or understanding of reality or phenomena in the world is based on their experience, thus, learning by participation (Maguire, 1987; Aigbavboa, 2014). The subjectivity stance of ontology opines that reality is not discovered but constructed by human beings as they engage the world they find themselves in (Maguire, 1987; Jean, 1992; Crotty, 1998). Vanson (2014) rejects absolute facts and maintains that facts are based on perception instead of objective truth. The subjectivity stance or subjectivism is also known as the interpretivist approach or interpretivism (Sarantakos, 2012; Saunders et al., 2007). It is inductive and adopts a qualitative method of enquiry (Sarantakos, 2005; Saunders et al., 2007; Vanson, 2014).

Epistemological Position or Epistemology

According to Crotty (1998), Aigbavboa (2014), and Vanson (2014), epistemology is the source of knowledge; it is concerned with the way we know things; the nature of knowledge, and the knowledge is abstract and interpreted. Epistemology also provides the basis for deciding on the kind of knowledge considered legitimate, appropriate, and adequate for the research being carried out and how to attain and interpret the knowledge (Crotty, 1998; Vanson, 2014). Saunders et al. (2007) state that epistemology is also concerned with what constitutes acceptable knowledge in a field of study. It considers aspects relating to the acceptability of knowledge in any discipline as a philosophical position of a research study (Vanderstoep and Johnson, 2008).

The causal relationship between the researcher and the phenomena being studied is unravelled by epistemology (Creswell, 2010). Nevertheless, Creswell (2010) and Bryman (2016) opine that there are two main epistemology positions: positivism and interpretivism. Amaratunga (2001) asserted that epistemologists recognise four sources of knowledge, namely authoritative knowledge – books, people, supreme being, empirical knowledge – based on objective facts which are determined by observation or experimentation or by both; intuitive knowledge – beliefs, faith, intuition, among others; and logical knowledge – arrived at when one reason from generally accepted knowledge to new knowledge.

When carrying out a study, once a researcher adopts an epistemological position, the researcher will adopt methods that are characteristic of the epistemological position (Vanson, 2014). However, all four sources of knowledge may be used simultaneously in a study (Vanson, 2014). Hence, Hill (1995) informs that research methodology applies epistemology since a research methodology ought to be supported by an epistemological position. Vanson (2014) asserts that it is of importance that epistemology informs the researcher's choice of the methodology used for justification. This should be based on the two stances of ontology that inform the two main epistemological positions in research, objectivity (positivist) epistemology and subjectivity (constructivist) epistemology (Vanson, 2014).

Quantitative Research

Nau (1995: 3) contends that quantitative research looks for "distinguishing characteristics, empirical boundaries and elemental properties" and tends to measure "how much" or "how often." Researchers use the quantitative research methodology to explain human behaviour and to discover scientific laws similar to those of the natural sciences (Sarantakos, 2005). Maguire (1987) and May (2001) opine that perception or conception of the individual, in this case, is excluded from the research process. Sarantakos (2012) and Vanson (2014) postulated that the positivist epistemology researchers advocated using quantitative research methodology in scientific enquiry into reality to explain and explore the relationship between attributes that constitute reality in the world.

According to Jean (1992), Crotty (1998), Saunders et al. (2007), and Aigbavboa (2014), positivist epistemology is based on empiricism; it is deductive in nature and devoid of intuitive and introspective knowledge. In discovering the true meaning of realities or phenomena in the world of positivist epistemology, the scientific method of enquiry is used (Saunders et al., 2007; Aigbavboa, 2014). Equally, in presenting research findings, positivist epistemology employs the use of statistical rhetorics such as correlation, validity, cause, reliability, and effect relationships, among others. Based on this, the findings from this philosophical stand tend to be more scientific, exact, and follow procedures (Sarantakos, 2005; Aigbavboa, 2014). Straub, Boudreau, and Gefen (2004) posited that certain principles guide a positivist's search for reality. They include the following:

* The process is neutral, and judgement-free observations are uncontaminated by the scientist's prediction. Thus, ethical issues can be included only if they are included as part of the research;
* Only phenomena that can be observed can be used to validate knowledge;
* Scientific knowledge is arrived at through the accumulation of verified facts derived from systematic observation or record-keeping; and
* Scientific theories are used to describe patterns of relationship between these facts to establish causal connections between them.

Jean (1992) and Aigbavboa (2014) suggest that one other important fact of positivist epistemology is that it uses scientific theories to describe the relationship pattern of the attributes that constitute reality. The process involved is devoid of predictions or the judgement of the researcher. Hence, when positivism is adopted for a study, there is limited room for bias, and the outcome seeks to predict and explain the causal relationship between the attributes or the elements that constitute reality. Data collection in positivism epistemology can be quantified and analysed using a mathematical formula (Maguire, 1987). Jean (1992) maintained that a positivist epistemology would result in using a scientifically guided research methodology. The object is to predict and explain causal relations among elements that constitute reality. This was considered in the current research. Also, as adopted for the current study, the success of positivist research depends on data collection that can be analysed and quantified using mathematical formulas (Maguire, 1987).

Consequently, Sarantakos (2005) and Aigbavboa (2014) postulated that positivist epistemology uses a quantitative research methodology in a scientific enquiry into a world reality. Nevertheless, in adopting a quantitative research methodology in a research study, some social phenomena that cannot be quantified are often ignored. Very important social phenomena are sometimes reduced to meaningless quantitative results (Maguire, 1987).

Nonetheless, Sarantakos (2012: 11) contends that in quantitative research, the researchers 'perceptions are detached from the research process since researchers who adopt a quantitative methodology are reduced to a research tool that does not have a mind while research respondents become research objectives and are treated as such. Conversely, May (2001) advocated that the relationship between the research and the researcher should not be disconnected because the connection to the flow of information will be affected once it is disconnected. Brieschke (1992) submitted that the subjectivity epistemology option for research is far better than a positivity epistemology.

Qualitative Research

According to Amaratunga (2001), qualitative research is a study conducted through prolonged contact with a life or "field" situation. The subjective epistemology stance is that reality is imposed on the object of the research by the subject of the research, and it is never discovered as advanced by positivist epistemology (Maguire, 1987; Jean, 1992). Jean (1992), Crotty (1998), and May (2001) posit that the researcher constructs or creates the phenomenon or meaning of reality in the world. This shows that researchers' input to the research process is not excluded; the object being studied even contributes less to the reality while meaning is created for the reality being studied by the researcher (Jean, 1992; Aigbavboa, 2014). As stated by Jean (1992) and Aigbavboa (2014), the subjectivist's research methodology is the qualitative research method. Jean (1992) maintains that qualitative research is a form of social interaction between the research subject and the research object, where the subject converses with the object and learns about the reality being studied (Jean, 1992).

What is noteworthy about subjectivity epistemology is that the meanings of realities are constructed (constructivism) and interpreted (interpretivism) by the researcher. This view is aligned with the concept of constructivism and interpretivism (Aigbavboa, 2014). Hence, Sarantakos (2012) states that the process through which necessary information is extracted through interpretation is called interpretivism. According to Crotty (1998), re-searchers under interpretivism pursue information relating to people's opinions, views, perceptions, and interpretations. Secondary data for research is also associated with interpretivism (Aigbavboa, 2014). This also applies in this current study.

Furthermore, Bryman (2016) found the qualitative research method more suitable for studying people and their environment (social science) than natural science. Sarantakos (2005) postulated that the qualitative research method is inefficacious because it cannot study the causal relationship between variables constituting a reality in the world. Sarantakos (2005) and Bryman (2016) opined that data mainly generated in qualitative studies are heavily impacted by the values and views of the researcher, and this undermines the reliability and generalisation of the findings. However, from the viewpoint of qualitative research advocates, Mitchell (1983) maintains that when qualitative research is based on sound theoretical reasoning, the weakness of non-generalisation of qualitative research will be overcome. Creswell and Clark (2017) emphasise that qualitative research findings are difficult to validate since it seeks to generate ideas instead of measuring the object of enquiry.

From the above, it can be deduced that positivists favour the quantitative research methodology (scientific and objective approach to enquiry), while interpretivists prefer qualitative research methodology (humanistic and subjective approach to enquiry). Amaratunga (2001) and Teddlie and Tashakkori (2003) posited that each research method has its weaknesses and strengths. In line with the aim and philosophical assumptions underpinning this current study, this book leans more towards the qualitative approach to research.

Data collection in a typical quantitative survey is through close-ended questionnaires, whereas an interview (open-ended and structured questions) assists in collecting data in a qualitative study. Teddlie and Tashakkori (2003) and Creswell and Clark (2017) opined that quantitative data typically are analysed with the aid of statistical instruments while qualitative data is analysed based on themes, which could also be transcribed into quantitative data. As shown in the present study, when the Delphi survey techniques were used to establish consensus in the responses of the Delphi survey participants, frequencies of measures of central tendencies were used. Aigbavboa (2014) specifies that data for a qualitative research study could be collected by gathering documents from private and public sources, observing the sites of a research study, and observing the participants of a research study, or from audio-visual materials such as videotapes or artifacts. Aigbavboa (2014) further states that in qualitative research, data analysis is largely done by organising the data into categories of information.

In data collection for this study, the qualitative method this study adopted used a structured interview (using an interview schedule) and a semi-structured interview (using an interview guide). This was done through the use of the Delphi technique. The Delphi survey findings assisted in refining the survey tool (structured questionnaire) and invalidating the qualitative findings by quantitative means. The analysis of the quantitative data as well as the development and validation of the integrated MM framework developed for the effective MM of municipal buildings in the South African education sector. Thus, the current study explored the extent to which the factors for effective MM identified in the qualitative study through the literature review and the Delphi technique contribute to the MM of municipal buildings in the South African education sector.

Hence, through the qualitative study, understanding of the factors for MM was enhanced, particularly when the Delphi technique was used to solicit views from experts regarding the issue being investigated.

Research Design

According to Churchill and Iacobucci (2004), in the successful pursuit of a study, the research design is the framework governing the research being carried out, serving as a guide to the research. Sarantakos (2012) posits that the research design influences the instrument for collecting data in a particular study. Likewise, Sekaran (2000) stresses that the research design helps the researchers justify the decisions and choices related to the research procedures. Accordingly, Sarantakos (2012) and Aigbavboa (2014) postulated that the choice of a research design is mostly influenced by the research methodology of the study (quantitative or qualitative research methods) as well as the philosophical position the study adopts. This shows that if a study is carried out from the standpoint of objectivist ontology. The researcher will adopt more of a positivist epistemology approach which culminates in the use of quantitative research methodology. As noted further by Sarantakos (2012) and Aigbavboa (2014), the survey method becomes the instrument for collecting data. The study that uses subjectivist ontology will be influenced to use constructivist epistemology, which culminates using a qualitative research methodology.

Hence, deciding on a research design for a particular study encompasses different connected steps (Sarantakos, 2012). The steps start with the philosophical assumptions guiding the study, which then suggest the research methodology of the study, guided by the research aim and questions, from where the research design is arrived at, which in turn predicts the instrument for collecting data in the research (see Figure 9.3).

According to Robson (2002), a research design for a particular study should show an appropriate interconnection between the research purposes, theoretical framework, research questions, research methods, and a sampling strategy. Thus, these aspects of research design were followed by this present study. Among others, the research design for the study was influenced by the research aim, objectives, and questions. On the other hand, the study's research method was influenced by three considerations. The study first considered a research method to identify various factors or attributes determining effective MM. Secondly, the study considered a research

	ONTOLOGY
	EPISTEMOLOGY
	METHODOLOGY
	DESIGN
	INSTRUMENTS

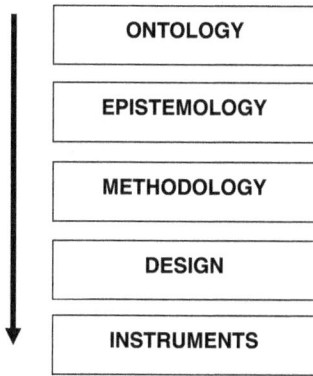

Figure 9.3 Steps in the research design process.
Source: Adapted from Sarantakos (2012).

method that could predict the pattern of relationships between the attributes of effective MM and the maintenance of municipal buildings.

Finally, a research method was selected that allows in-depth information to be collected and analysed regarding the views of some stakeholders in the education sector and the construction industry as to the extent to which the perceived attributes of effective MM influence the maintenance of municipal buildings. Hence, this current study adopted a qualitative research method as stated, discussed, and justified earlier. The qualitative approach to research was adopted to find answers to the research questions and objectives, thereby achieving the research aim of developing a MM framework for the municipal buildings of developing economies using the South African education sector as a case study.

According to Ahmad, Al-Shorgani, Hamid, Yusoff, and Daud (2013), a research design is necessitated by a research strategy, and the design leads to the research method one may employ for a study (Creswell, 2003). Consequently, the following strategies were adopted to meet the research objectives of the study:

RO1. to establish the factors that influence municipalities in the attainment of MM of the municipal buildings in the education sector.

To identify the general factors or attributes of effective MM literature was reviewed. Sources of literature reviewed included, among others, conference proceedings and journals. The expected outcome was information and a general picture of the attributes of effective MM globally.

RO2. to investigate and establish the current theories and literature that have been published on MM of buildings and to identify the gaps that need consideration.

A rigorous literature review on maintenance in general and the MM of educational buildings and other essential public facilities was conducted. A specific literature review was conducted on current theories on effective MM of public facilities, and research gaps that needed consideration were identified and incorporated as a construct of the integrated MM model that this study seeks to develop. The sources of the literature review included, among others, conference proceedings and journals.

Regarding this objective, the expected output was information on the current theories on MM. Thus, the review sought to determine gaps that have not yet been addressed by researchers, common themes and methodologies that have been adopted in MM research, and how terms have been defined. The information gathered through literature concerning this objective was important as it forms the core literature in the pursuit of this study. Further, the information is useful for the readers to understand how effective MM of buildings is developed and the extent of its relevance in meeting users' needs and expectations.

RO3. to determine the factors/attributes (main and sub-variables) that are perceived to be of importance and of major impact in obtaining effective MM of municipal buildings, and to determine whether the factors that have aided in obtaining effective MM in other geographical contexts are the same in the South African education sector.

This objective was achieved through the Delphi method. As this objective requires soliciting the views of experts, the Delphi method was the best method for the factors or attributes that determine the effective MM of municipal buildings and the relative contribution of each of the factors in the South African education sector. These types of questions could only be addressed through a focused group technique, experimental procedures, and Delphi techniques. Nevertheless, for this study, using an experimental procedure was not feasible because of challenges such as time and cost in bringing together experts under a common roof for the focus group discussions about these objectives. Thus, the purpose of conducting this study through a rigorous process would have also been defeated if the focus group method had been adopted.

On the other hand, the Delphi method was adopted for this present study to solicit experts' opinions concerning this objective. The merit of adopting the Delphi method is eradicating elements of bias among panel members. Moreover, the Delphi method also eliminates undue pressure among expert panellists because every member remains anonymous, removing undue influence from peers. However, the expected outcome is an estimation of the extent to which the established factors influence the effective MM of municipal buildings in the South African education sector to achieve consensus. Additionally, these factors and their inter-relationships developed a conceptual integration MM framework for municipal buildings in the South African education sector.

RO4. to develop a holistically integrated Maintenance Management framework (MMF) for municipal (educational) buildings in the South African education sector.

This objective was achieved by drawing upon the findings and conclusions from the literature review and the outcome of the qualitative Delphi study. Table 9.1 gives a summary of the research methods.

Research Methods

This section gives a detailed overview of the methods used to meet this study's objectives. Under this section, the methods described include the literature review and the Delphi technique.

Literature Review

A literature review examines what has been written on a particular subject and methodologies that have been used or adopted in investigating related realities or phenomena in the world (Aigbavboa, 2014). Boote and Beile (2005) postulated that the fundamental basis of the research is provided through literature. Additionally, Heppner and Heppner (2004) opine that the literature established trends in solutions that have already been advanced to solve a particular problem in the world. Consequently, it is of importance to embark on a review of literature for this present study to establish the following:

- Attributes or factors of effective MM from already existing literature,
- Theories and literature on MM, and
- Gaps in MM studies which could improve the effective MM of municipal buildings in the education sector.

However, a literature review, as noted by Aigbavboa (2014), is vital because the broad context for the study is set through it. Aigbavboa (2014) states further that it helps to highlight previous studies on the subject matter by relating the presenting study with what has been done previously. It gives the basis for comparing results (present findings with existing findings) and drawing lessons and conclusions. Literature can be obtained from published and unpublished sources for a robust study review. Sources of literature, among others, include theses, dissertations, books, and journals. Also, Boote and Beile (2005) posit that a literature review was cohesive and considered methods adopted in previous studies in relation to the current subject matter, which is done to ensure that the review does not only report

Table 9.1 Research procedure

Stage	Research Objective	Data collection method	Data analysis method	Output
1.0 Literature review	**RO1:** to establish the factors that influence municipalities in the attainment of MM of the municipal buildings in the education sector.	• Literature review		Information and a global picture of the determinants of effective MM.
	RO2: to investigate and establish the current theories and literature that have been published on MM of buildings and to identify the gaps that need consideration.	• Literature review		Information on the current theories on MM and especially to determine the gaps which other scholars have not addressed. Outline of constructs (factors) associated with effective MM that has not been considered in the previously developed models.
2.0 Delphi technique	**RO3:** to determine the factors/attributes (main and sub-variables) that are perceived to be of importance and of major impact in obtaining effective MM of municipal buildings, and to determine whether the factors that have aided in obtaining effective MM in other geographical contexts are the same in the South African education sector.	• Delphi technique	• Descriptive statistics	Consensus on the influence and impact level of the various attributes on effective MM.
	RO4: to develop a holistically integrated maintenance management framework (MMF) for municipal (educational) buildings in the South African education sector.	• Desk study • Delphi technique	• Theory	Holistic integrated MM framework.

findings or claims of earlier studies. In the subject area for this study, literature from the above sources, but not limited to those stated sources only, was accessed to establish the progression of research in the subject area, most importantly literature on the effective MM of buildings. This is in accordance with the description by Boote and Beile (2005), which states that a literature review should involve the following:

• Consulting a broad range of high-quality, specific books, dissertations, theses, reports, and reviews that relate to the study;
• Identifying methodologies adopted in previous studies;
• Reading and re-reading to establish progression and trends in the existing literature;
• Writing the literature review;
• Relating the present study to previous studies; and
• Summarising the studies read.

This present study's literature review approach is in line with the above description. Thus, the findings from the literature reviewed in this study revealed that several factors have contributed to the effective MM in different industries. Even in the facility management sector, several country-specific factors contribute to the effective MM of their facilities. Nevertheless, the literature findings affirmed that no universal factors or attributes contribute to the effective MM. Additionally, the literature reviewed shows that other factors should be considered key constructs in the "bundle of factors" which constitute attributes of effective MM studies and have been considered by previous or existing MM models. However, after missing factors that need consideration had been established, theories were developed about the influence of the factors and their interrelationship with the other factors advocated in the previous studies to determine the effective MM of buildings. These factors need to be tested to establish whether they will influence the effective MM of buildings and, if so, to what extent they would influence the effective MM of buildings. Because of this, the Delphi technique was employed. In the next section, the Delphi method is discussed.

Delphi Method

According to Sackman (1974) and Linstone and Turoff (1975), the Delphi method is a structured communication method developed as a systematic, interactive forecasting method that depends on a panel of experts. Rowe and Wright (2001) posit that the Delphi method is based on the belief that forecasts from a structured group of persons are more accurate than forecasts from unstructured groups. As Ameyaw, Hu, Shan, Chan, and Le (2016) noted, Delphi's output is based on believing that group judgement is more accurate than individual judgements. This is because the Delphi technique is grounded on the premise that "pooled intelligence" improves individual judgement and a shared view of the collective group of experts (Shariff, 2015). Furthermore, Tilakasiri (2015) postulated that the Delphi technique is recommended for developing standards, concepts, frameworks, and/or models for a study. Thus, the use of the Delphi technique was employed in the model it seeks to develop.

In the second stage of this study, the Delphi method was used to determine the main attributes or factors that determine the effective MM and to examine whether the factors that determine the effective MM of municipal buildings as identified in the literature in other countries are the same in the South African education sector. Also, the Delphi method was employed to explore the degree to which these main and sub-factors impact or /influence the effective MM of municipal buildings in the South African education sector.

However, the Delphi concept origin could be associated with the American defense industry, where experts' views were required on the enemy attack's frequency, probability, and intensity (Ameyaw et al., 2016). Moreover, in the 1950s, Helmer and Dalkey used Delphi as a tool to forecast and solve complex problems at the Rand Corporation (Buckley, 1995; Ogunbayo, Aigbavboa, Thwala, & Akinradewo, 2022). Buckley (1995) and Aigbavboa (2014) postulated further that the Delphi technique is informed by the premise that is soliciting the views or opinions of subject-matter experts aid in determining the possible occurrence of a future event. The forecasts (or decisions) from a structured group of persons are more accurate than forecasts from unstructured groups (Rowe & Wright, 2001). However, Keeney, McKenna, and Hasson (2011) noted that one merit of the Delphi process is that it is stopped after a predefined stop criterion, especially when there is an achievement of consensus or number of rounds, or the stability of the result is achieved.

However, as Keeney et al. (2011) noted, the classical Delphi process usually involves at least three survey rounds. Rowe and Wright (2001) and Aigbavboa (2014) state that the Delphi process is directed by a coordinator (researcher), and the median or mean scores of the final round determine the results of the Delphi survey. The breakdown of the process shows that Round 1 of the Delphi process is carried out to solicit experts' opinions on the issue being examined in an open-ended way. Round 2 of the process asks expert panellists to rate the statements in the questionnaire according to their opinions on the subject matter. The essence of round 3 is to ask expert panellists to re-assess the rankings in light of the consolidated results from round 2. This way, the rounds of the Delphi survey may continue until consensus among expert panellists is attained on some or all of the items. Keeney et al. (2011) and Hon et al. (2012) all stated that round 1 of the Delphi survey could be skipped when round 2 of the Delphi survey questionnaire can be developed through a literature review and/or interviews. Equally, Aigbavboa (2014) posits that during the survey, the expert panellists remain anonymous to one another; it is only their views that are known to each other to inform other experts on the panel in making a decision. The Delphi technique employs questionnaires, and a statistical estimate of the experts' opinions is calculated and analysed using percentages, median and interquartile ranges after each survey round. Nonetheless, Aigbavboa (2014) observed that during the feedback aspect of the Delphi technique, in some instances, arguments and counterarguments of extreme answers were also provided.

Furthermore, the Delphi technique is appropriate for studies that seek to solicit the opinions of a group of experts in an attempt to have a consensus on experts' opinions and to predict the timing and likelihood of a future event. According to Lang (2001), in instances where there is little data about the past on which a researcher can base his/her extrapolation, as well as instances where moral, ethical, economic, and social considerations are of significance, the Delphi technique is the most preferred group technique. The technique is relatively strong in a controlled group communication process (Chan, Yung, Lam, Tam & Cheung, 2001; Aigbavboa, 2014). Grisham and Walker (2008) pointed out that in analysing the information of each round of the Delphi survey, the experts' responses are processed with the main aim of looking for central and extreme tendencies and their validation. As emphasised by Miller (1993) and Hanson Creswell, Clark, Petska, and Creswell (2005), the Delphi method is a qualitative methodology that attempts to establish consensus in experts' opinions through successive survey rounds. In addition, Aigbavboa (2014) and Ameyaw et al. (2016) hypothesise that the Delphi method is based on a structured survey and employs the use of instinctively available information of the participating experts in the survey.

Additionally, Aigbavboa (2014: 261) asserts that the Delphi method provides both qualitative and quantitative results and has explorative, predictive, and even normative elements beneath its explorative. Hallowell and Gambatese (2010) refer to the Delphi method as an iterative process that provides a platform where consensus is often reached among experts on a particular subject. It is based on a number of rounds and the feedback of experts' opinions and judgements on a particular subject. Unlike interviews, Chan et al. (2001) postulated that the Delphi method provides a more reliable and efficient alternative for issues or problems with high levels of doubt and/or uncertainty. In addition, Aigbavboa (2014) and Ameyaw et al. (2016) contend that one of the aims of the Delphi technique is to eliminate likely bias usually related to techniques used in soliciting the opinions of a group of experts when the group of experts meets during a study.

The Delphi survey uses a nominal group technique (NGT) and interacting group method (IGM), which is part of the group decision or policy-making techniques that are individually based, anonymous and independent. Loo (2002) posits that in the Delphi process and feedback, the element of group interaction is eliminated from analysing the experts' responses gathered with the aid of questionnaires in written form. Loo (2002) further opines that when a research study is aimed at investigating or evaluating policies or setting policy direction for the future in private and public sectors alike or when experts whose views are being sought are dispersed geographically, the Delphi technique is an appropriate tool of enquiry. Moreover, the strength of the Delphi method also lies in its forecasting strength as well as in identifying issues in research that can provide the basis for the initiation of further studies (Aigbavboa, 2014; Ameyaw et al., 2016; Ogunbayo et al., 2022). Hence, the Delphi method was adopted for this current study to identify the main attributes or factors that determine the effective MM of buildings and to examine whether the factors that determine the effective MM of municipal buildings as identified in the literature in other countries are the same in the South African education sector.

In a diverse field of study, the use of the Delphi technique is increasingly gaining ground as a method of enquiry (Czinkota & Ronkainen, 1992). As noted by Linstone and Turoff (1975), available literature advocates its usage in medical fields, Buckley (1995) suggests its use for the library and information science field, with the Society of Actuaries (SOA) (1999) advocating it for communication science, actuarial science, and economics research, Chan et al. (2001) and Ameyaw et al. (2016) suggest it for engineering, and quasi-engineering disciplines, Aigbavboa (2014) advocates its use for built environment research, and Schoemaker (1993) also suggests it for the business field, among others. However, all studies generally that employed the use of Delphi techniques of enquiry show that it produces results that are acceptable, valuable, and supported by the majority of the expert community (Czinkota & Ronkainen 1992). Accordingly, Schoemaker (1993) categorically affirms that in the field of management and business research, the Delphi technique is highly rated as a systematic thinking tool, and it further assists as an identifier of strategic issues in research.

Contrariwise, Gupta and Clarke (1996) and Keeney et al. (2011) observed that owing to poor choice of experts, weak bias control, inappropriate design and execution of the Delphi study, and limited feedback during the Delphi study. Sometimes the reliability of the findings derived from a Delphi survey may raise some elements of controversy. Nevertheless, Chan et al. (2001) and Ameyaw et al. (2016) affirm that where objective data are unattainable when experimental research is unrealistic or unethical. Where there is a lack of empirical evidence or when the need to draw upon the collective knowledge and experience of selected experts in a given area is of paramount importance in consensus building, the Delphi method remains a predominantly useful alternative in situations (Hallowell & Gambatese, 2010). Careful measures were incorporated to avoid the weak bias control, poor choice of experts, and inappropriate design and execution of this Delhi study. Thus, making the findings of this present study more acceptable, credible, valuable, and able to be supported by the majority of experts in the community.

Subsequently, this present study might initiate a further study, and as the views of experts are scattered across various geographical locations in South Africa, it is of importance that this current study adopted the Delphi method of enquiry in soliciting the views of the experts. Moreover, little is known about using the Delphi method of enquiry in MM studies in the South African education sector. Hence, its application in a MM study in the South African education sector will be an innovation. This claim was supported by Loo (2002), namely that the Delphi method is an appropriate tool of enquiry when the research aims to evaluate or investigate policies to set policy direction for the future, especially in the public and private sectors. Hence, adopting the Delphi method in this present study was deemed appropriate since the study has implications for policymaking that concerns the MM of municipal buildings, especially in the education sector.

Moreover, Sackman (1974) observed that because the process involved in the Delphi technique is based more on opinions that do not follow any scientific procedure or psychometric validity, it has led to the Delphi method being more criticised. Nevertheless, Buckley (1995), Hasson et al. (2000), and Chan et al. (2001) all postulated that criticisms of the Delphi technique are erroneous because Delphi seeks to establish an agreement or achieve consensus in the opinions of experts. However, it is not an opinion poll among experts, nor does it seek to reach a majority of opinions among experts. Additionally, the Delphi method does not force consensus but discovers agreement and identifies differences (Buckley, 1995). Furthermore, Coates (1975) stated that it helps in the quest for deliberate judgements and provides a better understanding of an area that does not lend itself to traditional scientific approaches. Also, Helmer (1977) emphasised that judgmental information should be used to augment the standard operations research technique and that the Delphi technique cannot be condemned for using simple opinions and violating the rules of random sampling in the 'polling of experts. Hence, in this present study, the Delphi study (qualitative study) was adopted to augment the quantitative study. As Helmer (1977) noted, the Delphi technique has suffered a great deal of criticism owing to the gross misunderstanding of the Delphi method.

However, the Delphi technique is not without criticisms and limitations and is not suitable for all research. Brill, Bishop, and Walker (2006) and Ameyaw et al. (2016) suggest the use of Delphi techniques due to validation in literature and its acceptability in and across diverse fields of study as a comprehensively used method. Additionally, to build consensus among the experts, the Delphi technique is a reliable and credible means of soliciting their subjective views on specific issues. In summary, Lang (1995), Brill et al. (2006), Ameyaw et al. (2016) Ogunbayo et al. (2022) all concluded that the Delphi technique is a reliable technique for studies that aim to reach a consensus from the experts' subjective opinions to predict the possible occurrence of a future event.

An Epistemological Approach Towards the Delphi Design

According to Aigbavboa (2014), the definition of the Delphi method and the difference among the many group techniques, together with various criticisms of the Delphi method, form the epistemological basis for defining a typical Delphi design approach. Ogunbayo et al. (2022) state that this is achievable by assuring that all expert feedback is anonymous.

Scheele (2002) postulates that conducting a Delphi study through a Delphi design framework is vital as it aids the research of the overall objectives of the Delphi study. However, the basic premises of the Delphi research design for this study towards developing a MM framework for municipal buildings in the education sectors of developing countries are embedded in some form of general agreement and consensus regarding the core ingredients and components of the subsequent framework. Given the present status of the municipal buildings in the South African education sector and the absence of a generally agreed upon MM framework for the education sectors, the research for consensus on the factors that determine effective MM of municipal buildings in the education sector of the developing economies is justified.

Consequently, for this present study, the objective of the Delphi design is to obtain the most reliable consensus concerning the opinions of the group of experts in the field being studied. As Lang (1995) noted, the Delphi technique is appropriate for studies that seek to solicit the views/opinions of a group of experts to reach/build consensus and predict the timing and likelihood of a future event. Lang (1995) states further that in instances where there is little data about the past for a researcher on which to base an extrapolation, as well as instances where ethical, economic, social, and moral considerations are involved, the Delphi technique is the most widely preferred group technique.

Nevertheless, bearing in mind the outcome of the literature review that this current study undertook, it reveals that there is no existing structured research on MM that focuses on the maintenance of municipal buildings in the education sector of South Africa and other developing economies at large. Moreover, there is no existing study that developed a model to identify and predict the determining factors of MM of municipal buildings in the South African education sector. Also, the Delphi technique's definition, function, and nature are brought to the fore. However, this study justifies that the Delphi technique is one of the best methods to explore the subject of this research as well as exploring the aim and objectives of this research study.

When to Use the Delphi Technique

As Chan et al. (2001) stated, the Delphi technique is used to forecast the possibility of an event's occurrence. When look-term issues have to be assessed, the Delphi technique is employed as the subject of this present study. Additionally, Aigbavboa (2014) postulated that the Delphi technique helps reduce tacit and complex knowledge to make it possible to be judged. Thus, modelling the MM of the municipal buildings in the education sector of developing economies using the combination of the Delphi method with other methodologies, such as a survey design, is interesting. Chan et al. (2001) and Ameyaw et al. (2016) assert that the anonymity of the experts is to be preserved when Delphi is employed. In the structural group communication process, there is no incentive to be given to the participants involved (Ameyaw et al., 2016). Aigbavboa (2014) observed that the Delphi technique is used when experts whose views are being sought cannot meet because they are located in different geographic locations; bringing them together economically and timewise is impossible. Because of this, Adler and Ziglio (1996) posit that questionnaires were used to gather information from experts in a survey over several rounds. Furthermore, Jairath and Weinstein (1994) contend that in decision making, when a large number of individuals across diverse locations and areas of expertise can be included anonymously, the Delphi technique is ideal because it avoids the domination of the consensus process by one or a few experts.

Conversely, the Delphi technique is employed when the enquiry problem does not lend itself to an exact analytical technique but can benefit from collective subjective judgements (Aigbavboa, 2014). In a situation where more individuals are involved in a study but cannot meet, the Delphi technique is the best option as the element of group interaction is eliminated (Adler & Ziglio, 1996; Loo, 2002). When the study aims to solicit the opinions of a group of experts to reach a consensus and predict the timing and likelihood of a future event, the Delphi technique is used (Lang, 1995; Chan et al., 2001). Conversely, Lang (1995) believed that when there is little data about the past for a researcher (facilitator or leader) on which to base his/her extrapolation, as well as instances where ethical, social, economic, and moral considerations are involved, the Delphi technique is the most preferred group technique. Also, Chan et al. (2001) stressed that the Delphi technique provides a more efficient and reliable alternative for solving problems with high uncertainty levels.

Aigbavboa (2014) and Ameyaw et al. (2016) postulate that for soliciting the opinions of the group of experts when a group of experts meets, the Delphi technique helps to eliminate possible bias usually associated with techniques used. Loo (2002) asserts that the Delphi technique is an appropriate enquiry tool when research aims to investigate or evaluate policies and set policy direction for the future in private and public sectors alike. Hence, for this present study, the Delphi techniques were found suitable. A qualitative Delphi technique was used to solicit experts' opinions in developing a MM framework for municipal buildings in the South African education sector.

Components of the Delphi Technique

Components of the Delphi process should be followed during the execution of the Delphi technique for a study (see Figure 9.4). As stated by Loo (2002) and Aigbavboa (2014), five major components are involved in the Delphi survey. The five components, as noted by Loo (2002) and Aigbavboa (2014), include the following:

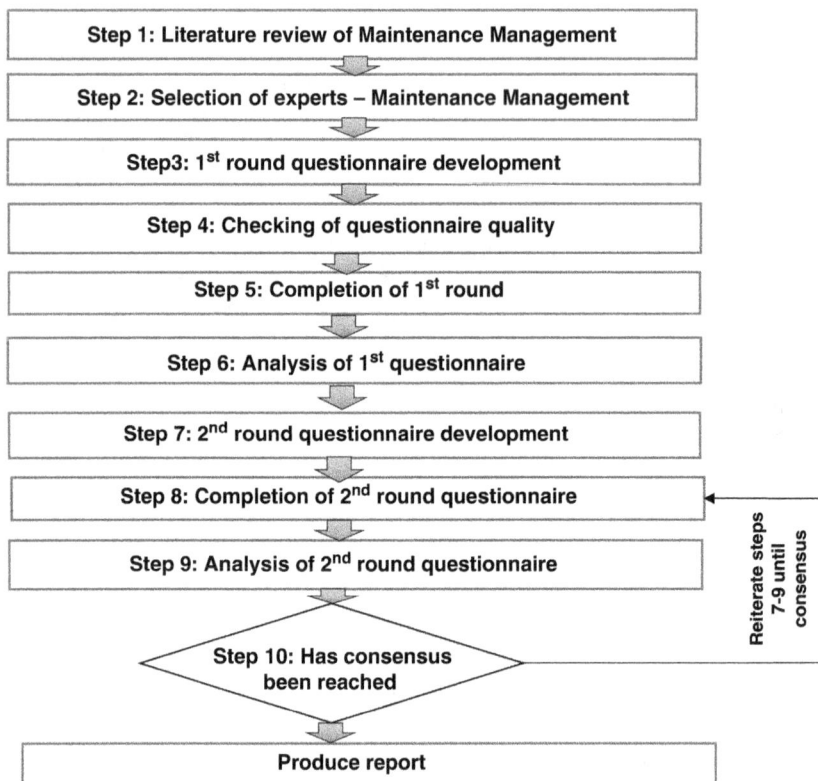

Figure 9.4 Diagram of the Delphi process.

Source: Adapted from Aigbavboa (2014).

- A carefully selected panel of experts should represent a broad spectrum of opinions on the issue being studied.
- Throughout the survey, the experts on the Delphi pane should remain anonymous to one another.
- To solicit experts' opinions on the issue under investigation, the researcher should construct a structured questionnaire and report on feedback to the experts throughout the Delphi process.
- The Delphi process should involve at least three or four rounds of questionnaires and feedback.
- The Delphi result should contain a research report showing forecasts (decisions), policy and programme options (with their strengths and weaknesses), recommendations to senior management, and a possible action plan for developing and implementing the policies' programmes.

It is noteworthy that these five components were adopted for the Delphi process for this current research as outlined by Loo (2002) and Aigbavboa

(2014). Nevertheless, Hasson, Keeney, and McKenna (2000) assert that the use of the Delphi technique in research should be guided by the following:

- **The research problem**: The use of the Delphi technique, as postulated by Reid (1988), should be informed by the appropriateness of available alternatives. Thus, Hasson et al. (2000) affirm that there is a need to identify the research problem the study seeks to solve, which informs the use of a particular research technique.
- **Understanding the Delphi process**: Ameyaw et al. (2016) postulated that the process involved in using the Delphi technique combines analysing solicited group opinions and giving feedback at each stage to establish consensus on the opinions of the experts. The Delphi process, as further informed by McKenna (1994), involves the following steps:

  i   Pilot testing of a small group,
  ii An initial questionnaire that solicits qualitative comments (not in all cases),
  iii Initial feedback – quantitative, after statistical analysis of the initial opinions,
  iv A subsequent questionnaire – which solicits qualitative comments again, and
  v   Subsequent feedback – quantitative after statistical analysis. However, McKenna (1994) emphasised that this process provides an opportunity for participants who may have to change their opinions. The above view, as promoted by McKenna (1994), is not too different from other opinions of Chan et al. (2001), Aigbavboa (2014), and Ameyaw et al. (2016), among others. Hence, Adler and Ziglio (1996) and Ameyaw et al. (2016) state that this stresses that the Delphi technique involves a process, and understanding this process is essential in the design of a Delphi survey.

- **Careful selection of experts**: The success of every Delphi study depends largely on an objective and careful selection of experts on the Delphi panel (Chan et al., 2001). Based on this, some studies have adopted a flexible system, while others have adopted some criteria for prequalifying experts on a Delphi panel (Ameyaw et al., 2016). In the study of Hallowell (2008), it was informed that experts on a Delphi panel should possess some special knowledge/experience, evidence of which could be professional qualifications, working experience, working appointments, and relevant publications. Additionally, in their study, Chan et al. (2001) and Manoliadis, Tsolas, and Nakou (2006) adopted the working experience and the involvement of the experts in a specific kind of project as key criteria to pre-qualify experts for a Delphi study. Nevertheless, Jairath and Weinstein (1994) advised that a Delphi panel should be carefully selected to avoid domination of the consensus process by one or a few experts. Aigbavboa (2014) affirms that the issue being examined should be of interest to the experts, who should be impartial.

Moreover, different opinions have been advanced regarding the size of a Delphi panel. Murphy et al. (1998) opined that the bigger the panel size, the more reliable the Delphi result is, a belief held by most scholars. On the other hand, some other scholars believe there is no significant correlation between the size of the Delphi panel and the effectiveness and accuracy of the Delphi method (Boje & Murnighan, 1982). Hasson et al. (2000), Manoliadis et al. (2006), and Hallowell and Gambatese (2010) concluded that the variation in the numbers of Delphi panel experts is influenced by factors such as available resources for the research in terms of time and money, the number of readily available experts and the scope or nature of the problem being investigated. Thus, the panel size for a typical Delphi survey, suggested by Aigbavboa (2014), could range from 15 to over 60 experts. Ameyaw et al. (2016) postulated that based on the majority of engineering studies they reviewed, a Delphi panel size of 8 to 20 experts is the most widely used size for a Delphi study.

- **Informing/invitations to experts**: In every Delphi process, experts should be invited to participate in a Delphi survey (Aigbavboa, 2014). The invited Delphi panellists should be informed about the study's objectives, what they will be required to provide during the study, how much time will be required regarding the survey, and what will be done with the information they will provide (Aigbavboa, 2014).
- **Data analysis:** According to Chan et al. (2001), Aigbavboa (2014), and Ameyaw et al. (2016), during the Delphi process, each round of a Delphi survey needs to be analysed with anonymous feedback given to the expert panellists for the experts to have the opportunity to change their former decisions once they now have access to the responses of other experts as well as their responses. Also, to quantify experts' opinions on the subjects under investigation, Ameyaw et al. (2016) informed that a centesimal system has been adopted in a few studies, whereas a Likert scale ranging from 3 to 12 has been adopted in most studies mentioned in the literature. Nevertheless, to maintain measurement accuracy, attitude scales employed in Delphi questionnaires should be in the range of over five (5) points since most Delphi sample sizes are small (Hsu & Sandford, 2007). Green, Jones, Hughes, and Williams (1999), Chan et al. (2001), and Ameyaw et al. (2016) observed that views have varied in the literature on how the Delphi data should be analysed, what constitutes consensus, and what the number of rounds should be. However, Aigbavboa (2014) suggested that 60% of consensus should be the goal, whereas Green et al. (1999) recommended that 80% should be the goal consensus. However, in opposing the usage of percentages Crisp, Pelletier, Duffield, Adams, and Nagy (1997) posited that the Delphi process should only stop when data stability occurs. Hence, the consensus in the Delphi technique is only achieved when stability occurs

in the data collected and analysed. Ameyaw et al. (2016) suggested that the analytical software could be used to analyse the Delphi data and provide feedback to expert panellists using central tendencies (median scores and interquartile range) to report on the level of dispersion standard deviation could be used. However, irrespective of the data analysis technique used, there should be credibility (truthfulness), fittingness (applicability), audit ability (consistency), and confirmatory ability in data analysis. Thus, this study adopted a 60% goal as the consensus achievement goal. In a Delphi study, there is no specific regulation on the number of rounds to achieve consensus (Chan et al., 2001; Ameyaw et al., 2016). After the second and/or the third round, studies adopting the Delphi technique usually reach a consensus (Chan et al., 2001; Ameyaw et al., 2016). Dalkey, Brown, and Cochran (1970), Chan et al. (2001), Rajendran and Gambatese (2009), and Xia, Chan, and Yeung (2011) all agreed that Delphi results are more accurate after two iterations; after these two rounds, experts will begin to drop out of the study. This, as noted by Hasson et al. (2000), can be attributed to factors such as participants' fatigue, cost, time, and attrition rates. Thus, this study set three rounds of Delphi survey to achieve consensus.
- **Presentation and interpretation**: Delphi results presentation can be done in different ways. The methods of presentation and interpretation, as noted by Aigbavboa (2014), include graphical and statistical methods. This present study employed graphical and statistical methods in presenting Delphi results.

Hence, given the nature of the present research, it is believed that the Delphi technique is well suited to obtain credible inputs and information from experts in the industry, government bodies, and academics, among others, in the South African education sector to serve as key inputs in the development of an integrated MM framework for municipal buildings. The next section provides an overview of how the Delphi technique was used in this research.

Design, Construction, and Execution of the Delphi Study

Available literature from the studies of Loo (2002), Aigbavboa (2014), and quite recently, Ameyaw et al. (2016) advocate that the design, construction, and execution of the Delphi technique should follow a sequential order. Hence, Loo (2002) pre-arranged the Delphi chronological order as problem definition, panel selection, panel size determination, and conduction of the Delphi iterations. Moreover, this sequential order has been widely adopted in Delphi studies across disciplines. The doctoral study on residential satisfaction carried out by Aigbavboa (2014) is a typical example of this. Likewise, in a recent qualitative study on the Delphi by Ameyaw et al. (2016), the design, construction, and execution of the Delphi survey were organised to include the selection of the panellists, the number of expert panellists, the number of rounds, and an anonymous feedback process. This showed that the Delphi survey's design, construction, and execution approach does not differ widely among researchers. Thus, the sequential order advocated by Loo (2002) is adopted for this current study. Also, it is in line with the study of Delbecq, Van de Ven, and Gustafson (1975), which suggested a basic methodology for a Delphi study that included Delphi questionnaire development (objective), expert panel selection, sample size, first questionnaire analysis, and follow-up questionnaires. However, for this present study, the basic Delphi methodology of Delbecq et al. (1975) was adopted and will be explained. In addition, Table 9.2 gives details of the design, construction, and execution of a Delphi study.

Table 9.2 Delphi questionnaire formulation

Key Delphi questions	Phrasing for this study
Why are you interested in this study?	This study was initiated based on the belief that there is a dearth of studies that holistically investigate the attributes of MM of municipal buildings in the South African education sector; hence this has led to abandonment, collapsing, and deterioration of municipal facilities among the educational institution. This assumption is concrete because diverse factors influence the attributes of effective MM. Thus, there is a lack of understanding of the features that determine effective MM of municipal buildings.
What do you need to know that you do not know now?	Despite the knowledge about the attributes that determine effective MM of buildings, they have not been placed together and put into a model to inform policymakers and predict MM of municipal buildings in the South African education sector. At the end of this study, it will be known what the attributes are that determine the effective MM of municipal buildings in the South African education sector.
How will the results from the Delphi study influence the effective MM of buildings?	The results from the Delphi study will aid in developing a conceptual framework for the MM framework for municipal buildings in the South African education sector that this present study seeks to develop. Thus, the attributes which would be collectively determined and predict effective MM of municipal buildings in the South African education sector will be established.

Phase 1 – Delphi Question Development

For the whole Delphi study, questionnaire development is vital. In achieving the objectives of the current study, some key questions were asked. The construction of the Delphi questions in this present study was based on the guidelines in Table 9.2 and the current study's corresponding phrasing.

Phase 2 – Delphi Expert Panel Selection

The success of a Delphi study depends upon the expert panellists (participants or respondents) for the study (Hasson et al., 2000). Thus, in selecting the experts to constitute a panel for the Delphi study, a researcher must be careful and objective (Chan et al., 2001; Ameyaw et al., 2016). Aigbavboa (2014) asserts that restraint should be exercised to avoid the "least resistance" syndrome (the selection of like-minded individuals/or cozy friends), which tends to negate the study's strength. Conversely, Hasson et al. (2000) advocated that for a particular study, the experts selected must be interested in the study and involved in the subject being investigated. In their view, Hasson et al. (2000) observed that when professionals become expert panellists, controversial debates usually occur. Based on this, McKenna (1994) definition of an expert was adopted for this study. McKennas (1994) defined "experts" as a panel of informed individuals (otherwise called "experts" hereafter).

In agreement with this, Goodman (1987: 731) postulated that the Delphi technique "... recommends the use of experts or at least of informed advocates ... but tends not to advocate a random sample of panellists." It is noteworthy that some criteria for prequalifying experts for a Delphi panel have been adopted for some studies. As shown in a doctoral thesis by Hallowell (2008), the researcher advocated that experts on a Delphi panel should possess some special experience/knowledge, evidence of which could be working experience, working appointments, professional qualifications, and relevant publications. Equally, Delphi expert panellists should display a high degree of knowledge on the subject being examined and being representatives of the profession for their suggestions to be adaptable to the population (Rogers & Lopez, 2002).

Adler and Ziglio (1996) postulated four criteria for pre-qualifying an expert for a Delphi study. Adler and Ziglio (1996) suggest that the four criteria include knowledge and experience of the subject being examined, sufficient time to participate in the study, capacity and willingness to participate, and effective communicative skills. However, in prequalifying experts for a Delphi survey, the criteria one could use vary among researchers. In choosing experts for this Delphi study, at least four of the following requirements were required to be met by each of the experts:

- Knowledgeable of MM of buildings at the educational institutions level;
- Academic qualification: Should hold at least a higher national diploma (HND);
- Employment: Should currently serve (or have previously served) in a professional or voluntary capacity (e.g., at a place of employment – institution, business, agency, department, company) as builders, quantity surveyors, architects, estate managers, engineers, maintenance managers or officers of an establishment that is involved in building development, maintenance and management for both public and private institutions in South Africa;
- Experience: Should have exhibited a high level of theoretical and/or practical experience in the subject being investigated over the years;
- Owner and/or top management member of a construction firm/physical planning department in the South African construction/real estate industry/educational institutions participating in the MM of both public and private buildings;
- An academic who is knowledgeable and teaches maintenance, MM, construction management, facility management, housing development, and management, within or as a model at educational institutions or a member of the education maintenance committee of educational institutions;
- Ability to communicate effectively;
- Have ample time to participate in the Delphi survey;
- Be a member of a professional body so that his or her views may be adaptable to that of the population; and
- Residency: Have lived in one of the provinces in South Africa for at least a year.

Chan et al. (2001) and Rogers et al. (2002) suggested that at least two criteria be considered for the Delphi study. This study adopts four of the above minimum criteria framed after the four recommendations by Adler and Ziglio (1996) with the inclusion of residency in South Africa, considered compulsory for all selected experts. This was considered important since experts were required to have a wide-ranging understanding of the attributes of effective MM of Municipal buildings in the South African education sector since the South African education sector is the setting for the study. Also, a minimum of four criteria was set for this study because of the suggestion from Keeney et al. (2011) that the Delphi method may be undermined if Delphi panellists are engaged who lack specialist knowledge qualifications and proven track records in the field.

Furthermore, it has been argued that one expert group representing valid experts' opinions has largely been criticised as overstated and scientifically untenable (Hasson et al., 2000; Aigbavboa, 2014). However, Hardy, O'Brien, and Gaskin (2004) and Aigbavboa (2014) observed that this is because expertise may come in many forms and may include those who are experts by experience. Because of this, the panel members for this present study were identified from four sources. The first source was South African institutes of higher learning faculties, departments, and research institutes, such as the Construction Industry Development Board (CIDB), which engages in teaching and/or research and development, including various aspects of the building and construction sectors of the economy.

The second source was owners/top management members of construction firms, as well as the head of the physical planning department in the South Africa construction/real estate industry/educational institutions who undertake/provide maintenance services on building and civil works. These persons are involved in the process of construction, management, and maintenance of the building and civil works for both public (government) and private organisations such as the City of Johannesburg and Engineering Centre of Excellence; hence, their expertise is worth tapping into. Thus, they were identified as part of the expert panellists for this Delphi study. The third source of experts were persons who belong to professional bodies and/or associations in the South African construction industry, such as the Association of South African Surveyors (ASAQS), the South African Council for the Quantity Surveying profession (SACQSP), South African Council for planners (SACPLAN), South African Council for the Property Valuers Profession (SACPVP), The South African Council for the Project and Construction Management Professions (SACPCMP), Association of Construction Project Managers (ACPM), Institute for Landscape Architecture in South Africa (ILASA), Consulting Engineers South Africa (CESA), and South African Institute of Architects (SAIA). The fourth source was individuals who have committed themselves to work in the area of maintenance and management of buildings in the construction industry.

For the recruitment process, expert panellists were recruited through e-mail containing a brief overview of the study's objective. Consequently, panellists who had assented to the preliminary invitation to participate in the Delphi study had a detailed description of the Delphi study sent to them. Finally, their demographics were well scrutinised to establish whether the panellists met at least four of the criteria that prequalified them to be part of the expert panel.

Thus, all the selected experts for the study met the four criteria set for the study. After the verification exercise had been carried out, selected experts were sent the first-round questionnaire survey, presented as both open and closed-ended questions. The panellists' qualifications based on their curriculum vitae that they had been requested to submit in response to the initial invitation were ascertained and included in the study. From all the sources mentioned above, 30 invitations were sent out. Out of 30 invitations, 18 responded to the invitation, 18 completed the first round, and 15 were retained throughout the study as three panellists could not meet the demands of the study. Thus, the Delphi study involved 30 invited panellists and 15 active members. According to recommendations based on scholars who have adopted the Delphi technique in previous studies, the number of panellists

(15) used for this study was considered adequate. For instance, Delbecq et al. (1975) postulated that if the background of the panellists is homogenous, 10–15 panellists could be sufficient, which was achieved in the current study.

Rowe and Wright (2001) posit that in peer review studies, the number of Delphi panellists' could range from three (3) to eighty (80), whereas Hallowell and Gambatese (2010) suggest a minimum of eight panellists for the Delphi because most studies incorporate between eight and 16 panellists. Hallowell and Gambatese (2010) further maintained that the size of a panel should be guided by the capacity of the facilitator, the study's characteristics, the desired geographical representation, and the number of available experts. Furthermore, in an engineering management study undertaken by Aigbavboa (2014), 15 experts were used for the Delphi study. However, based on the existing previous studies and the fact that the Delphi method does not depend on statistical power for arriving at consensus amongst experts but on group dynamics, the panel of 15 experts for the current Delphi study was considered suitable.

As Aigbavboa (2014) observed, the Delphi method is time-consuming and rigorous. This possibly explains the fallout of more than half of the potential experts who had consented to participate when they learned of their obligations. Three (3) members dropped out after the first round, while fifteen members (15) eventually completed the study was also a verification of the quality of the study and its engaging nature in the present South African construction industry. All (100%) panel members were from South Africa: three (3) are currently residing in Free State Province; two (2) reside in Eastern Cape Province; five (5) reside in Gauteng Province; while KwaZulu-Natal Province, Limpopo Province, North Cape Province, Mpumalanga Province, and Western Cape Province had one (1) each (Table 9.3).

Table 9.4 shows the highest qualification held by the experts. Five (5) of the experts had a Doctor of Philosophy (PhD) degree, seven (7) experts had a master's degree, and three (3) had a bachelor's degree. As per their curriculum vitae analysis, all the experts were theoretically and practically knowledgeable in the effective MM of buildings issues.

Table 9.3 Residential location of experts

Province of residence in South Africa	Number of experts
Free State	3
Eastern Cape	2
Gauteng	5
KwaZulu-Natal	1
Limpopo	1
North Cape	1
Mpumalanga	1
Western Cape	1
Total	**15**

Table 9.4 Expert panellists' qualifications

Highest qualification	Number of experts
Doctor of Philosophy (PhD)	5
Master's degree	7
Bachelor's degree	3
Total	**15**

Table 9.5 Expert panellists' designations

Designation	Number of experts
Lecturer	5
Member of maintenance committee of educational institutions	4
Maintenance manager of public buildings	3
Property manager/developer	1
Facility manager	1
Professional/research institution of buildings	1
Total	**15**

Conversely, in terms of their designations, as shown in Table 9.5, five (5) of the selected experts were lecturers, four (4) were a member of the maintenance committee of educational institutions, and three (3) were maintenance managers of public buildings, one (1) was a property manager/developer, one (1) was a facility manager, while one (1) was with a professional/research institution. The experts were either faculty members of educational institutions, public or private sector practitioners, or workers.

All the experts listed in Table 9.6 had at least one year of experience, whereas most were within the experience bracket of 21–30 years.

Table 9.6 indicates that one (1) expert panellist had 1–5 years of experience, one (1) had 6–10 years of experience, five (5) had 11–20 years of experience, and six had 21–30 years of experience. Only two had more than 31 years of experience. The experts were faculty staff of reputable educational

Table 9.6 Expert panellists' years of experience

Years of experience	Number of experts
1–5	1
6–10	1
11–20	5
21–30	6
Over 31 years	2
Total	**15**

institutions and registered members of professional bodies and associations in the South African built environments such as the Association of South African Surveyors (ASAQS), the South African Council for the Quantity Surveying profession (SACQSP), South African Council for planners (SACPLAN), South African Council for the Property Valuers Profession (SACPVP), the South African Council for the Project and Construction Management Professions (SACPCMP) among others.

Phase 3 – Determining the Panel Size

Chan et al. (2001) and Ameyaw et al. (2016) all postulated that researchers have no consensus regarding the optimal size for a Delphi panel. Because of this, there are different views in literature by researchers on the optimal size for a Delphi panel. For instant, Andranovich (1995) suggests that a panel size of 10–15 if the group of experts is homogenous (that is, shares similar opinions), whereas Zami and Lee (2009) and Aigbavboa (2014) posited that the panel size could be increased to ensure balance if diverse interests exist among the group of experts (heterogeneous experts' group). Philips (2000) states that the ideal panel size for a Delphi study ranges from 7 to 12, whereas Miller (1993) assumed that any additional response beyond the first 30 responses does not generate any new information. Equally, Dunn (1994) submitted that a panel size of 10 to 30 is ideal for a Delphi study. It should involve formal and informal stakeholders who understand and have vested interests in the subject. Thus, concerning the panel size for a Delphi study, it can be presumed that two views are maintained in the literature, as shared by Ameyaw et al. (2016). One school of thought shows that the larger the size, the more reliable and accurate the Delphi results or findings (Murphy et al., 1998).

However, the followers of this view usually prefer a larger panel size for Delphi studies. On the other hand, Boje and Murnighan (1982) believed that there is no correlation between the panel size and the accuracy and reliability of a Delphi survey. Supporting this view, Hasson et al. (2000), Manoliadis et al. (2006), and Hallowell and Gambatese (2010) postulated that many factors determine the panel size for the Delphi study. They include the number of available experts, the available resources in terms of money and time, and the nature of the problem under investigation. In alignment with this, Ameyaw et al. (2016) conducted a qualitative study on 67 papers that used the Delphi study and specified the panel size. The findings of the study revealed that high numbers (41/60) of studies that used the Delphi technique used a panel size of eight (8) to twenty (20), whereas quite a few studies used a panel size above 21.

Nevertheless, the panel size that has been used previously ranges from 3 to 91 (Ameyaw et al., 2016). Delbecq et al. (1975) and Flanagan, Jewell, Ericsson, and Henrics (2005) state that existing literature suggests that a total of five to ten experts is recommended for a heterogeneous panel. Hence, the small panel size approach was followed for this present Delphi study, and a panel size of 15 was adopted in light of money and time constraints, together with the expert panellists' conflicting schedules. Additionally, in adopting a panel size of 15 for the study, the following factors, as postulated by Rowe and Wright (2001), were considered in conjunction with the prequalification criteria:

- To guarantee a wide base of experience and knowledge, the panel size should involve participants of different backgrounds,
- Panellists in both categories (academics and practitioners) should have extensive knowledge and experience in the effective MM of buildings, and
- To solicit views that balance theoretical and practical opinions in the Delphi study, the expert panellists should be adequately constituted from academics and practitioners.

Phase 4 – Conducting the Delphi Iterations

DATA COLLECTION THROUGH THE DELPHI

In conducting the Delphi study, there is no specific guidance on the optimal number of rounds in a Delphi study (Ameyaw et al., 2016). The studies of Loo (2002) and Aigbavboa (2014) show that to achieve consensus, the number of rounds for the Delphi study should be at least three or four rounds of questionnaires and feedback. Also, Dalkey et al. (1970) posit that the results of the Delphi study are more accurate after two iterations. Nonetheless, Xia et al. (2011) observed that the number of experts usually starts dropping off immediately after the second round. On the other hand, Crisp et al. (1997) informed that when data stability occurs in the Delphi study, the process should be stopped. Hence, a consensus is only achieved when stability occurs in the data collected and analysed. Ameyaw et al. (2016) note that it usually ranges between two to six.

The Delphi method used in this study involved three rounds of an iterative process, intending to achieve consensus in the views of the expert panellists on the influence or impact of attributes of effective MM of municipal buildings in the South African education sector as well as the key factors to effective MM of public buildings in South Africa. The Delphi questionnaire was sent out to the experts via e-mail, and the experts were asked to respond to the questions, bringing their capabilities and expertise to bear. However, the development of the Delphi questionnaires was informed by the literature review.

Furthermore, the Delphi questionnaires aimed to achieve and address the Delphi objectives in this study.

The Delphi study for this current research consisted of three rounds of iteration. For each round, a questionnaire was designed based on the re- sponses of the previous one. The first-round questions were developed based on the literature review of this study, which, among others, highlighted the attributes and sub-attributes of effective MM of buildings. Also, the comprehensive literature review extracted issues relating to the critical fac- tors militating against the effective MM of public buildings. These were constructively and structurally organised to frame the first round of the Delphi study. As a result, round one of the Delphi surveys was intended to be a brainstorming exercise that was used to produce a list of empirical attributes/factors that determine the effective MM of municipal buildings in the South African education sector as well as determining the key factors of effective MM of municipal buildings in South Africa. In the Delphi study's first round, closed-ended and open-ended questions were used. The responses from round one of the studies were analysed, and the results formed the basis of rounds two and three. To measure the degree of consensus reached in the participants' opinions, frequencies were used concerning the attributes that determine the effective MM of buildings and other related questions in line with the Delphi-specific objectives of this study.

The second round of the Delphi questions aimed at allowing the expert panellists to review and comment on the factors/attributes that determine the effective MM of municipal buildings in the South African education sector, as well as the key factors of effective MM of municipal buildings in South Africa which the expert panellists proposed in the round one of the Delphi studies. In the second round of the Delphi study, close-ended questions were used to investigate the expert panellists' comments expressing agreement, disagreement, or clarification concerning proposed attributes/factors that determine the effective MM of municipal buildings in the South African education sector. The very nature of the close-ended questions stimulated participants' reactions. Frequencies were used to measure the degree of consensus reached in the responses of the expert panellists regarding the factors/attributes that determine the effective MM of buildings and other related questions.

The third round was the final round, and its purpose was the following:

- To inform the expert panellists about the results from the analysis conducted on round two; and
- To solicit their final views (affirmation or comments) on factors and issues on which consensus had not yet been reached, even after round two. The questionnaire for round three of the Delphi was designed based on the findings from round two. Likewise, close-ended questions were used, and frequencies were obtained to measure the consensus in the responses of the expert panellists in the third round on the factors/attributes of effective MM of buildings and other related questions.

Based on the results of the third round Delphi survey, a list of attributes/factors that determine effective MM of Municipal buildings in the South African education sector, as well as the key factors to effective MM of municipal buildings in South Africa. The list of attributes/factors that determine the effective MM of municipal buildings in the South African education sector was generated with the aid of the Delphi technique, informed by the conceptual framework for the broader study.

Regarding the Delphi study, participants were asked to rate the likelihood of a factor/attribute (main attribute) influencing the effective MM of municipal buildings in the South African education sector, as well as the impact of sub-attributes/factors in predicting the effective MM of municipal buildings in the South African education sector. A probability scale ranging from 0 to 10, representing 0 to 100%, was used. Interval ranges were also set at 10. Moreover, with the impact of the sub-attributes and the critical factors affecting the MM of buildings, a 10-point ordinal scale ranging from "no impact" to "very high impact" was used (Tables 9.7 and 9.8).

For each response, the group median was calculated on each attribute of effective MM. Regarding this study, the group median was deemed an appropriate measure of central tendency because it was found to be more suitable for the information being collected. Furthermore, using the median to measure central tendency considers outlier responses, and consensus-building makes the notion more reasonable and eliminates bias (Aigbavboa, 2014). On the other hand, using the mean instead would have only considered outlier responses in measuring central tendency (Aigbavboa, 2014). Conversely, the first round of Delphi survey results formed the first round of the Delphi survey.

Thus, the group median for each element was computed and sent back to expert panellists in the second round of the survey. In this round, the expert panellists were requested to either maintain their first-round responses or change their responses as informed by the group median of the first round. Finally, in the third round, the group's medians were recomputed, and the absolute deviation in building consensus in the respondents' opinions and sent back to the expert panellists to reconsider their initial responses. After the third round's responses were analysed and consensus had been determined, the Delphi process was terminated. Maintaining participants' anonymity is crucial to the Delphi result's credibility (Aigbavboa, 2014). Thus, in this study, the participants' anonymity was preserved throughout

Table 9.7 Probability scale (influence or likelihood scale)

0–10%	11–20%	21–30%	31–40%	41–50%	51–60%	61–70%	71–80%	81–90%	91–100%
1	2	3	4	5	6	7	8	9	10

Table 9.8 Impact scale

No impact		Low impact		Medium impact		High impact		Very high impact	
1	2	3	4	5	6	7	8	9	10

Source: Aigbavboa, 2014.

	Researcher's role	Expert Panelists' role
Round One	Circulates questionnaire to expert panelists	Rate the likelihood of an attribute contributing to effective maintenance management and the influence of other related issues
Round Two	Computes group median for round and re-circulate questionnaire	Review individual ratings in view of the group medians. Give reason if necessary
Round Three	Re-computes group medians standard deviation and compiles comment	Reconsider initial rating if necessary
Conclusion of Process	Determines consensus and terminates the Delphi process	

Figure 9.5 An adapted outline of the Delphi process.

Source: Adapted from Thangaratinam and Redman (2005); Aigbavboa (2014).

the Delphi process to avoid any undue influence on other panel members and ensure the study's credibility (Figure 9.5).

Delphi – Specific Objectives

From the literature reviewed in this study, it could be gathered that various attributes determine the effective MM of buildings as measured from different MM typologies. Nevertheless, the extent to which the identified attributes influence the effective MM of municipal buildings in the South African education sector was unclear from the literature reviewed. Although attempts have been made to determine the influence of the attributes in the effective maintenance of public buildings in developed countries, little is known about the industry in developing countries. Moreover, previous models of effective MM have not adequately organised the attributes of MM into models to form a holistic attribute that determines the MM of buildings. Therefore, a more reliable and holistic measure of the attributes that determine the effective MM of buildings was essential. Also, the extent to which each attribute influences the effective MM of municipal buildings in the South African education sector, together with the identified gaps of factors/attributes from the literature reviewed. It is noteworthy that the context of this study ordinarily would have called for an experimental research method. Nonetheless, experimental research was not feasible for the study due to practical ethical issues, the time frame for the study, and the willingness of the would-be participants.

Because of this, the Delphi method was considered the most suitable and appropriate method to determine the influence and impact of the identified attributes on the effective MM of municipal buildings in the South African education sector. As postulated by Tilakasiri (2015), Delphi is an appropriate and suitable technique for developing models, frameworks, standards, and principles. Hence, given the broader aim of this study, which is to develop the MM of municipal buildings of developing countries using the South African education sector as a case study, the Delphi method was chosen to formulate the conceptual framework at the first stage.

At the Delphi stage, attributes or factors identified from the literature review that defines and determines effective MM were formulated into questions. The expert panellists were expected to rate the factors individually as being influential or impacting the effective MM of municipal buildings in the South African education sector. However, the output of the Delphi process was a set of attributes or factors that determine the effective MM of Municipal buildings in the South African education sector. Hence, informing the education sector to maintain their buildings effectively, the successful attributes should be, and the interplay of the successful attributes should be given the necessary consideration and priority. This is because several studies have identified that different attributes or factors determine the effective MM of buildings, as learned from the literature review.

The literature review showed attributes such as the impact of organisational maintenance policy, maintenance budget factors, human resources management, training factors, monitoring and supervision, and MIS. However, this current study extends the attributes mentioned above by considering MM holistically by adding new constructs, namely communication among stakeholders and maintenance culture. To develop a model that predicts MM and their relative contribution (extent of influence or impact) in the MM of municipal buildings in the South African education sector.

The philosophy behind the Delphi specific objectives is to do away with the tendency of a noncoherent dialogue for effective MM of municipal buildings in the South African education sector. Therefore, achieving the above objectives resulted in the following outcomes:

- Determining the attributes/factors and constructs that have critical significance (influence) in determining the effective MM of municipal buildings in the South African education sector;
- Developing a holistic conceptual framework for MM of municipal buildings in the education sector.

Computation of Data From Delphi Study

Microsoft Office Excel spreadsheet software programs were used to compute the Delphi survey data. In the views expressed by the expert panellists, stage one analysis was aimed at determining consensus based on predetermined criteria. This involved the determination of the group median for the responses to each question. The median value for the responses was adopted as a measure of central tendency since it can ensure that the effect of potential bias in the responses of individual respondents was eliminated.

It is noteworthy that studies of this type fairly often adopt the median as a measure of central tendency instead of the mean and interquartile deviation (IQD). However, in some cases, the IQD and the mean could be used or a combination of two. According to Aigbavboa (2014), the IQD is based upon and related to the median. Hence, to determine consensus in this study, for every round of responses from the expert panellists. The respective IQD in the responses was computed as a measure of the central tendencies. As Aigbavboa (2014) notes, the interquartile range of deviation is a measure that shows the degree to which the central 50% of values within the dataset are dispersed.

Furthermore, Whitley and Ball (2002) inform that the interquartile range or deviation removes or ignores outlying values and is used to summarise variability in the data. Consequently, the IQD in this study helped identify which measures were most appropriate to influence the MM of municipal buildings. Similarly, a clearer picture of the overall dataset was provided through the use of the IQD as it ignores or removes outlying values.

Additionally, absolute deviations (*Di*) of the group medians *(m(X))* of each rating for the relevant questions as pre-determined were also computed using Equation (9.1) after the third round of the Delphi survey. The absolute deviation, as postulated by Aigbavboa (2014), is a measure of the absolute difference between a response within a data set and a given point. Hence, the point from which the deviation is measured in this study is based on the median (a measure of central tendency).

$$D_i = [x_i - m(x)] \tag{9.1}$$

where:

   Di = Absolute deviation
   *x*i = Panellise rating
   m (x) = Measure of central tendency

Computation for the likelihood and impact of each question element was completed. For example, the influence and impact of the attributes of the effective MM were computed. Furthermore, the impact of the barriers to the attainment of effective MM of municipal buildings and the influence of the key factors on the effective MM of municipal buildings was computed as well as other maintenance issues surrounding the effective MM of municipal buildings in the South African education sector. Results from the Delphi study are presented as percentages and numbers in tables, bar, and column charts.

Determination of Consensus From the Delphi Process

In the Delphi study, consensus needed to be reached on all questions. However, a measure of the central tendency of the various responses from the expert panellists was adopted to determine consensus. Conversely, the group median and the IQD were adopted in this study. For all responses, the group median and IQD were calculated. In line with achieving consensus, the deviation of all responses from the Delphi panellists about the group median was determined not to be more than one (1) unit, and the same was adopted for the IQD. This was considered appropriate as the scale used for both influence (probability), and the impact was 1 to 10. The deviation of all responses from the panel of experts was calculated using the absolute median (Equation 9.1), while the IQD was computed based on the recommended statistical process of the absolute value of the difference between the 75th (Q3 or upper quartile) and 25th (Q1 or lower quartile) percentiles. As postulated by Whitley and Ball (2002), a percentile is the value of a variable below which a certain percent of observations falls.

Additionally, Aigbavboa (2014) asserts that the percentile (or centile) is often used in reporting scores from norm-referenced tests, as is the case in this study. The 75th percentile is the third quartile (Q3), the 50th percentile is the second quartile or the median (Q2), and the 25th percentile is also known as the first quartile (Q1). Thus, Whitley et al. (2002) posit that the deviation between the 75th (Q3) and 25th (Q1) percentiles gives an absolute value which is referred to as the interquartile deviation or range. Nevertheless, Aigbavboa (2014) observed that the interquartile range is a statistic with a breakdown point of 25%. Based on this, it is often referred to as the total range. However, a smaller value in the IQD indicates a higher degree of consensus (agreement).

Nevertheless, in Delphi studies, the consensus is difficult to measure. Moreover, concerning a set of opinions, there is no consensus on determining consensus (Aigbavboa, 2014). Nonetheless, consensus, as suggested by Holey, Feeley, Dixon, and Whittaker (2007) and Aigbavboa (2014), is synonymous with the agreement, and the agreement could be determined through the following:

- By confirming stability in responses with the consistency of answers between successive rounds of the study;
- The aggregate of judgements; or
- A move to a subjective level of central tendency.

Over the years, different researchers have used frequency distribution to measure agreement, and the criterion of at least 51% responding to any given response category has been used to determine consensus (McKenna, 1994). Whereas mean and standard deviation were used by some other studies to measure consensus. For instance, Rayens and Hahn (2000) used standard deviations and means to measure consensus. The study showed a decrease in standard deviations between rounds indicates increased response agreement. Furthermore, Rayens and Hahn (2000) used IQD to determine consensus in their study. Therefore, this present study adopts the use of IQD in determining consensus in the responses of the expert panellists. To achieve stability in their study, Rayens and Hahn (2000) added another criterion to determine consensus in addition to the IQD. The criterion, by Rayens and Hahn (2000), was that the IQD should equal one (1) unit for which more than 60% of respondents should have responded either generally negatively or generally positively. Items with IQD ≠1, for which the percentage of generally negative or generally positive responses was between 40% and 60%, were deemed to indicate a lack of consensus or agreement.

Equally, Raskin (1994) acknowledged that an IQD of 1.00 or less indicates agreement or consensus. Additionally, Spinelli (1983) considered a change of more than 1.00 IQD point in each successive stage as the criterion for the convergence of opinion. However, in existing literature, there is a lack of consensus on using or interpreting IQD as a data analysis method for a Delphi process. Aigbavboa (2014) noted that adopting a range of IQD values to interpret a Delphi response is influenced by the number of response choices. Aigbavboa (2014) states further that larger IQDs require an increase in the number of response choices. Hence, Holey et al. (2007) and Aigbavboa (2014) claimed that using a particular IQD as a cut-off for consensus requires consideration of the number of response choices. Conversely, to determine consensus, the following criteria were used by Holey et al. (2007):

- Percentage response;
- Computation of the weighted kappa (k) values to compare the chance eliminated agreement between rounds;
- Computation of median, standard deviation, and their associated group rankings;
- Computation of the means, standard deviation, and their associated group rankings using the importance ratings; and
- Percentages for each level of agreement for each question to compensate for varying response rates.

Additionally, Holey et al. (2007) contended that consensus is reached when the following are present:

- An increase in the percentage of agreements;
- A decrease in comments as rounds progressed;
- Increase in kappa values;
- A smaller range of responses;
- Smaller values of standard deviations; and
- The convergence of importance rakings.

From the literature reviewed, it can be deduced that there is little agreement on measuring consensus in a Delphi study. However, there is a general view that for consensus to have been achieved, there must be a conjunction of ideas and reasoning towards a subjective central tendency measure. Based on this, in this study, if the following exists, a consensus was deemed to have been reached:

- The IQD was less than 1.00. This suggests that items with IQD = 0.00 were considered to have reflected high consensus;
- More than 60% of responses were generally positive or generally negative with certain questions; and
- The average absolute deviation was not more than one unit. The absolute deviation is computed using Equation (9.1).

As a result, the scales of consensus adapted for this current study are the following:

- Strong consensus – median 9–10, mean 8–10, IQD ≤1 and ≥80% (8–10);
- Good consensus – median 7–8.99, mean 6–7.99, IQD≥1.1≤2 and ≥60%≤ 79% (6–7.99); and
- Weak consensus – median ≤ 6.99, mean ≤5.99, and IQD≥2.1≤3 and ≤ 59% (5.99).

Reliability and Validity of the Delphi Method

Reliability is the extent to which a procedure produces similar results when conditions are constant (Els & Delarey, 2006). According to Aigbavboa (2014), this kind of statistical reliability is almost impossible to achieve in a typical Delphi study owing to the likelihood that another panel may reach a different conclusion in a Delphi study depending on their knowledge of the subject area and interest. Nevertheless, in this study, to ensure reliability, care was taken that credibility showed in truthfulness, fittingness exhibited in applicability, audit ability was shown in response consistency, and conformability was exhibited in the responses from all the expert panellists. Additionally, during the selection of the panel for this study, credibility was also ensured. They all distinguished themselves based on the criteria for selecting expert panellists and their experience and knowledge depth.

Similarly, validity was enhanced by maintaining the anonymity of all panellists. This ensures the Delphi study avoids influence or preconceptions from other experts, and this helps eliminate the "bandwagon" effect, which is one of the strong points of the Delphi method. Furthermore, internal validity was also boosted through the number of iterations implemented in the Delphi study. Hence, for dissenting views on the study, experts in the Delphi survey were given a chance to maintain or change their opinions with an argument or a written explanation. Moreover, another way of enhancing the study's internal validity is the constant feedback to the researchers and constant communication between the researcher and the individual panellists. Aigbavboa (2014) postulated that validity could be classified into two forms, namely internal validity and external validity. The external validity deals with the extent of generalizability of the results from the study regarding a larger view, while external validity is determined by how participants were selected for the study. Nevertheless, since the validation process of the conceptual model was done using the questionnaire survey, this process was not needed.

However, the selection process of the participants for the Delphi study has guaranteed external validity because pre-determined scientific criteria consistent with previous scholarly works were adopted. Additionally, the expert panel for the study comprised members from various sectors, all with in-depth knowledge of the maintenance of buildings. All resided in South Africa, were members of professional bodies in the built environment, and we're very experienced. Based on this, the study satisfied the requirements for external validity in line with standard research ethics.

Variables

The research instrument (see Table 9.9) was designed to measure the exogenous variables, namely organisational maintenance policy (OPY), maintenance budget factors (MBF), human resources management (HRM), training factor (TF), monitoring and supervision factor (MSN), maintenance information system (MIS), communication among stakeholders (CAS), and maintenance culture factors (MCF). These exogenous variables were hypothesised to be characterised by indicator variables, which collectively constituted the questionnaire items apart from the demographics, which are also measured by the questionnaire.

Ethical Consideration

The researcher assured the anonymity of research respondents in the survey, and their anonymity in the survey was maintained. Because of this, the respondents' names and any information that may reveal their identity were regarded as confidential. Also, no one was coerced to participate in the survey. They were asked to participate of their own free will. They were informed of their rights not to participate in the survey or to end their participation in the survey if they so desired. Also, they were told the purpose of the study, as well as how and why they had been chosen.

Table 9.9 Conceptual model indicator variables

Organisational maintenance policy (OPY)	
	Suitable maintenance procedures and process
	Organisation operational efficiency
	Optimisation of the maintenance policy
	Preparation of safety procedure
	Optimisation of the maintenance action plan
	Meeting maintenance objectives
	Optimisation of preventive maintenance design
	Preparation of safety procedure
	Well-defined priority system
	Assembling of maintenance organisation structure
	Maintenance strategies development
	Analysis of maintenance procedures
	Change in policy and its associated results
	Risk factor establishment
	Reduction in mean time to repair
Maintenance budget factors (MBF)	
	Corruption-free maintenance process
	Cost implication of maintained asset
	Audit of maintenance operational cost
	Cash flow indexing
	Valuation of maintenance operation budget
	Maintenance budget implementation
	Yearly maintenance budgets certainties
	Prioritisation of maintenance financing
	Incorporation of financial indicators
	Maintenance financial plan
	Market and financial terms of operations
	Optimisation of finance outsourcing
	Reduction in maintenance expenditure
	Maintenance operation financing
	Maintenance funding
	Optimising maintenance resources
Human resources management factors (HRM)	
	End-of-job documentation
	Recruitment of skilful and experienced personnel
	Interdepartmental conflicts resolution
	Developing the maintenance leader
	Resources management
	Personnel self-training encouragement
	Assessment of personnel performance
	Prioritisation of required resources
	Resources assignment
	Keeping records of training courses

(*Continued*)

Table 9.9 (Continued)

	Career path for maintenance personnel
	Roles for maintenance personnel
	Assigning personnel for specific tasks
	Counselling arrangements for personnel under stress
	Recognition of maintenance personnel effort
	Maintenance personnel Identification
	Institutional and in-house training for personnel
	Opportunities for growth
Training factor (TF)	
	Integration of new techniques
	Education on current maintenance knowledge
	Usage of proper tools
	Usage of proper procedures
	Understanding organisation maintenance policy
	Semi-skilled personnel can do better
	Safety of personnel
	Personnel training on maintenance Management
	Training of personnel on maintenance skills
	Time management
	Maintenance problem-solving skills development
	On-job training on maintenance
	Reduction in the scarcity of needed skills
	Reliability of personnel
Monitoring and supervision (MSN)	
	Maintenance performance reporting
	Zero-error tolerance
	Monitoring of safety procedure
	Maintenance operation inspection
	Value improvement
	Material wastage reduction
	Monitoring of maintenance operations
	Periodic maintenance planning
	Meeting maintenance target
	Maintenance process monitoring
	Risk reduction
	Maintenance personnel involvement
	Suitable replacement material usage
	Appropriate maintenance strategy
Maintenance information system (MIS)	
	Online maintenance monitoring systems
	Maintenance tools inventory
	Early maintenance warning guide
	Data collection
	Failure rates prediction

(Continued)

Table 9.9 (Continued)

	Data monitoring
	Analysis of maintenance process
	Maintenance descriptions
	Implementation action on data collected
	Balanced maintenance costs
	Implementation action on data collected
	Tracking maintenance performance indicators
	Integration of operation process
	Integration of management tools
	Data processing
	Maintenance activities record keeping
	Strategy for material cost
Communication among stakeholders (CAS)	
	Users' satisfaction
	Quality-driven service delivery
	Collaboration among stakeholders
	Identify gaps in maintenance process
	Early defects detection
	E-mails among stakeholders
	Users feedback
	Increase in users' participation
	Eradication of uncertainty
	Understanding of users' maintenance needs
	Addressing stakeholders' concerns
	Good working relationship among stakeholders
	Use of visualisation technique
	Understanding users' expectations
	Timely and accurate feedback
	Help desk
	Liaison with relevant government agencies
	Checking maintenance information with users
	Establishment of the need for change
	Web-based systems meetings and report
	Stakeholders' participative actions
	Liaison with maintenance institutions
	Telephone support
Maintenance culture factors (MCF)	
	Effective maintenance management style
	Effective decision-making capacity
	User change of attitude to maintained buildings
	Attitudinal change to maintenance operations
	Sustainability of infrastructure
	Use of right skills maintenance personnel
	Diligence in maintenance operations
	Enhance performance level of buildings

(*Continued*)

Table 9.9 (Continued)

Benchmarking for maintenance performance
Attitude of ensuring regular repairs
Proper maintenance system
Improve economic value of a country
Commitment to change initiatives
Building reliability
Attitude of ensuring regular servicing
Motivation of maintenance personnel
Maintenance decision based on societal value
Maximal utilisation of infrastructural

Summary

The methodology adopted for this current study is presented in this chapter. Additionally, it provided the justifications for the methods of data collection and the philosophical position, among others. Moreover, ethical considerations were also discussed. The following chapter presents the results of the data analysis from the Delphi study.

References

Adler, M., & Ziglio, E. (1996). *Gazing into the oracle: The Delphi method and its application to social policy and public health.* London and Philadelphia: Jessica Kingsley Publisher.

Ahmad, R., Al-Shorgani, N. K. N., Hamid, A. A., Yusoff, W. M. W., & Daud, F. (2013). Optimization of medium components using response surface methodology (RSM) for mycelium biomass and exopolysaccharide production by Lentinus squarrosulus. *Advances in Bioscience and Biotechnology, 4*(12), 1079–1085.

Aigbavboa, C. O. (2014). An integrated beneficiary centred satisfaction model for publicly funded housing schemes in South Africa. Published doctoral dissertation, University of Johannesburg. Johannesburg.

Amaratunga, R. G. (2001). *Theory building in facilities management performance measurement: Application of some core performance measurement and management principles.* Published doctoral dissertation, University of Salford.

Ameyaw, E. E., Hu, Y., Shan, M., Chan, A. P., & Le, Y. (2016). Application of Delphi method in construction engineering and management research: A quantitative perspective. *Journal of Civil Engineering and Management, 22*(8), 991–1000.

Andranovich, G. (1995). Achieving consensus in public decision making: Applying interest-based problem solving to the challenges of intergovernmental collaboration. *The Journal of Applied Behavioral Science, 31*(4), 429–445.

Boje, D. M., & Murnighan, J. K. (1982). Group confidence pressures decisions. *Management Science, 28*(1), 1187–1196.

Boote, D. N., & Beile, P. (2005). Scholars before researchers: On the centrality of the dissertation literature review in research preparation. *Educational Researcher, 34*(6), 3–15.

Brieschke, P. A. (1992). Reparative praxis: Rethinking the catastrophe that is social science. *Theory into practice, 31*(2), 173–180.

Brill, J., Bishop, M., & Walker, A. (2006). The competencies and characteristics required of an effective project manager: A web-based Delphi study. *Educational Technology Research and Development, 54*(2), 115–140.

Bryman, A. (2016). *Social research methods.* United Kingdom: Oxford university press.

Buckley, C. (1995). Delphi: A methodology for preferences more than predictions. *Library Management, 16*(7), 16–19.

Burrell, G., & Morgan, G. (1994). *Sociological paradigms and organizational analyses* (6th ed.). Portsmouth NH, Heinemann.

Chan, A. C., Yung, E. K., Lam, P., Tam, C. M., & Cheung, S. (2001). Application of Delphi method in selection of procurement systems for construction projects. *Construction Management and Economics, 19*(7), 699–718.

Coates, J. F. (1975). *Review of Sackman Report-Technological forecasting and social change* (Vol.7). New York: American Elsevier Publishing Co.

Creswell, J. W. (2010). Mapping the developing landscape of mixed methods research. *SAGE handbook of mixed methods in social & behavioral research, 2,* 45–68.

Creswell, J. W., & Clark, V. L. P. (2017). *Designing and conducting mixed methods research.* Thousand Oaks, CA: Sage.

Creswell, J. W., & Creswell, J. (2003). *Research design* (pp. 155–179). Thousand Oaks, CA: Sage publications.

Crisp, J., Pelletier, D., Duffield, C., Adams, A., & Nagy, S. (1997). The Delphi method? *Nursing Research, 46,* 116–118.

Crotty, M. (1998). *The foundations of social research: Meaning and perspective in the research process.* London: Sage Publications Limited.

Churchill, G., & Iacobucci, D. (2004). *Marketing research: Methodological foundations.* Ohio: Thomson South-Western.

Czinkota, M. R., & Ronkainen, I. A. (1992). Global marketing 2000: A marketing survival guide. *Marketing management, 1*(1), 36–43.

Dalkey, N., Brown, B., & Cochran, S. (1970). Use of self-ratings to improve group estimates. *Technological Forecasting, 1*(3), 283–291.

Delbecq, A. L., Van de Ven, A. H., & Gustafson, D. H. (1975). *Group techniques for program planning: A guide to nominal group and Delphi process.* Glenview, IL: Scott, Foresman, and Company.

Dunn, W. N. (1994). *Public policy analysis: An introduction.* Englewood Cliffs, NJ: Prentice Hall.

Easterby-Smith, M. (1991). *Management research: An introduction.* London: Sage Publications.

Ellram, L. M. (1996). The use of the case study method in logistics research. *Journal of Business Logistics, 17*(2), 93.

Els, D. A., & Delarey, R. P. (2006). Developing a holistic wellness model. *South African Journal of Human Resource Management, 4*(2), 46–56.

Fellows, R. F., & Liu, A. M. (2015). *Research methods for construction.* Hoboken, New Jersey: John Wiley & Sons.

Flanagan, R., Jewell, C., Ericsson, S., & Henrics, P. (2005). *Measuring construction competitiveness in selected countries*. Reading: The University of Reading. School of Construction Management and Engineering.

Goodman, C. M. (1987). The Delphi technique: A critique. *Journal of Advanced Nursing, 12*, 729–734.

Green, B., Jones, M., Hughes, D., & Williams, A. (1999). Applying the Delphi technique in the study of GP's information requirements. *Health and social care in the community, 17*(3), 198–205.

Grisham, T., & Walker, D. H. (2008). Intercultural leadership. *International Journal of Managing Projects in Business, 1*, 439–445.

Gupta, U. G., & Clarke, R. E. (1996). Theory and applications of the Delphi technique: A bibliography (1975–1994). *Technological Forecasting and Social Change, 53*(2), 185–211.

Hallowell, M. (2008). *A formal model of construction safety and health risk management*. Corvallis, Oregano, USA: Oregon State University.

Hallowell, M. R., & Gambatese, J. A. (2010). Qualitative research: Application of the Delphi method to CEM research. *Journal of Construction Engineering and Management, 136*(1), 99–107.

Hanson, W. E., Creswell, J. W., Clark, V. L. P., Petska, K. S., & Creswell, J. D. (2005). Mixed methods research designs in counseling psychology. *Journal of Counseling Psychology, 52*(2), 224.

Hardy, J. D., O'Brien, A. P., & Gaskin, C. J. (2004). Practical application of the Delphi technique in a bicultural mental health nursing study in New Zealand. *Journal of Advanced Nursing, 46*(1), 95–109.

Hasson, F., Keeney, S., & McKenna, H. (2000). Research guidelines for the Delphi survey technique. *Journal of Advanced Nursing, 32*(4), 1008–1015.

Helmer, O. (1977). Problems in futures research: Delphi and causal cross-impact analysis. *Future, 9*(1), 25–52.

Heppner, P. P., & Heppner, M. J. (2004). *Writing and publishing your thesis, dissertation, and research: A guide for students in the helping professions*. California: Thomson/Brooks/Cole.

Hill, S. (1995). The social organization of boards of directors. *British Journal of Sociology, 46*(2), 245–278.

Holey, E. A., Feeley, J. L., Dixon, J., & Whittaker, V. J. (2007). An exploration of the use of simple statistics to measure consensus and stability in Delphi studies. *BMC Medical Research Methodology, 7*(25), 1–10.

Hon, C. K., Chan, A. P., & Yam, M. C. (2012). Empirical study to investigate the difficulties of implementing safety practices in the repair and maintenance sector in Hong Kong. *Journal of Construction Engineering and Management, 138*(7), 877–884.

Hsu, C. C., & Sandford, B. A. (2007). The Delphi technique: Making sense of consensus. *Practical Assessment, Research, and Evaluation, 12*(1), 10.

Jack, E. P., & Raturi, A. S. (2006) Lessons learned from methodological triangulation in management research. *Management Research News, 29*(6), 345–357.

Jairath, N., & Weinstein, J. (1994). The Delphi methodology (Part one): A useful administrative approach. *Canadian Journal of Nursing Administration, 7*(3), 29–42.

Jean, K. (1992). *Livelihood strategies among farm youth in Rwanda*. Michigan: Michigan State University.

Jonker, J., & Pennink, B. (2010). *The essence of research methodology: A concise guide for master and PhD students in management science.* London: Springer Science & Business Media.

Keeney, S., McKenna, H., & Hasson, F. (2011). *The Delphi technique in nursing and health research.* Hoboken, New Jersey: John Wiley & Sons.

King, N. (1994). The qualitative research interview. In C. Cassell & G. Symon (Eds.), *Qualitative methods in organisational research.* London: Sage Publications.

Lang, T. (1995). An overview of four futures methodologies. *Manoa Journal of Fried and Half Fried Ideas, 7,* 1–43.

Lang, T. (2001). An overview of four futures methodologies (Delphi, environmental scanning, issues management, and emerging issue analysis). The Manoa Journal of Fried and Half-Fried Ideas (about the future) Hawaii Research Center for Futures Studies 7. March, S. T., & Smith, G. F. (1995). *Design and natural science research on information technology. Decision Support Systems, 15,* 251266.

Linstone, H. A., & Turoff, M. (Eds.). (1975). *The Delphi method* (pp. 3–12). Reading, MA: Addison-Wesley.

Loo, R. (2002). The Delphi method: A powerful tool for strategic management, policing. *An International Journal of Police Strategies and Management, 25*(4), 762–769.

Maguire, M. (1987). *Doing participatory research: A feminist approach.* Amherst, MA: The center of International Education, University of Massachusetts.

Manoliadis, O. G., Tsolas, I., & Nakou, A. (2006). Sustainable construction and drivers of change in Greece: A Delphi study. *Construction Management and Economics, 24*(2), 113–120.

May, T. (2001). *Social research: Issues, methods, and process* (3rd ed). Buckingham: Open University Press.

McKenna, H. (1994). The Delphi technique: A worthwhile research approach for nursing? *Journal of Advanced Nursing, 19*(6), 1221–1225.

Miller, B. A. (2007). *Assessing organizational performance in higher education.* USA: John Wiley and Sons, Inc.

Miller, M. M. (1993). Enhancing regional analysis with the Delphi Method. *Review of Regional Studies, 23*(2), 191–212.

Mitchell, J. C. (1983). Case and situation analysis. *The Sociological Review, 31*(2), 187–211.

Murphy, M. K., Black, N., Lamping, D. L., McKee, C. M., Sanderson, C. B., Askham, J., & Marteau, T. (1998). Consensus development methods and their use in clinical guideline development. *Health Technology Assessment, 2*(3), 1–88.

Nau, D. (1995). Mixing methodologies: Can bimodal research be a viable postpositivist tool. *The Qualitative Report, 2*(3), 1–5.

Ogunbayo, B. F., Ohis Aigbavboa, C., Thwala, W. D., & Akinradewo, O. I. (2022). Assessing maintenance budget elements for building maintenance management in Nigerian built environment: a Delphi study. *Built Environment Project and Asset Management*, Vol. ahead-of-print No. ahead-of-print. 10.1108/BEPAM-06-2021-0080

Pacitti, B. J. (1998). *Organisational Learning in R&D Organisations: A Study of New Product Development Projects.* Unpublished PhD Thesis. Manchester: University of Manchester.

Phillips, R. (2000). *New applications for the Delphi technique.* Annual San Diego: Pfeiffer and Company.

Rajendran, S., & Gambatese, J. A. (2009). Development and initial validation of sustainable construction safety and health rating system. *Journal of Construction Engineering and Management, 1067-1075,* 135.

Raskin, M. S. (1994). The Delphi study in field instruction revisited: Expert consensus on issues and research priorities. *Journal of Social Work Education, 30,* 75–89.

Rayens, M. K., & Hahn, E. J. (2000). Building consensus using the policy Delphi method. *Policy Politics Nursing Practice, 1*(2), 308–315.

Reid, N. G. (1988). The Delphi technique, its contribution to the evaluation of professional practices. In *Professional Competence and Quality Assurance in the Caring Professional* (Ellis, R. ed.). London: Chapman & Hall.

Remenyi, D., Williams, B., Money, A., & Swartz, E. (1998). *Doing research in business and management.* London: Sage Publications.

Robson, C. (2002). Real world research: A resource for social scientists and practitioner researchers (Vol. 2). Oxford: Blackwell.

Rogers, M. R., & Lopez, E. C. (2002). Identifying critical cross-cultural school psychology competencies. *Journal of School Psychology, 40*(2), 115–141.

Rowe, G., & Wright, G. (2001). Expert opinions in forecasting: The role of the Delphi technique. In *Principles of forecasting* (pp. 125–144). Boston, MA: Springer.

Sackman, H. (1974). *Delphi assessment: Expert opinion, forecasting, and group process* (No. RAND-R-1283-PR). Santa Monica CA: Rand Corp.

Sarantakos, S. (2012). *Social research.* London: Macmillan International Higher Education.

Saunders, M., Lewis, P., & Thornhill, A. (2007). *Research methods.* Business Students (4th ed.). England: Pearson Education Limited.

Saunders, M., Lewis, P., & Thornhill, A. (2006). *Understanding research approaches.* London: SAGE Publication.

Scheele, D. S. (2002). Reality construction as a product of Delphi interaction: The Delphi Method: Techniques and Applications. In H. A. Linstone & M. Turoff (Eds.). Available online at www.is.njit.edu/pubs/delphibook/ch2c.html; last accessed Feb. 27, 2007.

Schoemaker, P. H. (1993). Multiple scenario development: Its conceptual and behavioral foundation. *Strategic Management Journal, 14*(3), 193–213.

Sekaran, U. (2000). *Research methods for business: A skill-building approach.* New York: John Wiley & Sons.

Shariff, N. (2015). Utilizing the Delphi survey approach: A review. *Journal of Nursing Care, 4*(3), 246.

Society of Actuaries (SOA) (1999). *Final Report of the 1999 Delphi study.* Schaumburg, IL: Society of Actuaries.

Spinelli, T. (1983). The Delphi decision-making process. *Journal of Psychology, 113*(1), 73–80.

Straub, D., Boudreau, M. C., & Gefen, D. (2004). Validation guidelines for IS positivist research. *Communications of the Association for Information systems, 13*(1), 24.

Teddlie, C., & Tashakkori, A. (2003). Major issues and controversies in the use of mixed methods in the social and behavioral sciences. *Handbook of mixed methods in social & behavioral research.* Thousand Oaks, CA: Sage Publications.

Thangaratinam, S., & Redman, C. E. (2005). The Delphi technique. *The Obstetrician and Gynaecologist, 7*(2), 120–125.

Tilakasiri, K. K. (2015). Development of new frameworks, standards, and principles via Delphi data collection method. *International Journal of Science and Research, 4*(9), 1189–1194.

Vanderstoep, S. W., & Johnson, D. D. (2008). *Research methods for everyday life: Blending qualitative and quantitative approaches* (p. 32). New York: John Wiley & Sons.

Vanson, S. (2014). *What on earth are Ontology and Epistemology* (pp. 4–13). Rome: TPS the Performance Solution.

Whitley, E., & Ball, J. (2002). Statistics review 1: Presenting and summarising data. *Critical Care, 6*(1), 66–71.

Xia, B., Chan, A. C., & Yeung, J. Y. (2011). Developing a fuzzy multicriteria decision-making model for selecting Design Build operational variations. *Journal of Construction Engineering and Management, 137*(12), 1176–1184.

Zami, M. S., & Lee, A. (2009). A review of the Delphi technique: To understand the factors influencing adoption of stabilised earth construction in low-cost urban housing. *The Built and Human Environment Review, 2*(2), 37–50.

10 The Outcome of the Delphi Study

Introduction

The Delphi study sought to determine and solicit experts' opinions on the influence (probability) and the impact of attributes of maintenance management (MM) on municipal buildings in the South African education sector. Specifically, it identified the attributes (main and sub) that determined effective MM and examined whether the attributes that determine effective MM in other geographical contexts are the same in the South African education sector; to identify key factors to effective MM of municipal buildings and to determine the relative influence of each of the factors in the South Africa; and to identify the main measuring indicators (outcomes) of effective MM and to establish the relative influence of the indicators on the effective MM of municipal buildings in the South African education sector. Three rounds of the Delphi process were conducted before the panel of experts reached a consensus on the questions that were posed to them. The summary of results from the Delphi survey is presented in this section. The composition of the panel of experts and general background of the Delphi study are described, followed by the findings from the Delphi study. This chapter concludes with a summative discussion of the findings based on the Delphi-specific objectives for the Delphi study.

Background of the Delphi Survey

The Delphi-specific objectives that governed the Delphi study were the following:

DSO1. To identify the attributes (main and sub-) that determine effective MM of municipal buildings and to examine whether the attributes that influence the MM of municipal buildings in other geographical contexts are the same in the South African education sector.

DOS2. To identify the key factors to effective MM and determine the relative influence of each of the factors in the maintenance of municipal buildings in the South African education sector.

DOI: 10.1201/9781003344681-10

DOS3. To identify the measuring indicators (outcomes) of effective MM and to establish the relative influence of the indicators in the MM of municipal buildings in the South African education sector.

The philosophy behind the Delphi-specific objectives is to do away with the tendency of a noncoherent dialogue in the MM of municipal buildings in the South African education sector. Consequently, achieving the above objectives resulted in the following outcomes:

- Determining the attributes/factors and constructs that are of critical significance (influence) in determining the effective MM of municipal buildings in the South African education sector.
- A holistic conceptual model of MM of municipal buildings in the South African education sector.

A panel of 21 experts took part in the first round of the Delphi process, whereas 15 experts were retained who took part in the Delphi process from the second to the third round of the study. In selecting the panel of experts, particular attention was given to the expert's knowledge of the MM of buildings and the high level of theoretical and/or practical experience in the investigated subject. These attributes were considered compulsory for all selected experts. These were considered significant because experts were required to have had an in-depth understanding of the MM of buildings. Panel members were selected to achieve some level of balance between practitioners and theorists in the field of MM of buildings, particularly in terms of municipal buildings such as educational institution buildings. A questionnaire was designed for each round of the Delphi survey based on experts' responses from the previous round(s) of the Delphi study. However, the round one questionnaire of the Delphi study was designed based on a summary of the findings of the comprehensive literature review. This highlighted some sets of main and sub-attributes potentially relevant to achieving effective MM of municipal buildings in the South African education sector. Also, issues relating to the barriers to effective MM of municipal buildings, key factors to effective MM of municipal buildings, intentions of a maintenance policy/law, among others, were also extracted from the literature review. These were structurally and constructively put together to frame the first-round questionnaire of the Delphi survey. The intent of round one of the Delphi study was to be a brainstorming exercise in order to produce a list of empirical attributes that determine the effective MM of municipal buildings as well as issues about maintenance in the South African education sector. Open and close-ended questions were used in this round. Subsequently, the questions were analysed, and the results formed the basis of the questionnaires for rounds two and three of the Delphi study. Moreover, descriptive statistics were used to measure the degree of consensus reached among the survey's respondents on the questionnaire elements.

Additionally, content analysis methodology as postulated by Rubin, Pronovost, and Diette (2001: 491) and Aigbavboa (2014: 325) was adopted to analyse responses to the open questions to "minimise redundancy."

The goal of the second round of the Delphi study was to allow the experts to review and make comments on the attributes that determine the effective MM of municipal buildings in the South African education sector and other issues relating to the MM of municipal buildings in the South African education sector, which were proposed by the experts who had participated in round one of the Delphi studies. Closed questions were used in round two to investigate participants' comments expressing agreement, disagreement, or clarification concerning proposed attributes that determine the MM of municipal buildings in the South African education sector. The specific nature of the closed-ended questions stimulated the Delphi participants' reactions. Generally, the results of round two indicated that the expert panellists were generally in agreement.

Consequently, the third round of the Delphi study was a revision of round two of the Delphi survey. In round three, the statistical information calculated from the second round was presented to each expert. The experts were once again asked to give responses using the provided rating scale as applicable to the question. Round three successfully refined the discussion to the stage where clear points of agreement or disagreement could be determined. As a result, a fourth-round was not necessary. Accordingly, each question's descriptive statistics (the mean, median, percentages, standard deviation, and interquartile deviation [IQD] scores) were calculated. The expert panellists were requested to explain their responses in situations where the score was two points from the median score.

Assuming that an agreement was not reached at the third round. The data from the third round would have been analysed and sent again to the expert panellists for consideration; hence round four of the Delphi survey. However, this study predicted the use of three rounds in order to achieve a consensus. This prediction was correct. In view of this, a fourth-round was not necessary. The survey respondents were informed of this when the third round of questionnaires for the Delphi survey was sent out. The main aim of the Delphi research technique was to cycle the questions towards achieving a consensus among the panel of experts. During each round of questionnaires, the experts were given the results of the median of the previous round. Consensus is attained when 100% of the participants are in agreement. However, Stitt-Gohdes et al. (2004) and Aigbavboa (2014) informed that when two-thirds of the participants are in agreement, it is considered common consent. Thus, in this study, common consent was also acceptable. Common consent was attained when 60% of the experts agreed on each statement or question. All statements were examined individually for consensus. The results were statistically analysed after each round of questionnaires. This helped determine whether consensus was attained for each statement or question based on the provided scale for each question or statement.

Moreover, if consensus was attained before the final round, then that question or statement may no longer be required in the next rounds. After the third round of the Delphi survey, a consensus was attained with regard to most of the attributes that determine the MM of municipal buildings and on other MM-related issues in the South African education sector. Based upon the findings of the Delphi study, a list of attributes that determine MM was prepared, which informed the conceptual framework for the broader study, while issues surrounding the MM of municipal buildings in the South African education sector were also highlighted.

The results of the Delphi study were therefore presented in relation to the Delphi-specific objectives that guided the Delphi study.

Findings From the Delphi Study

DSO1. *To identify the attributes (main and sub-) that determine effective MM of municipal buildings and to examine whether the attributes that influence the MM of municipal buildings in other geographical contexts are the same in the South African education sector.*

From the summary of the comprehensive literature review, some attributes and sub-attributes potentially relevant to MM of municipal buildings in the South African education sector were identified. However, the reviewed literature was largely based on studies from developed countries. They were collectively used to examine the attributes of MM of municipal buildings in the South African education sector.

The level of influence of the main attributes of MM was established by assessing the extent to which the listed attributes determine the MM of municipal buildings in the South African education sector. Similarly, the sub-attributes impact in determining MM of municipal buildings was also evaluated. The rating was based on an ordinal scale of one (1) to ten, with one (1) being low influence or impact and the (10) being high influence or very high impact. The levels of influence and impact were then obtained as a product of the consensus attained based on the adopted scale for measuring consensus in the Delphi study.

From the eight (8) identified main attributes that determine the MM of municipal buildings, only six attributes, namely organisational maintenance policy, maintenance budget factors, training factors, monitoring and supervision, MIS, and maintenance culture, were considered by the experts to have a high influence. The attribute recorded a group's median score of 9.0 and IQD of between 0.00 and 1.00, signifying that there is a strong consensus in the expert's views and there is little deviation or inconsistency in the responses from the experts. Similarly, two other main attributes, as detailed in Figure 10.1, namely human resources management and communication among stakeholders, were scored as having an average influence on the MM of municipal buildings. The median scores were scored

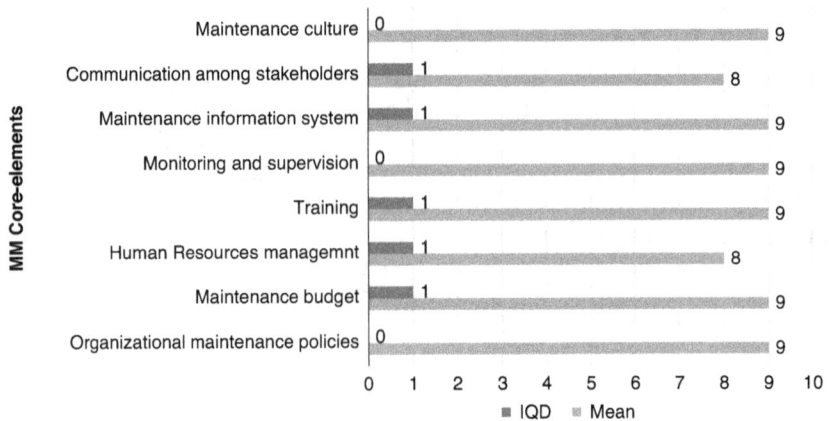

Figure 10.1 Influence of core attributes on maintenance management of municipal buildings.

Source: Author's review (2022).

as having an average influence on the MM of municipal buildings. The median scores were not more than 8.99, while the IQD was less than 1.00. The IQD values suggested a strong consensus was attained in the views of the experts regarding the five main attributes as a determinant of the MM of municipal buildings.

Additionally, the impact of the main attributes' sub-attributes was measured in the determination of MM of municipal buildings, and the results were presented in the form of tables (Tables 10.1–10.8).

Out of the twenty-three (23) listed variables for the organisational maintenance policy construct, five (5) of the items had a very high impact (VHI: 9.00–10.00) in determining the MM of municipal buildings, while ten (10) variables had a high impact (HI: 7.00–8.99) and eight (8) other variables scored a medium impact (MI: 5.00–6.99). Thus, none was found not to have had an impact on the determination of the MM of municipal buildings. Similarly, the IOD scores revealed that consensus was achieved for fifteen (15) items as they obtained scores ranging from 0.00 to 1.00. Likewise, their respective standard deviation (σx) revealed consistency in the responses of the experts as their respective (σx) values were at most (1). However, eight (8) of the items recorded good consensus, with scores ranging from 1.10 to 2.00. Their respective values for standard deviation (σx) revealed inconsistency and variability in the responses of the experts as their respective (σx) values were more than one (1). Additionally, in ranking the items by their respective mean scores, suitable maintenance procedures and process was ranked 1st out of the 23 variables. This was followed by organisation operational efficiency, while analysis of spare-part management was ranked 23rd.

Table 10.1 Organisational maintenance policy attributes

Sub-attributes	Median (M)	Mean ($\bar{x}$)	Standard deviation (σx)	Interquartile deviation (IQD)	Mean scores ranking (R)
Organisation operational efficiency	9	8.74	0.68	1.00	2
Suitable maintenance procedures and process	9	9.13	0.88	1.00	1
Assembling of maintenance organisation structure	8	7.80	0.91	1.00	10
Analysis of maintenance procedures	7	7.28	0.85	1.00	12
Preparation of maintenance operation	8	7.99	0.73	0.58	8
Preparation of safety procedure	8	8.34	0.87	1.00	4
Risk factor establishment	7	7.26	1.24	1.00	14
Target performance measurement	6	6.33	1.75	2.25	21
Routinisation of optimisation techniques	6	6.55	1.67	3.00	19
Maintenance strategies development	8	7.73	1.48	0.35	11
Spare-part management	6	6.28	1.53	2.00	23
Meeting maintenance objectives	8	8.00	1.65	1.00	6
Optimisation of the maintenance policy	9	8.40	1.68	1.00	3
Optimisation of the maintenance action plan	9	8.07	1.83	1.00	5
Well-defined priority system	9	7.93	2.46	1.00	9
Appropriate procurement strategy	6	6.33	1.88	3.00	21
Change in policy and its associated results	8	7.27	1.58	1.00	13
Working together with regulatory agencies	6	6.93	1.58	2.00	16
Maintenance contracts suitability	6	6.47	1.64	2.00	20
Timely identification of maintenance work	6	6.73	2.12	3.50	17
Reduction in mean time to repair	7	7.20	1.32	1.00	15
Design of a preventive plan	6	6.60	1.80	2.50	16
Optimisation of preventive maintenance design	8	8.00	1.25	0.50	6

Table 10.2 Maintenance budget factors

Sub-attributes	Median (M)	Mean (x̄)	Standard deviation (σx)	Interquartile deviation (IQD)	Mean scores ranking (R)
Prioritisation of maintenance financing	8	8.34	0.70	0.13	8
Valuation of maintenance operation budget	9	8.39	0.88	1.00	5
Maintenance operation financing	8	7.80	0.91	1.00	14
Optimising maintenance resources	7	7.28	1.53	1.00	16
Market and financial terms of operations	8	8.19	1.17	1.00	11
Maintenance budget implementation	8	8.36	1.14	1.00	6
Asset maintenance prioritisation	6	6.43	1.41	2.25	17
Optimisation of finance outsourcing	8	8.00	1.15	0.50	12
Audit of maintenance operational cost	8	8.46	0.81	1.00	3
Yearly maintenance budgets certainties	9	8.35	1.01	1.00	7
Cash flow indexing	8	8.43	1.09	1.00	4
Planning for future asset replacement	5	5.60	1.92	2.50	21
Cost implication of maintained asset	9	8.67	1.05	1.00	2
Reduction in maintenance expenditure	8	7.93	1.71	1.00	13
Maintenance financial plan	9	8.20	1.70	1.00	10
Maintenance operation system improvement	6	6.40	1.64	2.50	18
Incorporation of financial indicators	8	8.33	0.98	1.00	9
Optimisation of business profitability	6	6.40	1.30	2.00	18
Maintenance funding	8	7.73	1.03	0.50	15
Maintenance materials assessment	5	5.67	1.72	2.00	20
Corruption-free maintenance process	9	9.00	0.76	1.00	1

Table 10.3 Human resources management attributes

Sub-attributes	Median (M)	Mean (x̄)	Standard deviation (σx)	Interquartile deviation (IQD)	Mean scores ranking (R)
End-of-job documentation	9	8.86	0.50	0.05	1
Recruitment of skilful and experienced personnel	9	8.83	1.47	1.25	2
Equipment availability provision	5	4.91	1.61	2.00	25
Leadership responsibilities	5	5.15	1.34	2.00	24
Roles for maintenance personnel	8	8.07	0.68	0.03	12
Resources assignment	8	8.28	0.77	0.10	9
Resources management	9	8.60	1.02	1.00	4
Identifying external resources	6	6.35	1.45	3.00	21
Incentives for development	6	6.47	1.73	2.50	20
Developing the maintenance leader	9	8.60	0.74	0.50	4
Career path for maintenance personnel	8	8.13	1.25	0.50	11
Opportunities for growth	7	7.20	1.37	1.00	18
Interdepartmental conflicts resolution	9	8.80	0.68	0.00	3
Recognition of maintenance personnel effort	8	7.87	0.52	0.00	15
Prioritisation of required resources	8	8.33	1.11	1.00	8
Encouragement of creativity among personnel	6	5.47	1.73	2.00	23
Counselling arrangements for personnel under stress	8	8.00	0.76	1.00	14
Institutional and in-house training for personnel	7	7.33	0.62	0.50	17
Personnel self-training encouragement	8	8.40	0.83	1.00	6
Assessment of personnel performance	9	8.40	1.40	1.00	6
Keeping records of training courses	8	8.27	1.28	1.00	10
Developing the culture of continuous improvement	6	6.13	1.13	2.00	22
Maintenance personnel Identification	8	7.60	1.64	1.00	16
Assigning personnel for specific tasks	8	8.07	1.10	1.00	12
Evaluation of maintenance personnel task outcome	6	6.60	1.55	2.50	19

Table 10.4 Training factors attributes

Sub-attributes	Median (M)	Mean (x̄)	Standard deviation (σx)	Interquartile deviation (IQD)	Mean scores ranking (R)
Reliability of personnel	7	7.34	0.60	0.63	14
Safety of personnel	9	8.60	0.61	1.00	7
Priorities of the organisation objectives	6	5.97	1.25	2.00	17
Operation objectives improvement	5	4.83	1.47	3.00	20
Education on current maintenance knowledge	9	8.94	0.85	0.58	2
Training of personnel on maintenance skills	9	8.47	1.41	1.00	9
Maintenance training opportunities	6	5.88	1.36	2.00	19
On-job training on maintenance	8	7.81	0.75	1.00	12
Understanding organisation maintenance policy	9	8.73	0.77	1.00	4
Advancement in technology	6	6.06	1.34	2.00	16
Usage of proper procedures	9	8.73	0.85	1.00	4
Usage of proper tools	9	8.80	0.65	1.00	3
Semi-skilled personnel can do better	9	8.66	0.94	1.00	6
Maintenance problem-solving skills development	8	7.94	0.85	1.00	11
Time management	8	8.00	0.89	0.00	10
Interpersonal skills development	6	5.91	1.27	2.00	18
Personnel training on maintenance Management	9	8.48	0.89	1.00	8
Integration of new techniques	10	9.26	0.93	1.00	1
Strategy implementation improvement	6	6.28	1.39	2.00	15
Reduction in the scarcity of needed skills	8	7.69	1.89	0.33	13

Table 10.5 Monitoring and supervision attributes

Sub-attributes	Median (M)	Mean ($\bar{x}$)	Standard deviation (σx)	Interquartile deviation (IQD)	Mean scores ranking (R)
Determination of facilities performance	5	5.68	1.66	2.00	17
Maintenance operation inspection	9	8.66	0.70	1.00	4
Maintenance process monitoring	8	8.34	1.08	1.00	10
Maintenance performance reporting	9	8.79	0.54	0.00	1
Risk reduction	8	8.22	0.66	1.00	11
Diagnosis of maintenance task	6	6.17	1.63	2.25	15
Meeting maintenance target	9	8.39	0.88	1.00	9
Material wastage reduction	9	8.44	1.15	1.00	6
Maintenance personnel involvement	8	8.16	1.31	1.00	12
Value improvement	9	8.46	1.09	0.70	5
Zero-error tolerance	9	8.68	0.87	1.00	2
Suitable replacement material usage	8	7.87	1.19	1.00	13
Constant rescheduling of maintenance activities	6	5.93	1.53	2.00	16
Monitoring of maintenance operations	8	8.40	0.99	1.00	7
Monitoring of safety procedure	9	8.67	0.72	1.00	3
Observing faults trigger components	5	5.33	1.76	2.50	18
Periodic maintenance planning	9	8.40	1.40	1.00	7
Appropriate maintenance strategy	8	7.80	1.52	1.00	14

Table 10.6 Maintenance information system attributes

Sub-attributes	Median (M)	Mean (x̄)	Standard deviation (σx)	Interquartile deviation (IQD)	Mean scores ranking (R)
Maintenance descriptions	9	8.79	0.40	0.07	8
Data processing	8	8.14	0.34	0.00	14
Data monitoring	9	8.93	0.77	0.03	4
Maintenance activities record keeping	8	8.07	1.28	1.00	15
Maintenance specifications detailing	6	5.80	1.37	2.00	22
Data collection	9	8.93	1.03	1.50	4
Implementation action on data collected	9	8.60	0.63	1.00	9
Integration of operation process	8	8.47	0.99	1.00	11
Integration of management tools	8	8.27	1.10	1.00	13
Early maintenance warning guide	9	9.00	0.53	0.00	2
Reliability strategy results	6	6.60	1.59	2.50	17
Derived strategy for equipment usage	6	6.27	1.62	2.00	20
Strategy for material cost	8	7.87	1.51	1.00	16
Maintenance tools inventory	9	9.00	0.65	0.00	2
Analysis of maintenance process	9	8.80	0.86	1.00	7
Balanced maintenance costs	8	8.60	0.91	1.00	9
Benefits information channel	6	6.33	1.68	2.50	19
Failure rates prediction	9	8.93	0.70	0.50	4
Maintenance performance indicators	6	6.13	1.30	2.00	21
Tracking maintenance performance indicators	9	8.47	1.55	1.00	11
Integrating the maintenance information system	6	6.40	1.50	2.00	18
Online maintenance monitoring systems	10	9.13	1.25	1.00	1

Table 10.7 Communication among stakeholders' attributes

Sub-attributes	Median (M)	Mean (x̄)	Standard deviation (σx)	Interquartile deviation (IQD)	Mean scores ranking (R)
Collaboration among stakeholders	9	8.80	0.86	1.00	3
Addressing stakeholders' concerns	9	8.46	0.88	1.00	11
Liaison with maintenance institutions	8	7.80	0.75	1.00	22
Liaison with relevant government agencies	8	8.13	0.88	1.00	17
Stakeholders' participative actions	8	7.86	0.81	0.40	21
Establishment of the need for change	8	8.01	0.73	0.57	19
Quality-driven service delivery	9	8.86	0.50	0.05	2
Evaluation of maintenance performance	6	6.11	1.58	2.25	28
Understanding of users' maintenance needs	9	8.53	0.89	1.00	9
Understanding users' expectations	8	8.26	0.77	1.00	14
Increase in users' participation	9	8.60	0.61	1.00	8
Identify gaps in services quality	6	6.36	1.35	3.00	25
Identify gaps in maintenance process	9	8.79	0.40	0.07	4
Technical information documentation	6	6.09	1.88	2.00	29
Users satisfaction	9	9.19	0.65	1.00	1
Bias-free maintenance operations	6	6.23	2.14	2.50	27
Good working relationship among stakeholders	8	8.39	0.61	1.00	12
Timely and accurate feedback	8	8.15	0.62	0.55	15
Organisation loyalty	6	5.79	1.58	2.00	35
Eradication of uncertainty	9	8.53	0.89	1.00	9
Operational strategies support	6	6.29	1.39	2.25	26
Early defects detection	9	8.79	0.83	0.00	4
Timely correction of defects	6	5.84	1.61	2.00	34
Identification of suitable approach to maintenance	5	5.43	1.89	2.25	36
Meetings with line management	5	5.04	1.42	2.00	38

(*Continued*)

Table 10.7 (Continued)

Sub-attributes	Median (M)	Mean ($\bar{x}$)	Standard deviation (σx)	Interquartile deviation (IQD)	Mean scores ranking (R)
E-mails among stakeholders	9	8.67	0.79	1.00	6
Help desk	8	8.14	1.09	0.05	16
Telephone support	8	7.59	1.36	1.00	23
Verbal discussion	5	5.42	1.41	2.25	37
Web-based systems meetings and report	8	8.00	0.37	0.00	20
Checking maintenance information with users	8	8.12	1.03	1.03	18
Signs and symbols	6	5.99	1.41	2.00	32
Letter writing	6	6.02	1.68	1.08	31
Straightforward maintenance requests	6	6.03	1.55	2.00	30
Management suggestion matching interests of users	6	6.57	1.51	2.25	24
Use of visualisation technique	8	8.34	1.08	1.00	13
Users feedback	9	8.66	0.87	1.00	7
Maintenance schedule	6	5.88	1.67	1.28	33

Table 10.8 Maintenance culture factors

Sub-attributes	Median (M)	Mean ($\bar{x}$)	Standard deviation (σx)	Interquartile deviation (IQD)	Mean scores ranking (R)
Diligence in maintenance operations	9	8.73	0.85	0.48	7
Commitment to change initiatives	9	8.59	0.80	1.00	13
Attitude change to maintenance	6	6.24	1.80	2.00	26
Maintenance decision based on societal value	8	7.93	0.96	0.00	17
Maintenance decision based on environment effect	6	6.13	1.41	2.00	28
Proper maintenance system	9	8.60	1.06	1.00	11
Adequate knowledge on maintained building	6	6.73	1.10	1.50	21
Improve economic value of a country	9	8.60	0.83	1.00	11
Use of right skills maintenance personnel	9	8.87	0.52	0.00	6
Benchmarking for maintenance performance	9	8.67	0.72	0.00	9
Attitudinal change to maintenance operations	9	9.00	0.85	0.50	4
User change of attitude to maintained buildings	9	9.07	0.80	0.50	2
Maintenance equipment availability	6	6.20	1.78	2.00	27
Building reliability	8	8.33	0.82	0.50	14
Building maintainability	6	6.13	1.13	2.00	28
Maintenance values and norms	6	6.27	1.39	2.00	25
Maintenance policies provision	6	6.33	1.23	2.00	23
Prioritisation of maintenance economic value	6	5.73	2.02	2.50	32
Attitude of ensuring regular servicing	8	8.33	0.62	0.50	14
Attitude of ensuring regular repairs	9	8.67	0.90	1.00	9
Sustainability of infrastructure	9	9.00	0.65	0.00	4
Maximal utilisation of infrastructural	8	7.67	1.50	2.00	18
Investment returns guarantee	6	6.33	1.50	2.00	23
Enhance performance level of buildings	9	8.73	1.22	1.00	7
Viable maintenance system	6	7.00	1.20	2.00	19
Motivation of maintenance personnel	8	8.20	0.94	1.00	16
Provision of user guide within building	6	5.80	1.52	1.50	31
Effective maintenance management style	9	9.13	0.74	1.00	1
Prioritisation of maintenance activities	6	6.07	1.22	2.00	30
Management confidence in the employee capacity	6	6.93	1.62	2.00	20
Effective decision-making capacity	9	9.07	0.70	0.50	2
Early response to stakeholders' maintenance needs	6	6.67	1.29	2.00	22

Similarly, out of the twenty-one (21) listed variables for the maintenance budget factors construct, five (5) of the items had a very high impact (VHI: 9.00–10.00) in determining the MM of municipal buildings, while eleven (11) variables had a medium (HI: 7.00–8.99) and five (5) other variables were scored as having a medium impact (MI: 5.00–6.99). Thus, none was found not to have had an impact on the determination of the MM of municipal buildings. Also, the IQD scores revealed that strong consensus was achieved for sixteen (16) items as they obtained scores ranging from 0.00 to 1.00. Moreover, their respective standard deviations (σx) revealed consistency in the responses of the experts as their respective (σx) values were at most (1). However, five (5) of the items recorded good consensus, with scores ranging from 1.10 to 2.00. Their respective values for standard deviation (σx) revealed inconsistency and variability in the responses of the experts as their respective (σx) values were more than one (1). Likewise, when the 21 variables of maintenance budget were ranked by their respective mean scores, the corruption-free maintenance process ranked 1st while planning for future asset replacement was ranked 21st.

Furthermore, out of the twenty-five (25) listed variables for the human resources management attributes construct, six (6) of the items had a very high impact (VHI: 9.00–10.00) in determining the MM of municipal buildings, while twelve (12) variables had a high impact (HI: 7.00–8.99) and seven (7) other variables were scored as having a medium impact (MI: 5.00–6.99). Thus, none was found not to have had an impact on the determination of MM of municipal buildings. Likewise, the IQD scores revealed that strong consensus was achieved for eighteen (18) of the items as they obtained scores ranging from 0.00 to 1.00, while eight (8) of the items recorded good consensus with scores ranging from 1.10 to 2.00. Their respective values for standard deviation (σx) revealed inconsistency and variability in the experts' responses. Additionally, the 1st ranked variable of human resources was the end-of-job documentation, whereas equipment availability provision emerged as 25th.

Furthermore, out of the twenty (20) listed variables for the training factors attributes construct, nine (9) of the items had a very high impact (VHI: 9.00–10.00) in determining the MM of municipal buildings, while five (5) variables had a high impact (HI: 7.00–8.99) and six (6) other variables were scored as having a medium impact (MI: 5.00–6.99). However, none was found not to have had an impact on the determination of the MM of municipal buildings. Similarly, the IQD scores revealed that strong consensus was achieved for fourteen (14) of the items as they obtained scores ranging from 0.00 to 1.00, while six (6) of the items recorded good consensus with scores ranging from 1.10 to 2.00. Their respective values for standard deviation (σx) revealed inconsistency and variability in the experts' responses. Likewise, when the 20 variables of training factors attributes were ranked by their respective mean scores, integration of new techniques ranked 1st while operation objectives improvement was ranked 20th.

Moreover, out of the eighteen (18) listed variables for the monitoring and supervision attributes construct, eight (8) of the items had a very high impact (VHI: 9.00–10.00) in determining the MM of municipal buildings, while six (6) variables had a high impact (HI: 7.00–8.99) and four (4) other variables were scored as having a medium impact (MI: 5.00–6.99). However, none was found not to have had an impact on the determination of MM of municipal buildings. Similarly, the IQD scores revealed that strong consensus was achieved for fourteen (14) of the items as they obtained scores ranging from 0.00 to 1.00, while four (4) of the items recorded good consensus with scores ranging from 1.10 to 2.00. Their respective values for standard deviation (σx) revealed inconsistency and variability in the experts' responses. Likewise, when the 18 variables of monitoring and supervision were ranked by their respective mean scores, Maintenance performance reporting ranked 1st while observing faults trigger components was ranked 18th.

Moreover, out of the twenty-two (22) listed variables for MIS attributes construct, ten (10) of the items had a very high impact (VHI: 9.00–10.00) in determining the MM of municipal buildings, while six (6) variables had a high impact (HI: 7.00–8.99) and six (6) other variables were scored as having a medium impact (MI: 5.00–6.99). Hence, none was found not to have had an impact on the determination of the MM of municipal buildings. Similarly, the IQD scores revealed that strong consensus was achieved for sixteen (16) of the items as they obtained scores ranging from 0.00 to 1.00, while six (6) of the items recorded good consensus with scores ranging from 1.10 to 2.00. Their respective values for standard deviation (σx) revealed inconsistency and variability in the experts' responses. Furthermore, when the twenty-two (22) variables of the MIS were ranked by their respective mean scores, the online maintenance monitoring systems ranked 1st, while the maintenance specifications detailing was ranked 22nd.

Moreover, out of the thirty-eight (38) listed variables for the communication among stakeholders' attributes construct, eleven (11) of the items had a very high impact (VHI: 9.00–10.00) in determining the MM of municipal buildings, while twelve (12) variables had a high impact (HI: 7.00–8.99) and fifteen (15) other variables were scored as having a medium impact (MI: 5.00–6.99). Therefore, none was found not to have had an impact on the determination of MM of municipal buildings. Equally, the IQD scores revealed that strong consensus was achieved for twenty-three (23) of the items as they obtained scores ranging from 0.00 to 1.00, while fifteen (15) of the items recorded good consensus with scores ranging from 1.10 to 2.00. Their respective values for standard deviation (σx) revealed inconsistency and variability in the experts' responses. Also, when the thirty-eight (38) variables of communication among stakeholders were ranked by their respective mean scores, users' satisfaction ranked 1st while meetings with line management was ranked 38th.

Similarly, out of the thirty-two (32) listed variables for maintenance culture factors construct, thirteen (13) of the items had a very high impact

(VHI: 9.00–10.00) in determining the MM of municipal buildings, while five (5) variables had a high impact (HI: 7.00–8.99) and fourteen (14) other variables were scored as having a medium impact (MI: 5.00–6.99). Hence, none was found not to have had an impact on the determination of the MM of municipal buildings. Likewise, the IQD scores revealed that strong consensus was achieved for eighteen (18) of the items as they obtained scores ranging from 0.00 to 1.00, while fourteen (14) of the items recorded good consensus with scores ranging from 1.10 to 2.00. Their respective values for standard deviation (σx) revealed inconsistency and variability in the experts' responses. Further, when the thirty-two (32) variables of maintenance culture attributes were ranked by their respective mean scores, effective MM style ranked 1st while prioritisation of maintenance economic value was ranked 38th.

DOS2. *To identify the key factors to effective MM and determine the relative influence of each of the factors in the maintenance of municipal buildings in the South African education sector.*

In order to achieve this objective, literature was reviewed, and some key factors likely to lead to the effective MM of municipal buildings in the South African education sector were identified. Consequently, experts' views on the relative influence of each key factor were sought, and the qualitative survey's analysis was presented in Table 10.9.

From the eighteen (18) identified key factors to effective MM, twelve (12) of the items had a very high impact (VHI: 9.00–10.00) on the key factors to effective MM, while six (6) of the items had a high impact (HI: 7.00–8.99). None of the key factors identified was found not to have had an impact on the effective MM of buildings. Moreover, the IQD values revealed a strong consensus in the views of the expert panellists as all the IQD values for all the eighteen (18) items were at most one (1). Hence, the standard deviation values affirmed a high degree of consistency and non-variability in the views of the experts as respective standard deviation values recorded were less than one (1). In terms of ranking by respective mean scores, the Maintenance budget emerged 1st while the motivation of maintenance personnel emerged 18th among the eighteen (18) key factors to effective MM identified.

DOS3. *To identify the measuring indicators (outcomes) of effective MM and to establish the relative influence of the indicators in the MM of municipal buildings in the South African education sector.*

According to Table 10.10, all twenty-four (24) outcomes were outcomes of effective MM. Sixteen (16) of the outcome indicators were found to be of a very high impact, while the remaining eight (8) were of high impact. However, in terms of ranking by the mean scores, maintenance work well supervised was the topmost ranked outcome, while MIS is effectively used was ranked 24th among the outcome of effective MM of municipal buildings.

Table 10.9 Key factors to effective maintenance management

Key factors	Median (M)	Mean ($\bar{x}$)	Standard deviation (σx)	Interquartile deviation (IQD)	Mean scores ranking (R)
Policy development and organisation	9	8.99	0.82	0.25	2
Human resources management	9	8.81	0.75	0.78	6
Maintenance budget	9	9.00	0.63	0.00	1
Continuous improvement	8	8.40	0.80	1.00	12
Monitoring and supervision	9	8.74	0.68	0.80	7
Task planning and schedule	8	8.25	1.00	1.25	16
Computerised MM system	9	8.66	0.79	1.00	10
User satisfaction	8	8.14	0.34	0.00	17
Maintenance approach/strategy	9	8.67	0.70	1.00	9
Education and training	8	8.35	0.94	1.00	14
Spare part management	9	8.38	1.02	1.00	13
Outsourcing strategy	8	8.33	1.07	1.00	15
Attitudinal change to maintenance	9	8.88	0.81	0.00	5
Communication among stakeholders	9	8.94	0.68	0.25	3
Motivation of maintenance personnel	8	7.90	0.97	2.00	18
Performance measurement	9	8.73	0.77	1.00	8
Quality management	9	8.94	0.85	0.50	3
Users requirement	9	8.66	1.08	1.00	10

Table 10.10 Outcomes of effective maintenance management of municipal buildings

Outcomes	Median (M)	Mean (x̄)	Standard deviation (σx)	Interquartile deviation (IQD)	Mean scores ranking (R)
Maintenance budget is adequately planned for	9	8.73	0.77	1.00	9
Maintenance organisation structure is effective	9	8.94	0.57	0.00	2
Strategy for maintenance is better selected	9	8.59	0.71	1.00	14
Policies for maintenance is better enacted	8	8.27	0.93	0.50	22
Maintenance operations is planned appropriately	9	8.60	0.71	1.00	13
MIS is effectively used	8	8.26	0.85	1.00	24
Maintenance work complying with standard expected	9	8.58	0.61	1.00	15
Change in attitude to maintenance operations	9	8.93	0.57	0.03	3
Replaceable material used for maintenance is of quality	9	8.53	0.88	1.00	16
Technical support to maintenance is adequate	8	8.34	0.79	1.00	21
Maintenance work is well supervised	9	9.14	0.34	0.00	1
Maintenance work is better monitored	9	8.67	0.60	1.00	11
Maintenance procedure is better understood	9	8.73	0.77	1.00	9
Users' attitude to maintained building is improved	9	8.40	0.80	1.00	19
Users' maintenance needs and expectation are met	8	8.27	0.44	0.50	22
Delay to users' maintenance request is avoided	9	8.51	0.89	1.00	17
Users' performance is enhanced	9	8.82	0.66	1.00	6
Users' productivity is enhanced	9	8.67	0.79	1.00	11
Users' well-being is improved	9	8.93	0.44	0.00	3
Buildings' life cycle performed optimally	9	8.79	0.54	0.50	7
Deterioration within the buildings is avoided	9	8.74	0.68	1.00	8
Buildings and their support components run effectively	8	8.40	0.80	1.00	19
High output in the use of assets is provided	9	8.44	0.81	1.00	18
Improved and sustainable building performance is achieved	9	8.86	0.34	0.00	5

Discussion of the Delphi Results

Objective DSO1

The first objective of the Delphi study was to identify the attributes (main and sub-) that determine the effective MM of municipal buildings and to determine whether the attributes that influence the MM of municipal buildings in other geographical contexts are the same as in the South African education sector. Findings emanating from the study revealed that organisational maintenance policy, maintenance budget factors, human resources management, training factors, monitoring and supervision, and MIS significantly influence the MM of municipal buildings in the South African education sector. This is consistent with earlier studies' findings that these attributes have influenced the MM in their geographical setting (Pintelon & Gelders, 1992; Campbell, 1998; Wireman, 2005; Marques et al., 2009). Likewise, the two additional constructs in the study included communication among stakeholders and maintenance culture, which were also found to influence the MM of municipal buildings significantly. They confirmed previous studies' findings that communication among stakeholders and maintenance culture factors have significantly influenced the MM of buildings (Bothma & Cloete, 2000; Campbell & Jardine, 2001; Wilkinson & Townsend, 2012; Suwaibatul et al., 2012). Thus, a strong consensus was attained with respect to organisational maintenance policy, maintenance budget factors, human resources management, training factors, monitoring and supervision, MIS, communication among stakeholders, and maintenance culture factors as the main attributes influencing the MM of municipal buildings as they all recorded a high influence on MM.

Also, the assessment of the sub-attributes of the eight (8) main attributes of MM reveals that the sub-attributes that determine the MM of municipal buildings in the South African education sector were largely similar to those in other geographical contexts. Nonetheless, a consensus was not reached for some of the sub-attributes found to have strongly determined the MM of municipal buildings in other geographical contexts. Among these sub-attributes were spare-part management, appropriate procurement strategy, and target performance measurement, which were under the main attribute, namely organisational policy factors. Likewise, planning for future asset replacement and maintenance materials assessment was under the maintenance budget main attributes. Also, equipment availability provision and leadership responsibilities were under human resources management attributes.

Similarly, operation objectives improvement and maintenance training opportunities were under training factor attributes. Equally, observing faults trigger components and determination of facilities' performance were under monitoring and supervision attributes. Moreover, maintenance specifications detailing and maintenance performance indicators results were under MIS attributes. Furthermore, meetings with line management and verbal discussion were under Communication among stakeholders. While prioritisation of maintenance economic value and provision of user guide within the building were under maintenance culture. These sub-attributes had IQD scores ranging from 1.1 to 2.0, an indication that there was variability in the views of the experts and that strong consensus was not achieved.

In conclusion, the results suggested that the attributes (main and sub-) that determine the MM of municipal buildings in the South African education sector are largely similar to the determinants in other geographical contexts. Additionally, the effective MM of municipal buildings is assured if there is a consideration of these factors in the development of the MM model for the municipal buildings in the South African education sector, namely organisational maintenance policy, maintenance budget factors, human resources management, training factors, monitoring, and supervision, MIS, communication among stakeholders and maintenance culture factors, which have all been described as being of significant influence and having a high impact on determining the MM of municipal buildings.

Objective DSO2

The second objective of the Delphi survey sought to identify the key factors to effective MM and determine the relative influence of each factor in the MM of municipal buildings in the South African education sector. In all, eighteen (18) key factors were identified. Out of the eighteen (18) identified key factors to the effective MM, twelve (12) of the key factors were found to have had a very high impact of (VHI: 9.00–10.00) on the effective MM of municipal buildings. These key factors were maintenance budget with a mean score of 9.00, policy development and organisation with a mean score of 8.99, communication among stakeholders and quality management both with a mean score of 8.94, attitudinal change to maintenance with a mean score of 8.88, human resources management with a mean score of 8.81, monitoring and supervision with a mean score of 8.74, performance measurement with a mean score of 8.73, maintenance approach/strategy with a mean score of 8.67, computerised MM system and users requirement both with a mean score of 8.66, as well as continuous improvement with a mean score of 8.40. This was found to be consistence with the view of Wardhaugh (2004), Söderholm (2005), Hull, Jackson, and Dick (2005), Omar et al. (2016), and Alzaben (2015), whose earlier studies found these factors as key factors to the effective MM of municipal buildings.

In addition, five (5) factors also achieved a high impact (HI: 7.00–8.99) among the factors for effective MM of municipal buildings. These factors were spare part management with a mean value of 8.38, education and training with a mean score of 8.35, outsourcing strategy with a mean score of 8.33, task planning and schedule with a mean value of 8.25, user satisfaction with a mean value of 8.14 and as well as the motivation of maintenance personnel with a mean value of 7.90. They were found to be in alignment with the view of Kaplan and Norton (1996), Wardhaugh (2004), Omar et al. (2006), and Alzaben (2015), whose earlier studies also found these factors as key factors to the effective MM of buildings. Moreover, all the key factors identified were found to have had an influence on the effective MM of municipal buildings.

Likewise, the IQD values revealed a strong consensus in the views of the expert panellists as all the IQD values for all the eighteen (18) key factors regarding the effective MM of municipal buildings were at most one (1). Thus, the standard deviation values affirmed a high degree of consistency and non-variability in the views of the experts as respective standard deviation values recorded were less than one (1).

Objective DSO3

The third objective of the Delphi survey was to identify the measuring indicators (outcomes) of effective MM and to establish the relative influence of the indicators on the MM of municipal buildings in the South African education sector. The result of the Delphi study suggested 24 outcomes of effective MM of buildings. The outcomes of effective MM of buildings were maintenance budget is adequately planned for, maintenance organisation structure is effective, strategy for maintenance is better selected, policies for maintenance is better enacted, maintenance operations is planned appropriately, and MIS is effectively used. The outcomes further include maintenance work complying with standard expected, change in attitude to maintenance operations, replaceable material use for maintenance is of quality, technical support to maintenance is adequate, maintenance work is well supervised, maintenance work is better monitored, maintenance procedure is better understood, users' attitude to maintained building is improved, and users' maintenance need and expectation is met. Similarly, it also includes a delay to users' maintenance request is avoided, users' performance is enhanced, users' productivity is enhanced, users' well-being is improved, buildings life cycle performed optimally, deterioration within the buildings is avoided, buildings and its support components run effectively, high output in the use of assets is provided and improved and sustainable building performance is achieved.

The outcomes of the effective MM of municipal buildings in the South African education sector were consistent with existing literature as they shared much resemblance with the existing body of knowledge. Maintenance work is well supervised, with a mean value of 8.34 ranked first among the outcomes. Accordingly, Amaratunga, Baldry, and Sarshar (2000), Amaratunga (2001), Atkin and Brooks (2000), Márquez (2007), and Price and Pitt (2012) state that effective MM planning helps the maintenance process to be well supervised in order to obtain maximum returns for the investment on the building maintained. The maintenance organisation structure is effective ranked 2nd with a mean value of 8.94. This affirms the views of Cholasuke, Bhardwa, and

Antony (2004), Karia et al. (2014), and Pinjala et al. (2006) that effective MM guides a maintenance organisation to achieve an effective operational maintenance structure and meet their organisational maintenance policies and set goals. The 3rd ranked outcome was the change in attitude to maintenance operations with a mean value of 8.93, and users' well-being is improved with a mean value of 8.93. This supports the assertion of Eti, Ogaji, and Probert (2006) that effective MM brings to bear the adoption of the attitude of ensuring regular servicing, repairs, and maintenance of working assets. Also, it supports the suggestion of Atkin and Brooks (2000) that effective MM will enhance individuals' well-being, especially users of the maintained buildings. The 5th ranked outcome was improved, and sustainable building performance is achieved with a mean value of 8.86. This aligns with Swanson (2001) and Elsevier (2008) that effective MM is the significant driving force to improved and sustainable building performance. The 6th ranked outcome was users' performance is enhanced with a mean value of 8.82. This is consistent with the argument by Barrett and Baldry (2003), Zimring and Rashidi (2008), and Joe (2009) that effective MM is a positive influence that enhances users' performance and productivity.

The 7th outcome of effective MM was the building's life cycle performed optimally with a mean score of 8.79. This supports the assertion of Lateef (2009) and Ogunbayo et al. (2022) that effective MM will ensure optimal performance of the building life cycle. The 8th ranked outcome was deterioration within the buildings is avoided, with a mean value of 8.74. This is in alignment with Chew, Tan, and Kang (2004), Sherwin (2000), Too (2012), and Olanrewaju, Fang, and Tan (2018) that one important output expected from effective MM is that it delays and avoids decay and deterioration within the building maintained. The 9th ranked outcome of effective MM was the maintenance budget is adequately planned for with a mean value of 8.73, and the maintenance procedure is better understood with a mean value of 8.73. This agrees with Yaacob (2006) that effective MM guides towards adequate maintenance budgeting planning and cost control. Also, Visser (2002) affirmed that effective MM is a positive force for a better understanding of the procedure suitable for maintenance activities. The 11th ranked outcome of effective MM was maintenance work is better monitored with a mean score of 8.67, and users' productivity is enhanced with a mean score of 8.67. This affirmed the findings of Manaf and Alias (2005) and Ahzahar et al. (2011) that effective MM helps the maintenance managers better monitor maintenance work towards identifying gaps between present building performances and expected performance that could guide the organisation to prospect for maintenance process improvement. Likewise, it also supported the assertion of Zimring (2001), Barrett and Baldry (2003), Joe (2009), and Bateman et al. (2009) that effective MM will lead to achieving a suitable building performance and boost users' productivity. The 13th ranked outcome, maintenance operations, is planned appropriately, with a mean score of 8.60. This affirms the view of Hauer, Bombach, Mohr, and Masse

(2000) and Olatubara and Adegoke (2007) that effective MM will lead to better outcomes for preventive maintenance operations. The 14th ranked outcome of effective MM was the strategy for maintenance is better selected with a mean score of 8.59. This is in alignment with Márquez (2007) and Lee and Scott (2009) that effective MM will help in the formulation of maintenance policy through a better-selected maintenance strategy that will follow standard organisational planning methods.

The 15th ranked outcome was maintenance work complying with the standard expected with a mean score of 8.58. This agrees with Yacob (2006) and Ogunbayo and Aigbavboa (2019) that effective MM will provide the standard expected for maintenance operations. The 16th ranked outcome was replaceable material use for maintenance is of quality with a mean score of 8.53. The assertions align with the suggestion of Cholasuke et al. (2004) and Ogunbayo and Aigbavboa (2019) that effective MM will also encourage the use of high-quality materials and components as good workmanship. The 17th ranked outcome was the delay to users' maintenance request is avoided with a mean score of 8.51. This supports the view of Lateef (2009) and Okolie (2011) that effective MM helps to prevent delays in response to users' maintenance requests (effectiveness) with fewer resources (efficiency). The 18th ranked outcome of effective MM was high output in the use of assets is, provided with a mean score of 8.44. This supports the view of Amaratunga et al. (2000), and Atkin and Brooks (2000) that effective MM will lead to high output in the use of assets is maintained. The 19th ranked outcomes were buildings, and their support components run effectively with a mean score of 8.40, and users' attitude to the maintained building is improved with a mean score of 8.40. This aligns with Seeley (1987) and Ogunbayo et al. (2022) that effective MM will ensure that buildings run efficiently towards supporting the delivery of a wide range of services within the maintained building. It is in line with Heitor (2005), and Okolie (2011) that effective MM of buildings and their auxiliary services will improve users' attitudes and performance.

The 21st ranked outcome was technical support to maintenance is adequate, with a mean score of 8.34. This supports the view of Lateef (2009) that effective MM will help provide technical support to the functional requirement of the maintained building and available space. Policies for maintenance are better enacted with a mean score of 8.27, and users' maintenance needs and expectations are met with a mean value of 8.27, ranked 22nd among the outcomes of effective MM. This affirms the findings of Lee and Scott (2009) that effective MM will lead to the formulation of maintenance policies that will guide maintenance managers in preparing maintenance programmes and strategy choices. The 24th ranked outcome was MIS is effectively used with a mean value of 8.26. This agrees with Trappey Sun, Trappey, and Ma (2009) and Ko (2009) that effective MM will lead to the effective use of MISs in the maintenance process.

In conclusion, this study's outcomes of effective MM are similar to those of previous MM studies. However, the relative influence of each outcome differed even though the expert panellists ranked all the twenty-four (24) outcomes to have influenced the MM of municipal buildings in the South African education sector. From the above, many factors considered important in determining the effective MM of buildings have been identified and amplified by the Delphi study. The factors considered to be paramount determinants of the effective MM of municipal buildings include organisational maintenance policy, maintenance budget factors, human resources management, training factors, monitoring and supervision, MIS, communication among stakeholders, and maintenance culture. These factors have been collectively considered for the development of a holistic MM model in this study. Six of the factors have been previously considered in the development of the MM model in other cultural contexts. However, none of the existing models to date have included both communications among stakeholders and maintenance culture as inclusive factors to develop a model to guide maintenance managers of municipal buildings to maintain their buildings in the South African education sector effectively.

Summary

This chapter presented a summary of findings and a discussion of the results from the Delphi study. Each element was computed for the influence and impact of the attributes that determine the effective MM of municipal buildings in the South African education sector. Additionally, issues relating to key factors to effective MM buildings, among others, were addressed in this chapter. The chapter concluded with a summative discussion of the findings based on the objectives of the Delphi study. The findings from the expert participants revealed a coherent dialogue on the attributes of MM municipal buildings in the South African education sector, with the consensus being reached in most cases and others with a discrete conclusion. The result of the Delphi study assisted in the determination of key factors and constructs that are of critical significance (influence) to determine the MM of municipal buildings in the South African education sector, and which led to the development of the holistically integrated conceptual MM framework for municipal buildings in the South African education sector. The evaluation of these factors and their interrelationships are presented in the next chapter.

References

Ahzahar, N., Karim, N. A., Hassan, S. H., & Eman, J. (2011). A study of contribution factors to building failures and defects in construction industry. *Procedia Engineering, 20*, 249–255.

Aigbavboa, C. O. (2014). *An integrated beneficiary centred satisfaction model for publicly funded housing schemes in South Africa*. Published doctoral dissertation. Johannesburg: University of Johannesburg.

Alzaben, H. (2015). *Development of a MM framework to facilitate the delivery of healthcare provisions in the Kingdom of Saudia Arabia.* Published doctoral dissertation. Nottingham Trent University.

Amaratunga, D., Baldry, D., & * Sarshar, M. (2000). Assessment of facilities management performance – what next? *Facilities, 18*(1/2), 66–75.

Amaratunga, R. G. (2001). *Theory building in facilities management performance measurement: Application of some core performance measurement and management principles.* Published doctoral dissertation. University of Salford.

Atkin, B., & Brooks, A. (2000). *Total facilities management.* London: Blackwell Science.

Barrett, P., & Baldry, D. (2003). *Facilities management: Towards best practice.* Oxford: Blackwell.

Bateman, T. S., & Snell, S. (2009). *Administración: Liderazgo y colaboración en un mundo competitivo* No. Sirsi) a 458252.

Bothma, B., & Cloete C. (2000). A facilities management system for the maintenance of government hospitals in South Africa. *Acta Structilia, 7*(1&2), 1–21.

Campbell, J. D., & Jardine, A. K. (2001). *Maintenance excellence: Optimizing equipment life-cycle decisions.* New York: Marcel Dekker Inc.

Campbell, J. D. (1998). *Uptime, strategies for excellence in maintenance management.* Portland, OR: Productivity Press.

Chew, M. Y. L., Tan, S. S., & Kang, K. H. (2004). Building maintainability—Review of state of the art. *Journal of Architectural Engineering, 10*(3), 80–87.

Cholasuke, C., Bhardwa, R., & Antony, J. (2004). The status of maintenance management in UK manufacturing organisations: Results from a pilot survey. *Journal of Quality in Maintenance Engineering, 10*(1), 5–15.

Crespo Márquez, A. Moreu de León P., Gómez Fernández J. F., Parra Márquez C., & González V. (2009). *The maintenance management framework: A practical view to maintenance management, Safety, Reliability and Risk Analysis: Theory, Methods, and Application–Martorell et. al., 2009.* Taylor & Francis Group.

Crespo Márquez, A., Moreu de León, P., Gómez Fernández, J. F., Parra Márquez, C., & López Campos, M. (2009). The maintenance management framework: A practical view to maintenance management. *Journal of Quality in Maintenance Engineering, 15*(2), 167–178.

Elsevier, B. (2008). User driven innovation in building process. *Tsinghua Science and Technology, 13*(1), 248–254.

Eti, M. C., Ogaji, S. O. T., & Probert, S. D. (2006). Development and implementation of preventive-maintenance practices in Nigerian industries. *Applied Energy, 83*(10), 1163–1179.

Hauer, J., Bombach, V., Mohr, C., & Masse, A. (2000). *Preventive maintenance for local government buildings: A best practices review.* USA: ERIC Clearinghouse.

Heitor, T. (2005). Potential problems and challenges in defining international design principlesfor school. *Evaluating Quality in Educational Facilities, 48,* 44–54.

Hull, E., Jackson, K., & Dick, J. (2005). *Requirements engineering in the solution domain.* London: Springer. 109–129.

Joe, C. T. (2009). CLP experience on condition monitoring and condition-based maintenance. In *8th International Conference on Advances in Power System Control, Operation, and Management,* London, UK, 8–11 November 2009, pp.36–43.

Kaplan, R. S., & Norton, D. P. (1996). *The balanced score card.* Massachusetts, Boston: Harvard Business School Press.

Karia, N., Asaari, M. H. A. H., & Saleh, H. (2014). Exploring maintenance management in service sector: A case study. In Proceedings of *International Conference on Industrial Engineering and Operation Management*, Bali, 7–9 January 2014, pp. 3119–3128.

Ko, C. H. (2009). RFID-based building maintenance system. *Automation in Construction, 18*(3), 275–284.

Lateef, O. A. (2009). Building maintenance management in Malaysia. *Journal of Building Appraisal, 4*(3), 207–214.

Lee H. H. Y., & Scott D. (2009). Overview of maintenance strategy, acceptable maintenance standard and resources from a building maintenance operation perspective. *Journal of Building Appraisal, 4*(4), 269–278.

Manaf, Z., & Alias, A. (2005). Training needs in facilities management a study among local authority offices in Malaysia. In *International Real Estate Research Symposium (IRERS)*, Kuala Lumpur, Malaysia, 11–13 April, pp.1–14.

Márquez, A. C. (2007). *The maintenance management framework: Models and methods for complex systems maintenance*. London: Springer-Verlag.

Ogunbayo, B. F., & Aigbavboa, O. C. (2019). Maintenance requirements of students' residential facility in higher educational institution (HEI) in Nigeria. In *Proceedings of 1st International Conference on Sustainable Infrastructural Development*, Ota, 24–28 June 2019, IOP Conference Series: Materials Science and Engineering, 640, (1) 012014. IOP Publishing.

Ogunbayo, B. F., Aigbavboa, C. O., Thwala, W., Akinradewo, O., Ikuabe, M., & Adekunle, S. A. (2022). Review of culture in maintenance management of public buildings in developing countries. *Buildings, 12*(5), 677.

Ogunbayo, B. F., Ohis Aigbavboa, C., Thwala, W. D., & Akinradewo, O. I. (2022). Assessing maintenance budget elements for building maintenance management in Nigerian built environment: A Delphi study. *Built Environment Project and Asset Management, 12*(4), 649–666.

Okolie, K. C. (2011). *Performance evaluation of buildings in Educational Institutions: A case of Universities in South-East Nigeria*. Published doctoral dissertation. Port Elizabeth, South Africa: Nelson Mandela Metropolitan University.

Olanrewaju, A., Fang, W. W., & Tan, Y. S. (2018). Hospital building maintenance management model. *International Journal of Engineering and Technology, 2*(29), 747–753.

Olatubara, C. O., & Adegoke, S. A. O. (2007). Housing maintenance. In *Housing Development and Management. A book of readings* (pp. 391–318). Nigeria: Department of Urban and Regional Planning, Faculty of Social sciences, University of Ibadan.

Omar, M. F., Ibrahim, F. A., & Omar, W. M. S. W. (2017). Key Performance Indicators for Maintenance Management Effectiveness of Public Hospital Building. In *MATEC Web of Conferences*, Ho Chi Minh City, Vietnam, 5–6 August 2016, pp 01056. (Vol.7). EDP Sciences.

Omar, M. F., Ibrahim, F. A., & Omar, W. M. S. W. (2016). An assessment of the maintenance management effectiveness of p ublic hospital building through key performance indicators. *Sains Humanika, 8*(4-2), 51–56.

Pinjala, S. K., Pintelon, L., & Vereecke, A. (2006). An empirical investigation of the relationship between business and maintenance strategies. *International Journal of Production Economics, 104*(1), 214–229.

Pintelon, L. M., & Gelders, L. F. (1992). Maintenance management decision making. *European Journal of Operational Research, 58*(3), 301–317.

Price, S., & Pitt, M. (2012). The influence of facilities and environmental values on recycling in an office environment. *Indoor and Built Environment, 21*(5), 622–632.

Rubin, H. R., Pronovost, P., & Diette, G. B. (2001). Methodology matters. From a process of care to a measure: The development and testing of a quality indicator. *International Journal for Quality in Health Care, 13*(6), 489–496.

Seeley, I. H. (1987). *Building maintenance.* London: Macmillan Publishers Limited.

Sherwin, D. (2000). A review of overall models for maintenance management. *Journal of Quality in Maintenance Engineering, 6*(3), 138–164.

Söderholm, P. (2005). *Maintenance and continuous improvement of complex systems: Linking stakeholder requirements to the use of built-in test systems.* Published doctoral dissertation, Luleå tekniska universitet.

Stitt-Gohdes, W., &. Crews, T. B. (2004). The Delphi technique: A research strategy for career and technical education. *Journal of Career and Technical Education, 20*(2), 55–67.

Suwaibatul Islamiah, A. S., Abdul Hakim, M., Syazwina, F. A. S., & Eizzatul, A. S. (2012). An overview development of maintenance culture. In *Proceedings of 3rd International Conference on Business and Economic Research.* Bandung Indonesia. 12-13 March 2012. 2206–2217.

Swanson, L. (2001). Linking maintenance strategies to performance. *International Journal of Production Economics, 70*(3), 237–244.

Too, E. (2012). Infrastructure asset: Developing maintenance management capability. *Facilities, 30*(5/6), 234–253.

Trappey, A. J., Sun, Y., Trappey, C. V., & Ma, L. (2011). Re-engineering transformer maintenance processes to improve customized service delivery. *Journal of Systems Science and Systems Engineering, 20*(3), 323.

Visser, J. K. (2002). Maintenance management – A neglected dimension of engineering management. In *Proceedings IEEE Africon, Vols 1 and 2: Electro technological Services for Africa, IEEE,* New York, 2–4 October 2002. pp. 479–484.

Wardhaugh, J. (2004). Useful key performance indicators for maintenance. In *Singapore IQPC Reliability and Maintenance Congress.* Singapore, 2004, Lifetime Reliability Maintenance – the best Practices, www.lifetime-reliability.com

Wilkinson, A., Johnstone, S., & Townsend, K. (2012). Changing patterns of human resource management in construction. *Construction Management and Economics, 30*(7), 507–512.

Wireman, T. (2005). *Developing performance indicators for managing maintenance.* New York: Industrial Press Inc.

Yacob, S. (2006). *Maintenance management system through strategic planning for public school in Malaysia* (Doctoral dissertation, Tesis Master: Universiti Teknologi Malaysia, Skudai).

Zimring, C. (2001). Post-Occupancy Evaluation and Organizational Learning. In Counci, F. F. (Ed). Learning from Our Buildings: A State-of-the-Practice Summary of Post-Occupancy Evaluation. Washington: National Academy Press.

Zimring, C., & Rashidi, M. (2008). *Facility Performance Evaluation* [Online]. Available from www.wbdg.org/Resources/Fpe.Php. Accessed 7, October 2020.

11 An Integrated Maintenance Management Conceptual Framework for Municipal Buildings in the Developing Economies

Introduction

This chapter presents the discussion of the literature review findings as well as the Delphi study findings. This discussion forms the basis of the conceptual framework's theory. The hypothesised integrated, holistic maintenance management (MM) framework is also presented in this chapter based on an in-depth review of the previous models as presented earlier in this book. This chapter also describes the integrated, holistic MM framework and the variables of the framework in detail, except the communication among stakeholders and maintenance culture, which were discussed in this book as variable constructs identified as gaps in MM research. Furthermore, this chapter presents the framework identification and justification for the selected variables, and a conclusion is drawn for the chapter thereafter.

Selection of Variables for Maintenance Management

Most MM frameworks have combined objectives and subjective attributes to assess the MM of buildings. Accordingly, Campbell (1998) identified four key factors for effective MM. They were the development of strategies for assets, maintenance performance measures, value-driven maintenance, implementation of a maintenance function measurement system, and rivalry as exogenous factors (Campbell, 1998; Wireman, 2005; Marquez & Gupta, 2006). These four key constructs were adapted in this current study. Moreover, Campbell (1998) and Wireman (2005) identified attributes of MM to include maintenance objective, monitoring and supervision, MIS, maintenance strategy, task planning, and scheduling, organisational maintenance policy, continuous improvement, human resources management, and maintenance budget. The attributes identified by Campbell (1998) and Wireman (2005) share resemblances with those of Marques et al. (2006). The common attributes were organisational maintenance policy, maintenance budget, human resources management, training, continuous improvement, planning and scheduling, maintenance techniques/approach, monitoring and supervision, and MIS.

DOI: 10.1201/9781003344681-11

In related studies, Pintelon and Gelders (1992) found the attributes of MM to include organisational task planning, monitoring and supervision, MIS, maintenance budget, and human resources management. However, Márquez et al. (2009) opined that the attributes of MM comprise four main variables: effectiveness, efficiency, assessment, and continuous improvement. The core elements in the model include organisational maintenance policy, maintenance objective, monitory and supervision, maintenance budget, human resources management, and continuous improvement. Moreover, Takata et al. (2004) posit that the key attributes of MM include planning and scheduling, organisational maintenance strategy, monitory and supervision, human resources management, and maintenance budget. Similarly, Márquez (2007) posited that effective MM could be achieved through a maintenance strategy-setting process that should follow standard organisational planning methods based on the elements that include organisational maintenance strategy, task and planning schedule, monitory and supervision, maintenance objective, organisational maintenance policy, continuous improvement, education and training, maintenance budget, and human resources management. Likewise, Fernández and Márquez (2009) theorised that effective MM would reduce maintenance and other costs through attributes that include maintenance objective, task planning, organisational policy development, maintenance strategy, approach to maintenance, maintenance budget, and user satisfaction.

Furthermore, Lynch and Cross (1995) assert that maintenance attributes should include maintenance strategy, organisational policy, human resources management, maintenance budget, training, monitoring and supervision, and user satisfaction. Vanneste and Van Wassenhove (1995) also theorised that the attributes of MM effectiveness and efficiency include maintenance planning and scheduling, monitoring and supervision, organisational maintenance policy, continuous improvement, maintenance budget, training, and MIS. Equally, Hassanain, Froese, and Vanier (2001) found attributes of MM to include task planning and scheduling, human resources management, organisational maintenance policy, monitory and supervision, and a maintenance budget. Additionally, Sharma (2013) theorised that effective MM results from organisation strategy, productivity, and maintenance performance management. This can be achieved through MM attributes that include maintenance planning, outsourcing strategy, human resources management, maintenance budget, maintenance approach, organisational maintenance policy, continuous improvement, training, and education.

Hence, common attributes of MM espoused in the literature include organisational maintenance policy, maintenance budget factors, human resources management, training factors, monitoring and supervision, and MIS. These attributes were adopted in this study. Thus, the present study considers the MM of municipal buildings in the South African education sector to contain organisational maintenance policy with 15 variables, maintenance budget factors with 16 variables, human resources management with

18 variables, training factors with 14 variables, monitoring and supervision with 14 variables, and MIS with 16 variables. All these are the constructs that have frequently been conceptualised in most MM studies. Nonetheless, this present thesis brings into focus the impact of communication among stakeholders with 23 variables and maintenance culture factors with 18 variables. The two additions are the gaps identified from the literature review, which were found peculiar to situations in developing countries.

The next section of this chapter gives a detailed explanation of the six different constructs influencing MM. The explanation for the two newly added constructs (communication among stakeholders and maintenance culture) has been discussed comprehensively.

Organisational Maintenance Policy (OPY)

An organisational maintenance policy is one of the basic and integral parts of the MM function towards achieving set maintenance goals and objectives (Karia, Asaari, & Saleh., 2014). Ali and Mohamad (2009) opine that the policy contains a written plan, policies, and procedures which describe how the organisation will manage each specific component of buildings and their auxiliary facilities and services. Management of a maintenance organisation with an organisational maintenance policy will be able to effectively manage resources such as hardware, material selection, usage, assets, capital, and personnel, among others (Cholasuke, Bhardwa, & Antony, 2004). As noted by Lind and Muyingo (2012), once an organisational maintenance policy is in place in a maintenance organisation, it guides the management to ensure that maintenance task is executed based on the laid down maintenance policies towards achieving the set maintenance goals and objectives of the organisation. This is evident in some previous MM studies, especially the study by Vanneste and Van Wassenhove (1995) and Campbell (1998), which have formed the basis of most MM studies, even in present times. Accordingly, Vanneste and Van Wassenhove (1995) informed that the organisational maintenance policy is essential for any maintenance organisation desiring to have an effective MM system. The organisational maintenance policy influences its maintenance operations towards achieving set maintenance goals and objectives (Campbell, 1998). Over the years, some researchers have used some variables to measure organisational maintenance policies in some industries. For instance, Lind and Muyingo (2012) identified a clear mission, an implementation strategy for setting goals and objectives, corporate culture, job clarifying, delegation of authority, and chain of command as variables for measuring an organisational maintenance policy. On the other hand, roles and responsibilities, procedures, strategy, process, line of operational information, supported personnel training, optimising techniques, motivation, and organisation culture statement were identified by Ali and Mohamad Nasbi Bin Wan Mohamad (2009) as specified variables in an organisational maintenance policy. Albert and Brownsword (2002) also identified maintenance operations options, structuring of the maintenance function and tasks of the organisation, identifying equipment that supports the maintenance process, and understanding the procedure suitable for maintenance as variables in measuring organisational maintenance policy.

Thus, in this current study, organisational maintenance policy refers to policy, planning, and procedures that contribute to the effective MM system of a maintenance organisation. They include the suitable maintenance procedures and process, organisation operational efficiency, optimisation of the maintenance policy, preparation of safety procedure, optimisation of the maintenance action plan, and meeting maintenance objectives, as summarised in Table 11.1. These specific variables have already been considered in previous studies in measuring organisational maintenance policies in MM studies, except that the combination of the variables varied from one study to another. Accordingly, in this present study, organisational maintenance policy variables that have been hypothesised for the development of a holistic MM framework have been summarised in Table 11.1.

Table 11.1 Conceptual model indicator constructs

Latent variable constructs	Measurement variables
Organisational maintenance policy (OPY)	
	Suitable maintenance procedures and process
	Organisation operational efficiency
	Optimisation of the maintenance policy
	Preparation of safety procedure
	Optimisation of the maintenance action plan
	Meeting maintenance objectives
	Optimisation of preventive maintenance design
	Preparation of safety procedure
	Well-defined priority system
	Assembling of maintenance organisation structure
	Maintenance strategies development
	Analysis of maintenance procedures
	Change in policy and its associated results
	Risk factor establishment
	Reduction in mean time to repair
Maintenance budget factors (MBF)	
	Corruption-free maintenance process
	Cost implication of maintained asset
	Audit of maintenance operational cost
	Cash flow indexing
	Valuation of maintenance operation budget
	Maintenance budget implementation
	Yearly maintenance budgets certainties
	Prioritisation of maintenance financing

(*Continued*)

Table 11.1 (Continued)

Latent variable constructs	Measurement variables
	Incorporation of financial indicators
	Maintenance financial plan
	Market and financial terms of operations
	Optimisation of finance outsourcing
	Reduction in maintenance expenditure
	Maintenance operation financing
	Maintenance funding
	Optimising maintenance resources
Human resources management factors (HRM)	
	End-of-job documentation
	Recruitment of skilful and experienced personnel
	Interdepartmental conflicts resolution
	Developing the maintenance leader
	Resources management
	Personnel self-training encouragement
	Assessment of personnel performance
	Prioritisation of required resources
	Resources assignment
	Keeping records of training courses
	Career path for maintenance personnel
	Roles for maintenance personnel
	Assigning personnel for specific tasks
	Counselling arrangements for personnel under stress
	Recognition of maintenance personnel effort
	Maintenance personnel Identification
	Institutional and in-house training for personnel
	Opportunities for growth
Training factor (TF)	
	Integration of new techniques
	Education on current maintenance knowledge
	Usage of proper tools
	Usage of proper procedures
	Understanding organisation maintenance policy
	Semi-skilled personnel can do better
	Safety of personnel
	Personnel training on maintenance management
	Training of personnel on maintenance skills
	Time management
	Maintenance problem-solving skills development
	On-job training on maintenance
	Reduction in the scarcity of needed skills
	Reliability of personnel

(*Continued*)

Table 11.1 (Continued)

Latent variable constructs	Measurement variables
Monitoring and supervision (MSN)	
	Maintenance performance reporting
	Zero-error tolerance
	Monitoring of safety procedure
	Maintenance operation inspection
	Value improvement
	Material wastage reduction
	Monitoring of maintenance operations
	Periodic maintenance planning
	Meeting maintenance target
	Maintenance process monitoring
	Risk reduction
	Maintenance personnel involvement
	Suitable replacement material usage
	Appropriate maintenance strategy
Maintenance information system (MIS)	
	Online maintenance monitoring systems
	Maintenance tools inventory
	Early maintenance warning guide
	Data collection
	Failure rates prediction
	Data monitoring
	Analysis of maintenance process
	Maintenance descriptions
	Implementation action on data collected
	Balanced maintenance costs
	Implementation action on data collected
	Tracking maintenance performance indicators
	Integration of operation process
	Integration of management tools
	Data processing
	Maintenance activities record keeping
	Strategy for material cost
Communication among stakeholders (CAS)	
	Users' satisfaction
	Quality-driven service delivery
	Collaboration among stakeholders
	Identify gaps in maintenance process
	Early defects detection
	E-mails among stakeholders
	Users feedback
	Increase in users' participation
	Eradication of uncertainty
	Understanding of users' maintenance needs
	Addressing stakeholders' concerns

(*Continued*)

Table 11.1 (Continued)

Latent variable constructs	Measurement variables
	Good working relationship among stakeholders
	Use of visualisation technique
	Understanding users' expectations
	Timely and accurate feedback
	Help desk
	Liaison with relevant government agencies
	Checking maintenance information with users
	Establishment of the need for change
	Web-based systems meetings and report
	Stakeholders' participative actions
	Liaison with maintenance institutions
	Telephone support
Maintenance culture factors (MCF)	
	Effective maintenance management style
	Effective decision-making capacity
	User change of attitude to maintained buildings
	Attitudinal change to maintenance operations
	Sustainability of infrastructure
	Use of right skills maintenance personnel
	Diligence in maintenance operations
	Enhance performance level of buildings
	Benchmarking for maintenance performance
	Attitude of ensuring regular repairs
	Proper maintenance system
	Improve economic value of a country
	Commitment to change initiatives
	Building reliability
	Attitude of ensuring regular servicing
	Motivation of maintenance personnel
	Maintenance decision based on societal value
	Maximal utilisation of infrastructural

Maintenance Budget Factors (MBF)

According to Lee and Scott (2009), among the attributes and resources influencing the MM process of an organisation are the maintenance budget factors. This view is affirmed by the views of some researchers, including Alexander (1996), Wireman (2005), Yacob (2005), and Cloete (2014). Similarly, a study was carried out by Alzaben (2015), which aimed at developing a MM framework to facilitate the delivery of healthcare provisions in the Kingdom of Saudi Arabia. The researcher applied the Campbell and Wireman models, among other validated models. The study analysis was mainly based on questionnaire surveys designed and distributed to the staff, including contractors in the maintenance department within the study area. The study concluded that the attributes accounting for the development of a MM framework to facilitate the delivery of healthcare provisions in the Kingdom of Saudi Arabia include the maintenance budget factors (Alzaben, 2015). This affirms that the influence of a maintenance budget on MM has been ascertained in previous studies, including those of Sherwin (2000), Parida and Kumar (2009), Van Horenbeek et al. (2010), and Mekasha (2018) in some national contexts. However, it must be admitted that specific variables that constituted maintenance budget factors and the relative influence of each of the variables differed from one national context to the other. Thus, this study seeks to develop the inclusion of maintenance budget factors as one of the main constructs of the MM model to ascertain its relative contribution to other constructs in the South African education sector.

Maintenance budget factors are key factors of progression that influence the financial planning of maintenance operations in a maintenance organisation (Wireman, 2005; Ahzahar et al., 2011). Therefore, Wireman (2005) identified variables for measuring the maintenance budget factors to include cost control for labour, cost of monitoring the contractor, market and financial terms of operations, prioritisation of maintenance financing, valuation of a maintenance operation budget, maintenance operation financing, optimising maintenance resources as well as maintenance budget implementation. Subsequently, Omar et al. (2017) further maintained that the maintenance budget factors include financial planning, an annual budget, the achievement of expenses, analysis of damage cost, maintenance operation system improvement, a corruption-free maintenance process, maintenance materials assessment, maintenance funding, optimisation of business profitability and the incorporation of financial indicators. According to Ahzahar et al. (2011), the maintenance budget factors include a financial maintenance plan, an annual maintenance budget, reduction in an organisation's maintenance expenditure, better asset replacement planning, certainties in yearly maintenance budgets, operational maintenance auditing, and the optimisation of maintenance financing outsourcing. Marquez and Gupta (2006) further argued that among the specific variables for measuring maintenance budget factors were the cost of lost production, labour cost, cost of spare parts, cost of providing information systems, cost of human resources to support the programme, and equipment/line/plant production loss cost. Mekasha (2018) identified the cost of an asset for its entire life span, failure rate cost, cost of spares, personnel cost, repair times, and components costs as the variables for maintenance budget factors.

Thus, it is obvious from the previous studies that no single study is exhaustive of all the maintenance factors of MM. Moreover, there is much resemblance and commonness in the specific variables identified in previous studies for measuring maintenance factors. Hence, this study synthesised the specific variables and adapted them in measuring maintenance budget factors in relation to the MM of municipal buildings in the South African education sector, thereby having more comprehensive specific variables for measuring maintenance budget factors. Moreover, in this current study, maintenance budget factors refer to the elements of MM that deal with financial planning for maintenance operations and execution within a maintenance organisation. These include the corruption-free maintenance process, the cost implication of maintained assets, audit of maintenance operational cost, cash flow indexing, valuation of maintenance operation budget, and, among others, as summarised in Table 11.1.

Human Resources Management (HRM)

According to Amaratunga et al. (2000), a maintenance organisation is more proficient when there are suitable and efficient human resources that support the personnel towards the effective MM through the best combination of cost, efficiency, and quality of the organisation. This view is affirmed by Campbell (1998) and Sharma (2013) in their respective MM studies. Atkin and Brooks (2000), Amaratunga (2001), and Omar et al. (2017) affirmed human resources factors as variables that influence the MM though as sub-variables for measuring organisation facilities. Human resources management involves the management of the organisation's resources and the processes it supports, operating within the realm of available resources based on the organisational corporate culture. The human resources factors sub-variables vary from one organisation or industry to another, though there are instances when the criteria are invariably not different.

Previous studies have identified variables for measuring human resources management in some national contexts. According to Pintelon et al. (1990), specific variables in measuring human resources include maintenance cost/budget, equipment performance, personnel performance, materials management, work order control, end-of-job documentation, equipment availability provision, leadership responsibilities, and the recruitment of experienced personnel. Moreover, Wireman (2005) identified staffing, identification of maintenance personnel, training for personnel, encouragement of creativity among personnel, personnel task outcome evaluation, assigning tasks to personnel, and record-keeping of personnel activities. Campbell (1998) identified maintenance strategies for the asset, resources management, resources assignment, incentive for good effort, and career path for maintenance personnel. Omar et al. (2017) identified human resources management, staff competency, attending training, job scope, duty list, working spirit, staff performance, a total balanced staff, and continuous training programme.

Therefore, since in this present study, the human resources factors are organisationally driven, it only considered the specific variables that would be relevant to the South African education sector in measuring human resources factors. Also, the relative contribution of human resources factors amid other constructs was ascertained. Thus, in this study, human resources factors refer to the element of MM that deals with resources management and planning of maintenance operations in order to meet the optimum performance of the buildings being maintained. It includes end-of-job documentation, recruitment of skilful and experienced personnel, inter-departmental conflict resolution, developing the maintenance leader, and resources management, among others. A comprehensive summary of the specific variables for measuring human resources management in this study is presented in Table 11.1.

Training Factors (TF)

It is evident in the existing MM literature that another important element of MM is the training of maintenance personnel (De Groote, 1995; Wireman, 2005; Noe, 2010; Brinia & Efstathiou, 2012; Fatoni & Nurcahyo, 2018). According to Schreiber (2007), with enhanced training among personnel, maintenance errors will be better managed during maintenance operations. The main goal of training the maintenance personnel is applying knowledge gained in work practice and equipping personnel with additional skills to enable them to undertake new tasks (Kempton, 1996). Nonetheless, as to specific variables that constitute training, researchers have expressed varied views depending on the nature of the maintenance operation and set orga-nisational maintenance objectives (Fatoni & Nurcahyo, 2018). Moreover, Nikandrou, Brinia, and Bereri (2009) assert that the specific variables that were of importance to training factors in the maintenance organisation will depend on the training objectives and the extent of training and training equipment, as well as the training methods and means.

According to Brinia and Efstathiou (2012), specific variables in measuring training factors include motivation to learn, motivation to transfer training, opportunity to use training, personal career goals, motivation from work content of training, organisational commitment to development, colleagues support, and superior support. Moreover, Wilson (2005) identified person-nel's reliability, safety, organisational objectives' priorities, operational ob-jectives' improvement, and current maintenance knowledge. Fatoni and Nurcahyo (2018) identified the training of personnel on maintenance skills, training opportunities, on-job training on maintenance, and understanding organisational maintenance policies. Mekasha (2018) also states that better understanding of organisation policy, advancement in technology, usage of proper procedures, proper tools, personnel development, problem-solving skills development, time management, interpersonal skills development, and integration of new techniques all influence the MM of an organisation.

It could be deduced from the literature reviewed that researchers have a consensus that training factors influence MM (Vanneste & Van Wassenhove, 1995; Campbell, 1998; Wireman, 2005; Wilson, 2005; Nikandrou et al., 2009; Usanmaz, 2011; Mekasha, 2018). Nonetheless, there are differences in opinion as to specific variables that constituted training factors within maintenance organisations or countries. This could be attributed to the varying nature of policies in the various maintenance organisations or industries. The various studies focused on the training objectives and extent of training as well as training methods and means within the maintenance organisations and industries (Campbell, 1998; Nikandrou et al., 2009; Usanmaz, 2011). Thus, all the relevant factors that have been considered in the previous MM studies to have influenced training factors in other national contexts were also considered for this study. Thus, in this study, training factors refer to the elements of MM that deal with on-job knowledge improvement for maintenance personnel towards continuous improvement and maintenance task efficiency. It includes integration of new techniques, education on current maintenance knowledge, usage of proper tools, usage of proper procedures, and understanding of organisation maintenance policy, among others. A comprehensive summary of the specific variables for measuring training factors is presented in Table 11.1.

Monitoring and Supervision (MSN)

According to Duffuaa (2000) and Wordsworth and Lee (2001), monitoring and supervision are the strategic ways of assessing and ascertaining the performance of maintenance organisations towards implementing maintenance plans, objectives, policies, and procedures related to MM. This view is sustained by Hassanain et al (2001), Takata et al. (2004), and Márquez (2007). Moreover, Omar et al. (2017) advanced that the need for organisational maintenance success has compelled the establishment of monitoring and supervision in the maintenance organisations and industries supported by well-guided monitoring and supervision. Monitoring and supervision is an important aspect of maintenance operations that guides maintenance managers to identify gaps between present building performance and expected performance (Manaf & Alias, 2005; Ahzahar et al., 2011). Similarly, Jonsson (1997) stresses that the specific variables related to monitoring and supervision differ from organisation to organisation and from industry to industry. However, some previous studies have identified some variables for measuring monitoring and supervision in some national contexts.

According to Márquez et al. (2006), the specific variables in measuring monitoring and supervision include failure analysis, reliability analysis, risk analysis of the system's operation, design of the maintenance plan, employees' involvement in the maintenance process, continuous improvement, and the management of maintenance resources. Moreover, Omar et al. (2017) identified maintenance work complying with standard specifications and the internal audit international organisation for standardisation. Hassanain et al. (2001) identified suitable replacement material usage, constant rescheduling of maintenance activities, monitoring of maintenance operations, and monitoring safety and procedures. Takata et al. (2004) identified variables such as periodic maintenance planning, appropriate maintenance strategy, and observing faults trigger components. Additionally, Márquez et al. (2009) identified factors such as the diagnosis of maintenance tasks, meeting maintenance targets, material wastage reduction, maintenance personnel involvement, value improvement, and zero-error tolerance. Pintelon et al. (1992) identified factors such as the determination of facilities performance, maintenance operation inspection, maintenance process monitoring, and performance reporting.

Hence, since in this study, the monitoring and supervision factors are institutionally driven, it only considered the specific variables relevant to the South African education sector in measuring monitoring and supervision factors. Also, the relative contribution of monitoring and supervision factors amid other constructs was ascertained. Hence, in this study, monitoring and supervision refer to the element of MM that deals with coordinating maintenance activities through effective supervision and proper feedback towards achieving organisational set goals. It includes maintenance performance reporting, zero-error tolerance, monitoring of safety procedures, maintenance operation inspection, and value improvement, among others. A comprehensive summary of the specific variables for measuring monitoring and supervision factors in this study is presented in Table 11.1.

Maintenance Information System (MIS)

According to Marquez and Gupta (2006), maintenance information systems (MIS) guide maintenance managers and personnel to access maintenance equipment data. MISs within a maintenance organisation provide critical tools for three maintenance activities: effective information processing capability, the collaboration between expert systems, and communication tools and maintenance decision support (Marquez & Gupta, 2006). This view is affirmed by Yu et al. (2003), Garcia et al. (2006), Lee et al. (2006), Neelamkavil (2009), and Trappey et al. (2011). Additionally, Campbell (1998) and Wireman (2005) maintained that using MISs such as CMMS helps support communication and improve coordination among different functions in the maintenance organisation. In maintenance organisations, data collected can be transformed into information that will take greater decisions within the maintenance process and procedures used to prioritise maintenance actions (Lynch & Cross, 1995). Marquez and Gupta (2006) theorised that the provision of MISs would lead to proper monitoring and control of organisational assets, especially when the number of items to maintain is high. Various studies have identified some variables for measuring MIS in some national contexts.

Trappey et al. (2011) identified specific variables in measuring MISs, including checking specifications during building maintenance supervision, identifying and correcting faulty construction and methods, eliminating less durable materials and retaining the durable ones, evaluating maintenance running costs, and predicting a maintenance guide for the maintenance process. Moreover, Cato and Mobley (2002) identified invoices matching and accounts payable, human resources, purchasing of maintenance components, a PM plan, development, and scheduling and receiving, work order creation, scheduling, execution and completion, equipment/asset bill of materials creation and maintenance, equipment/asset and work order history and inventory control. Marquez and Gupta (2006) identified maintenance tools inventory, analysis of maintenance process, balanced maintenance costs, benefits information channel, failure rates prediction, maintenance performance indicators, tracking maintenance performance indicators, integration of the MIS and online maintenance monitoring systems. Pintelon and Gelders (1992) identified material cost strategies, a strategy for equipment usage, an early maintenance warning guide, integration of management tools, and operation processes. Vanneste and Van Wassenhove (1995) identified maintenance descriptions, data processing, data monitoring, maintenance activities record keeping, maintenance specifications detailing, data collection, and implementation action on data collected.

However, in this study, the MISs factors are institutional driven. It will only consider the specific variables relevant to the South African education sector in measuring MIS factors. Likewise, the relative contribution of MIS factors amid other constructs was ascertained. Thus, in this study, MISs refers to an effective tool that can report the analysis that relates to overall MM aspects accurately and more quickly compared to the usage of the manual technique. It also referred to CMMS tools used to control maintenance operations, inventories, material purchasing, and coordination of maintenance activities. It includes online maintenance monitoring systems, maintenance tools inventory, early maintenance warning guide, Data collection, and prediction of failure rates, among others. A comprehensive summary of the specific variables for measuring MIS factors in this study is presented in Table 11.1.

Model Specification and Justification

This research aimed to build a conceptual MM framework for the municipal buildings in the South African education section. The theoretical, conceptual framework for the current research builds on the work of Vanneste and Van Wassenhove (1995), Campbell (1998), and Wireman (2005) and their models of MM, as discussed in an earlier chapter in the book. The model of Vanneste and Van Wassenhove (1995) conceptualised that the MM structure should include two management processes for effectiveness (buildings) and efficiency (resources) of the maintenance processes. The two management processes include analysis of process effectiveness and analysis of process efficiency. As noted by Vanneste and Van Wassenhove (1995), the effective management process seeks to identify the most important problems in maintenance activities and their potential solutions, while the efficiency management process focuses more on identifying suitable procedures for maintenance operations.

Similarly, Campbell (1998) model also conceptualised that developing strategies for assets and human resources is necessary for an effective MM structure through implementing CMMS, MIS, and planning and scheduling for the maintenance process. Campbell (1998) model further conceptualised that eight different tactics should support MM for effective maintenance operations depending on the value these assets represent. These include redundancy, run to failure, scheduled overhaul, scheduled replacement, ad-hoc maintenance, preventive maintenance (use-based or age-based), condition-based maintenance, and redesign, if necessary. Likewise, the Wireman (2005) model also conceptualised that a preventive maintenance programme should be in place before advancing to the next level of maintenance activities. Wireman's (2005) model conceptualised further that before considering the implementation of RCM and predictive maintenance programmes, CMMS implementation is necessary with a suitable "work order release system" (to schedule appropriate prioritised tasks), the provision of spare parts, and the training of maintenance personnel (maintenance resources management system) that will be based on a preventive maintenance programme. For effectiveness and efficiency of MM structure, Wireman (2005) further conceptualised that there is a need for TPM towards operators' involvement and routinising the use of optimisation techniques and guides for configuring MIS and the application of statistical tools for financial optimisation.

Thus, the conceptual framework for this research is primarily based on the approach used by Vanneste and Van Wassenhove (1995) when they view MM as a criterion for an evaluation of maintenance while simultaneously as a criterion variable predicting MM. In this regard, MM was treated as a criterion and dependent variable. This approach which was also used by Pintelon and Gelders (1992), Lynch and Cross (1995), Hassanain et al (2001), Takata et al. (2004), Marquez and Gupta (2006), Márquez et al. (2009), Fernández and Márquez (2009) and Sharma (2013) has been adopted in the current study.

Based on the fundamental factors and constructs associated with all the earlier models, this present model or the conceptual framework model for this current study looks at the relationship of the organisational maintenance policy, maintenance budget, human resources management, training factors, monitoring and supervision, and the MIS, which are the essential variables that the majority of the previous studies have measured. However, this study has included consideration of the impact of communication among stakeholders and maintenance culture. These have been classified as the exogenous variables in predicting an organisation's overall MM, which is the endogenous variable. These will, in turn, predict the MM of municipal buildings in the South African education sector. The study aims to forecast the relative predictive power of these different variables for MM in order to test or determine whether the MM of municipal buildings in the South African education sector depends on the supposed features of the variables, considering the effects of communication among stakeholders and maintenance culture.

Some of the variables discussed above should be measured by objective means, some by subjective means, and better still, some by a combination of objective and subjective measurement forms. The combination of both subjective and objective indicators within the proposed model is supported by Campbell and Finch (2004) and Fallah-Fini et al. (2010); they opined that objective indicators, by themselves, are often misleading and will remain so until indicators that human beings attached to them are obtained. Similarly, subjective indicators, by themselves, are insufficient as guides to policy.

The conceptual model theorised that MM is established by the relationship between the exogenous variables, including the basic elements linked by the subjective and objective measurements. These variables identified from the generic literature review and the Delphi survey findings are considered the major determinants of the MM of municipal buildings in the education sector. These have been adapted to fit the peculiar MM characteristics in maintaining municipal buildings in the South African education sector. Hence, the combination of the objective and subjective measures will produce a measure of MM of municipal buildings in the South African education sector as defined in the previous sections (Table 11.2).

Structural Component of the Model

The present conceptual framework hypothesised that the MM of municipal buildings in the South African education sector (developing economies) is derived from the organisational maintenance policy (OPY), maintenance budget factors (MBF), human resources management (HRM), training factors (TF), monitoring and supervision (MSN), maintenance information system (MIS), communication among stakeholders (CAS), and maintenance culture factors (MCF). The model to be tested in the hypothesis postulates a priori that the MM of municipal buildings (maintenance management) is a multidimensional structure composed of OPY, MBF, HRM, TF, MSN, MIS, CAS, and MCF. This is presented schematically in Figure 11.1 (Model 1.0). The theoretical underpinning of this priority is derived from the works of Vanneste and Van Wassenhove (1995), Campbell (1998), and Wireman (2005), whose models of MM and the approach as adopted by Pintelon and Gelders (1992), Lynch et al. (1995), Hassanain et al. (2001), Takata et al. (2004), Márquez (2007), Márquez et al. (2009) and Fernández et al. (2009) have been discussed earlier. Inherent in the conceptualised framework is the notion that MM is related to evaluating many variables. While the principal variable under consideration is the organisational maintenance policy, it is difficult to discuss it without reference to variables of maintenance budget factors, human resources management factors, and the inclusion of the other exogenous variables. The outcome of the MM of buildings is expressed by the educational institutions' subjective evaluation of their MM system for

Table 11.2 Factors of maintenance management

Maintenance management elements	Organisational maintenance policy	Maintenance budget	Monitoring and supervision	Task planning and scheduling	Maintenance information system	Maintenance method /approach	Training	Spare part management	Human resources management	Outsource strategy	Users' needs and expectations	Continuous improvement
Pintelon et al. (1992) – Decision Making MM model	X	X	X	X	X	X			X			
Lynch et al. (1995) – The "Smart" Pyramid model	X	X				X	X		X		X	
Vanneste and Van Wassenhove (1995) – MM process model	X		X	X	X	X						X
Campbell (1998) – Asset strategy MM model	X		X		X		X		X			X
Hassanain et al. (2001) – Integrated MM model.	X	X	X				X		X			
Takata et al. (2004) – Life Cycle MM model		X	X				X	X	X	X		
Wireman (2005) – Preventive MM model	X	X			X		X	X	X			

(Continued)

Table 11.2 (Continued)

Maintenance management elements	Organisational maintenance policy	Maintenance budget	Monitoring and supervision	Task planning and scheduling	Maintenance information system	Maintenance method /approach	Training	Spare part management	Human resources management	Outsource strategy	Users' needs and expectations	Continuous improvement
Marquesz et al. (2006) – Supporting structure pillars MM model	X	X	X		X			X	X	X		
Márquez (2007) – Complex systems maintenance model	X	X	X		X		X		X			X
Márquez et al. (2009) – Generic MM model	X	X	X		X		X		X			
Fernández et al. (2009) – Risk assessment MM model	X	X	X						X		X	

Source: Researcher's literature review (2022).

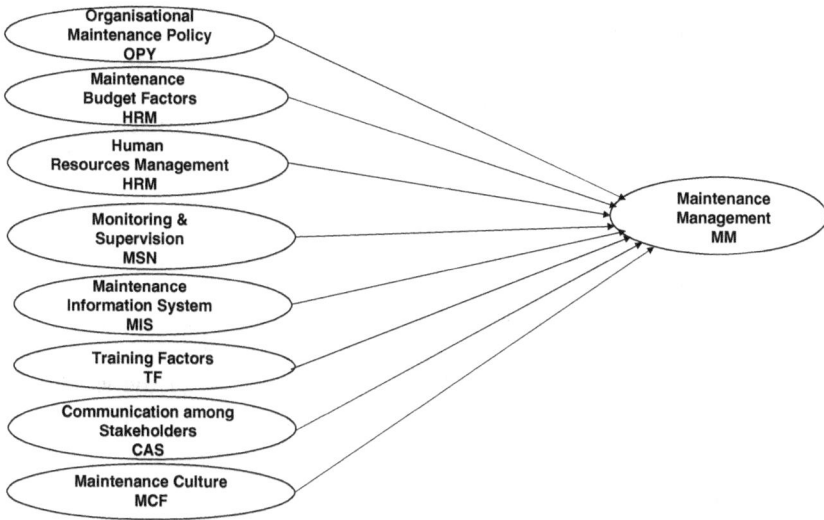

Figure 11.1 An integrated conceptual framework of maintenance management (Model 1.0).

Source: Author's review (2022).

their municipal buildings. The evaluation will depend on the educational institutions' assessment of several indicator variables under each exogenous variable. It is an empirical question regarding which attributes are most relevant and may differ under different circumstances. How educational institutions maintained their municipal buildings in a particular education sector, for instance, is considered dependent on specific characteristics of the education sector. This is meant to include all characteristics and stakeholders such as the maintenance units, educational institutions' management, and users that influence their evaluations. In this study, the objective evaluation of MM of municipal buildings is assessed by measuring the actual performance of municipal buildings on the MM outcomes.

Measurement Component of the Model

The hypothesised measurement component of the model comprises the following MM factors: OPY=15 measurement variables; MBF=15 measurement variables; HRM=17 measurement variables; TF=16 measurement variables; MSN=16 measurement variables; MIS=17 measurement variables; CAS=20 measurement variables; MCF=18 measurement variables and MM outcomes=24 measurement manifest variables. In the present model, it is theorised that MM is to be considered a sufficient indication to show the effective performance of the municipal buildings in the education sector.

Summary

In this chapter, a conceptual framework was theorised. The model postulates a priori that MM is a multidimensional structure composed of eight latent variables of organisational maintenance policy (OPY), maintenance budget factors (MBF), human resources management (HRM), training factors (TF), monitoring and supervision (MSN), a maintenance information system (MIS), communication among stakeholders (CAS), and maintenance culture factors (MCF). These factors were derived from the literature review this study conducted and the findings from the concurrent Delphi study. Likewise, this chapter highlights the theoretical framework for explaining the variables selected for the construction of the integrated conceptual MM framework for municipal buildings in the South African education sector. Future studies will present the findings for the validation of the conceptual framework developed in this chapter.

References

Ahzahar, N., Karim, N. A., Hassan, S. H., & Eman, J. (2011). A study of contribution factors to building failures and defects in construction industry. *Procedia Engineering, 20*, 249–255.

Albert, C., & Brownsword, L. (2002). Evolutionary Process for Integrating COTS-Based Systems (EPIC): An Overview. Key Elements in Building, Fielding, and Supporting Commercial-off-the-Shelf (COTS) Based Solutions.

Alexander, K. (1996). Facilities management: A strategic framework. *Facilities Management: Theory and Practice*, London: Routledge (Taylor &Francis group).

Ali, M., & Mohamad Nasbi Bin Wan Mohamad, W. (2009). Audit assessment of the facilities maintenance management in a public hospital in Malaysia. *Journal of Facilities Management, 7*(2), 142–158.

Alzaben, H. (2015). *Development of a MM framework to facilitate the delivery of healthcare provisions in the Kingdom of Saudia Arabia.* Published doctoral dissertation, Nottingham Trent University.

Amaratunga, R. G. (2001). *Theory building in facilities management performance measurement: Application of some core performance measurement and management principles.* Published doctoral dissertation, University of Salford.

Amaratunga, D., Baldry, D., & Sarshar, M. (2000). Assessment of facilities management performance - what next? *Facilities, 18*(1/2), 66–75.

Atkin, B., & Brooks, A. (2000). *Total facilities management.* London: Blackwell Science.

Brinia, V., & Efstathiou, M. (2012). Evaluation of factors affecting training transfer on safety in the workplace: A case study in a big factory in Greece. *Industrial and Commercial Training. 44*(4), 223–231.

Campbell, J. D. (1998). *Uptime, strategies for excellence in maintenance management.* Portland, OR: Productivity Press.

Campbell, L., & Finch, E. (2004). Customer satisfaction and organisational justice. *Facilities, 22*(7/8), 178–189.

Cato, W. W., & Mobley, R. K. (2002). *Computer-managed maintenance systems in process plants: A step-by-step guide to effective management of maintenance, labor, and inventory in your operation.* Houston: Gulf Publishing Company.

Cholasuke, C., Bhardwa, R., & Antony, J. (2004). The status of maintenance management in UK manufacturing organisations: Results from a pilot survey. *Journal of Quality in Maintenance Engineering, 10*(1), 5–15.

Cloete, N. (2014). The South African higher education system: Performance and policy. *Studies in Higher Education, 39*(8), 1355–1368.

Crespo Márquez, A., de León, P. M., Gomez Fernandez, J. F., Parra Marquez, C., & López Campos, M. (2009). The maintenance management framework: A practical view to maintenance management. *Journal of Quality in Maintenance Engineering, 15*(2), 167–178.

De Groote, P. (1995). Maintenance performance analysis: A practical approach. *Journal of Quality in Maintenance Engineering, 1*(2), 4–24.

Duffuaa, S. O. (2000). Mathematical models in maintenance planning and scheduling. In *Maintenance, modeling and optimization* (pp. 39–53). Boston, MA: Springer.

Fallah-Fini, S., Rahmandad, H., Triantis, K., & de la Garza, J. M. (2010). Optimizing highway maintenance operations: Dynamic considerations. *System Dynamics Review, 26*(3), 216–238.

Fatoni, Z. Z. Z., & Nurcahyo, R. (2018). Impact of training on maintenance performance effectiveness. In Proceedings of the *International Conference on Industrial Engineering and Operations Management*, Paris. 26–27 July 2018, pp. 619–628.

Fernández, J. F. G., & Márquez, A. C. (2009). Framework for implementation of maintenance management in distribution network service providers. *Reliability Engineering & System Safety, 94*(10), 1639–1649.

Garcia, M. C., Sanz-Bobi, M. A., & Del Pico, J. (2006). SIMAP: Intelligent system for predictive maintenance: Application to the health condition monitoring of a wind turbine gearbox. *Computers in Industry, 57*(6), 552–568.

Hassanain, M. A., Froese, T. M., & Vanier, D. J. (2001). Development of a maintenance management model based on IAI standards. *Artificial Intelligence in Engineering, 15*(2), 177–193.

Jonsson, P. (1997). The status of maintenance management in Swedish manufacturing firms. *Journal of Quality in Maintenance Engineering, 3*(4), 233–258.

Karia, N., Asaari, M. H. A. H., & Saleh, H. (2014). Exploring Maintenance Management in Service Sector: A Case Study. In Proceedings of *International Conference on Industrial Engineering and Operation Management*, Bali, 7–9 January 2014, pp. 3119–3128.

Kempton, G. E. (1996). Training for organizational success. *Health Manpower Management, 22*(6), 25–30.

Lee, H. H. Y., & Scott D. (2009). Overview of maintenance strategy, acceptable maintenance standard and resources from a building maintenance operation perspective. *Journal of Building Appraisal, 4*(4), 269–278.

Lee, J., Ni, J., Djurdjanovic, D., Qiu, H., & Liao, H. (2006). Intelligent prognostics tools and e-maintenance. *Computers in industry, 57*(6), 476–489.

Lind, H., & Muyingo, H. (2012). Building maintenance strategies: Planning under uncertainty. *Property Management, 30*(1), 14–28.

Lynch, R. L., & Cross, K. F. (1995). *Measure up: How to measure corporate performance*. Malden, MA: Blackwell.

Manaf, Z., & Alias, A. (2005). Training needs in facilities management a study among local authority offices in Malaysia. In *International Real Estate Research Symposium (IRERS)*, Kuala Lumpur, Malaysia, 11–13 April, pp. 1–14.

Márquez, A. C. (2007). *The maintenance management framework: Models and methods for complex systems maintenance.* London: Springer-Verlag.

Marquez, A. C., & Gupta, J. N. (2006). Contemporary maintenance management: Process, framework, and supporting pillars. *Omega, 34*(3), 313–326.

Mekasha, E. (2018). *Maintenance management framework development for competitiveness of food and beverage industry:* A case study on Asku Plc. Thesis Draft. Addis Ababa Institute of Technology, Addis Ababa University.

Neelamkavil, J. (2009). *A review of existing tools and their applicability to facility maintenance management.* Canada: Institute for Research in Construction Report # RR-285.

Nikandrou, I., Brinia, V., & Bereri, E. (2009). Trainee perceptions of training transfer: An empirical analysis. *Journal of European Industrial Training, 33*(3), 255–270.

Noe, R. A. (2010). *Employee training and development* (5th ed.). New York: McGraw-Hill.

Omar, M. F., Ibrahim, F. A., & Omar, W. M. S. W. (2017). Key Performance Indicators for Maintenance Management Effectiveness of Public Hospital Building. In *MATEC Web of Conferences*, Ho Chi Minh City, Vietnam, 5–6 August 2016, pp. 01056. (Vol.7). EDP Sciences.

Parida, A., & Kumar, U. (2009). Maintenance productivity and performance measurement. In *Handbook of maintenance management and engineering* (pp. 17–41). London: Springer.

Pintelon, L. M., & Gelders, L. F. (1992). Maintenance management decision making. *European Journal of Operational Research, 58*(3), 301–317.

Pintelon, L. M., & Van Wassenhove, L. N. (1990). A maintenance management tool. *Omega, 18*(1), 59–70.

Schreiber, F. (2007). *Maintenance briefing notes: Human performance error management.* Toulouse, France: Airbus Customer Service.

Sharma, S. K. (2013). Maintenance reengineering framework: A case study. *Journal of Quality in Maintenance Engineering, 19*(2), 96–113.

Sherwin, D. (2000). A review of overall models for maintenance management. *Journal of Quality in Maintenance Engineering, 6*(3), 138–164.

Takata, S., Kirnura, F., van Houten, F. J., Westkamper, E., Shpitalni, M., Ceglarek, D., & Lee, J. (2004). Maintenance: changing role in life cycle management. *CIRP Annals, 53*(2), 643–655.

Trappey, A. J., Sun, Y., Trappey, C. V., & Ma, L. (2011). Re-engineering transformer maintenance processes to improve customized service delivery. *Journal of Systems Science and Systems Engineering, 20*(3), 323.

Usanmaz, O. (2011). Training of the maintenance personnel to prevent failures in aircraft systems. *Engineering Failure Analysis, 18*(7), 1683–1688.

Van Horenbeek, A., Pintelon, L., & Muchiri, P. (2010). Maintenance optimization models and criteria. *International Journal of System Assurance Engineering and Management, 1*(3), 189–200.

Vanneste, S. G., & Van Wassenhove, L. N. (1995). An integrated and structured approach to improve maintenance. *European Journal of Operational Research, 82*(2), 241–257.

Wilson, J. P. (Ed.). (2005). *Human resource development: Learning & training for individuals & organizations.* New York: Kogan Page Publishers.

Wireman, T. (2005). *Developing performance indicators for managing maintenance.* New York: Industrial Press Inc.

Wordsworth, P., & Lee, R. (2001). *Lee's building maintenance management.* London: Blackwell Science.

Yacob, S. (2005). *Maintenance management system through strategic planning for public school in Malaysia.* Published doctoral dissertation, Universiti Teknologi Malaysia.

Yu, R., Iung, B., & Panetto, H. (2003). A multi-agents-based E-maintenance system with case-based reasoning decision support. *Engineering Applications of Artificial Intelligence, 16*(4), 321–333.

12 Conclusion and Recommendations

Introduction

This study aimed to develop a maintenance management (MM) framework for municipal buildings in developing economies, using the South African education sector as a case study. The framework's usefulness includes assessing the MM of educational buildings and aiding educational institutions to obtain effective MM of municipal buildings in the education sector.

Hence, an extensive literature review and a Delphi study were conducted to achieve the aim. Conclusions regarding the study were presented concerning the objectives of the study in the next sections.

Research Objective RO1

The study's first objective was to establish the factors that influence municipalities in the attainment of MM of the municipal buildings in the education sector. Concerning this objective, the relevant factors were identified through an extensive literature review of the study conducted. The literature findings also aided in designing the questionnaire that guided the Delphi study. Findings were that the MM of municipal buildings was not a product of one or two attributes but a multi-faceted construct. Additional findings revealed that the MM of buildings had been a major topic in various fields such as manufacturing industries, facility management, production companies, construction organisations, and agencies, among others.

Additionally, it was found that the attributes that influence the effective MM of buildings differed from one industry to another as well as from one nation to another. The difference in the attributes could be due to cultural, social, political, laws, and economic climates that may be prevailing in an organisation, industry, or a nation at any point in time, the nature of the MM arrangement within the organisation or industry, as well as the dynamic and subjective nature of the MM concept. Likewise, it was found that the factors which aid in the attainment of effective MM encompass policy, budget, spare part management, monitoring, and supervision, among others. Findings from the literature were that more effort and research are

DOI: 10.1201/9781003344681-12

essential to try and address the attributes of MM of municipal buildings in the education sector of developing economies as they are unknown.

Research Objective RO2

The second objective of the research was to establish the current theories in the literature that have been advanced on the MM of buildings and to identify the gaps that need consideration. An extensive literature review was carried out to achieve this objective. The findings revealed that the MM of buildings' research had not been studied with an all-inclusive construct in developing the previous models and theories. From the extensive literature review, the identified gaps were the influence of the communication among stakeholders and the maintenance culture factors in the MM of municipal buildings. The identified gaps formed the new constructs in this study's conceptual framework (Model 1.0). These gaps were considered essential because the appearance of South Africa's municipal buildings, such as hospitals, airports, roads, and especially educational buildings, indicated that the society lacks a cultural behaviour that ensures the effective and efficient functioning of the buildings fostering national development.

Similarly, provision for adequate care of the hard-earned municipal buildings has not gained ground in the consciousness of users and maintenance managers of municipal buildings in educational institutions over the years because of the absence of a maintenance culture. Also, poor communication and the dearth of knowledge sharing between the maintenance team and building users have been the main cause of problems, specifically on technical and documentation aspects for the failed sections of the building being maintained. Nonetheless, in educational institutions in the South African education sector, little is known of communication among stakeholders in the maintenance process and their impact on MM efficiency.

This study offers a synthesised classification of the constructs which should be collectively considered to predict the MM of municipal buildings. From the synthesised literature, this study argued that MM was an eight-factor construct.

Research Objective RO3 and RO4

The third (RO3) objective of the study was to determine the factors/attributes (main and sub-variables) that are perceived to be of importance and of major impact in obtaining effective MM and to determine whether the factors that have aided in obtaining effective MM in other geographical contexts are the same in the South African education sector. The fourth research objective (RO4) was to develop a holistically integrated Maintenance Management framework (MMF) for municipal buildings in developing economies using the South.

A Delphi study was conducted in the African education sector to achieve these objectives. Findings were that several factors that were significant in determining the MM of Municipal buildings were identified and amplified by the Delphi study. The factors considered to be paramount determinants of effective MM of municipal buildings were organisational maintenance policy, maintenance budget factors, human resources management, training factors, monitoring and supervision, maintenance information system, communication among stakeholders, and the maintenance culture factors. The findings suggested that the attributes that bring about effective MM of municipal buildings in the education sector in South Africa are similar to the determinants in other geographical contexts. Additionally, the effective MM of Municipal buildings is assured if these factors are considered in the maintenance processes and operations of municipal buildings in the South African education sector. These factors were collectively considered for the development of the all-inclusive (integrated holistic) MM framework.

The MM framework that this book has developed was based on both the literature and the Delphi Study findings. Hence, to achieve the fourth objective, a synthesis of the reviewed literature together with the findings from the Delphi study was used. The conceptual framework theorised that the MM of municipal buildings is an eight-factor construct. These factors are an organisational maintenance policy, maintenance budget factors, human resources management, training factors, monitoring and supervision, maintenance information system, communication among stakeholders, and the maintenance culture factors, which jointly predict the MM of municipal buildings in the South African education sector.

Contribution and Value of the Book

The value and contribution of this book are described at four levels. These are the integrated MM framework, theoretical, methodological, and practical levels of the research findings. Nevertheless, it is appropriate to note that the outstanding contribution of the book is the revelation and validation of the influences of communication among stakeholders and maintenance culture factors in predicting the MM of municipal buildings.

Integrated Maintenance Management Conceptual Framework

The uniqueness of this study also lies in the integrated maintenance management framework (MMF) that has been developed to aid educational institutions in achieving effective MM of municipal buildings in the South African education sector. The literature findings in relation to the MM studies in the South African education sector revealed that there is a lack of studies that developed a framework to aid educational institutions in achieving the effective MM of municipal buildings in the education sector. Thus, the integrated MMF this current study has developed contributes to knowledge as it addresses the lack of a framework to guide educational institutions to achieve the effective MM of their municipal buildings.

Furthermore, the framework informs as to the attributes that determine the MM of municipal buildings in the South African education sector. Likewise, the latent variables which led to the effective MM outcome variables could be used for the MM of municipal buildings in the South African education sector.

The general hypothesis was therefore sustained, namely that the MM of municipal buildings is directly related to the influence of the exogenous variables in predicting the overall MM of municipal buildings in the education sector of developing economies using South Africa as a case study. The findings supported previous MM studies, which informed that the MM of municipal buildings is a multi-dimensional construct (Pintelon & Gelders, 1992; Lynch et al., 1995; Vanneste & Van Wassenhove, 1995; Campbell, 1998; Hassanain et al., 2001; Takata et al., 2004; Wireman, 2005; Marquez and Gupta, 2006; Márquez, 2007; Crespo Márquez et al., 2009). It also contrasts the stance of Sharma's (2013) model on MM studies that goals and strategy, human aspects, support mechanisms, tools and techniques, and organisation are the only attributes that influence the MM of buildings.

Moreover, the endogenous variable was made up of nine latent variables. The nine latent variables of MM were that educational institutions with an effective MM system would achieve adequate planning for a maintenance budget, enjoy an effective maintenance organisation structure, achieve change in the attitude of their personnel to maintenance operations, maintain well-supervised maintenance work, ensure better-monitored maintenance work, improve users' attitude to the maintained building, enhance users' performance, allow buildings and their support components to run effectively, and ensure improved and sustainable building performance.

Theoretical Contribution and Value

The results of the Delphi study indicated that the MM of municipal buildings is an eight-factor framework. The researcher could not find evidence of a similar study that has been conducted in the education sector of a developing economies' context. The study is also significant because it addresses the lack of theoretical information about the factors most significantly predict the effective MM of municipal buildings.

The Delphi study's results also showed that the factors – organisational maintenance policy, maintenance budget factors, human resources management, training factors, monitoring and supervision, a maintenance information system, communication among stakeholders, and the maintenance culture factors were found to have a significant influence in determining the MM of municipal buildings in the South African education sector. The findings enforced the theory that the MM of municipal buildings is multi-faceted.

Nonetheless, the literature review did not reveal evidence of a similar study to the current one. Consequently, this suggests that this type of research has not yet been conducted in MM studies, especially in South Africa. Therefore, this study may offer a base for further follow-up studies for other researchers. Likewise, this study modelled the MM of municipal buildings as an eight-factor construct with the inclusion of two new variables: communication among stakeholders and maintenance culture factors. Previous studies have tried to model MM using other variables without the inclusion of these two additional constructs.

Also, this study has shown that more than one factor influences the MM of municipal buildings. As advanced by Omar et al. (2017) and validated through the Delphi result, the study suggested that KPIs for the effectiveness of MM of municipal buildings could be subdivided into four main factors (see Figure 12.1). They include administrative and organisational factors, individual factors, technical factors, and behavioural factors. This shows that MM is moving from the strategic approach to a new dimension (behavioural change to maintenance). Apart from the study contributing to theoretical knowledge, it also contributed to the methodological advances in terms of the approach used in conducting the research.

Methodological Contribution and Value

There was no evidence during the literature review that suggested that a Delphi study had been used in MM studies in South Africa. More so, the questionnaire survey instrument had high internal reliability values and could therefore be used in similar studies to validate this study or for similar purposes. Apart from this contribution and value to the body of knowledge in terms of the methodological approach, a contribution to practice and the education sector was also achieved.

Practical Contribution and Value

Educational institutions, especially in South Africa, have not realised the significance of communication among stakeholders in relation to the MM of their municipal buildings. However, the Delphi results have shown that the communication among stakeholders in the maintenance process significantly influences the MM of municipal buildings, especially in the education sector in South Africa.

Knowledge of the influence of communication among stakeholders in the maintenance process and operation could be developed by municipalities and educational institutions desirous of having an effective MM system. In particular, the influence of the following factors is significant; namely users' satisfaction, quality-driven service delivery, collaboration among stakeholder(s), identify gaps in the maintenance process, early defects detection, e-mails among stakeholder(s), users feedback, increased in users' participation, eradication of uncertainty, understanding of users' maintenance need(s), addressing stakeholders' concern(s), good working relationship

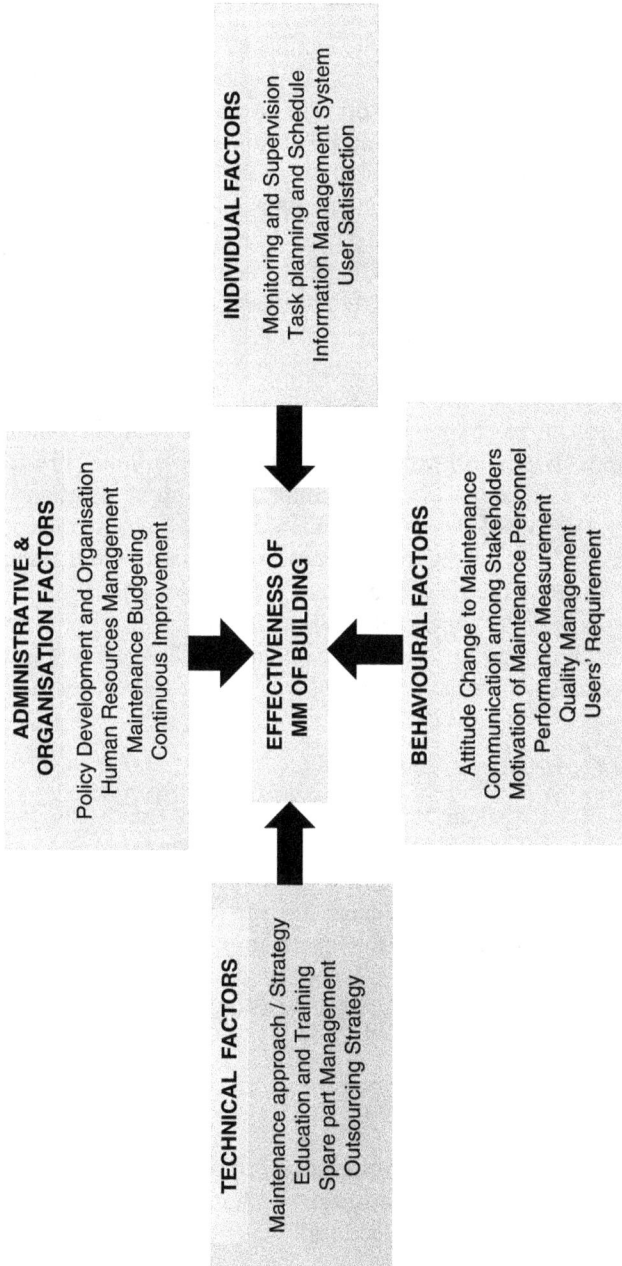

Figure 12.1 Key factors for the effectiveness of maintenance management of buildings.
Source: Author's review (2022).

among stakeholder(s), use of visualisation technique, understanding users' expectation(s), timely and accurate feedback(s), help desk, liaison with relevant government agencies, checking maintenance information with user(s), establishment of the need for change, web-based systems meetings and report, stakeholders' participative action(s), liaison with maintenance institutions and telephone support.

Similarly, maintenance culture factors significantly influence the overall MM of municipal buildings in the South African education sector. Knowledge of the influence of maintenance culture factors will also aid maintenance managers and administrators of municipalities and educational institutions in decision-making relating to the MM of municipal buildings in the education sector. In particular, the influence of the following factors is significant, namely effective maintenance management style, effective decision-making capacity, user change of attitude to maintained buildings, attitudinal change to maintenance operation(s), sustainability of infrastructure, use of right skills maintenance personnel, diligence in maintenance operation(s), enhance performance level of building(s), benchmarking for maintenance performance, attitude of ensuring regular repairs, proper maintenance system, improve economic value of a country, commitment to change initiative(s), building reliability, attitude of ensuring regular servicing, motivation of maintenance personnel, maintenance decision based on societal value, and maximal utilisation of infrastructural.

Furthermore, the knowledge of the influence of the eight-factor construct could help stakeholders in the education sector in curriculum design and delivery, policy formulation, a maintenance strategy for educational institutions, and capacity development of their maintenance managers and personnel in matters relating to the MM of municipal buildings in the education sector. The practical significance of the study is further elaborated as follows:

Significance to Curriculum Design and Delivery

Having the right curriculum that addresses current trends and needs in maintenance will enable teachers to teach subjects relevant to MM modules for which they are producing graduates. Hence, they will prepare graduates equipped with the requisite competencies for the job market. Therefore, the current findings of this study could help design curricula for teaching and learning in educational institutions, especially concerning the maintenance and management of public buildings. Furthermore, it will provide the basis for curriculum review.

Significance to Academia

This study's findings contribute to the existing body of knowledge in MM studies. It will form the basis for future research in MM studies. It will also provide relevant information for lecturers who teach maintenance-related courses in educational institutions.

Significance to Maintenance Personnel and Management

The knowledge advanced in this study will inform maintenance personnel and maintenance organisations, as well as the management of municipalities and educational institutions, of the factors that significantly influence the MM of their organisations. It will guide them in making decisions that impact the MM of municipal buildings within their organisations or institutions.

Significance to Capacity Development of Maintenance Personnel of Educational Institutions

The knowledge espoused by this study will help municipalities and educational institutions (both public and privately owned) in providing relevant training programmes that aim to develop the capacity of maintenance personnel of their maintenance department or units to enhance their knowledge of current trends on the job.

Significance to Policy Formulation in the Education Sector

Policy is an important tool for the fulfilment of a goal. Policies informed by knowledge provide the relevant policy framework for achieving institutional goals. Hence, the knowledge advanced by this study will help policymakers in the municipalities and education sector formulate policies geared towards assisting educational institutions in attaining the effective MM of their municipal buildings.

Significance to the South African Built Environment

The study results have also demonstrated the influence of the eight constructs on the MM of buildings in the South African built environment. The output of the study will help the Association of South African Surveyors (ASAQS), the South African Council for the Quantity Surveying profession (SACQSP), the South African Council for planners (SACPLAN), South African Council for the Property Valuers Profession (SACPVP), The South African Council for the Project and Construction Management Professions (SACPCMP), Association of Construction Project Managers (ACPM), Institute for Landscape Architecture in South Africa (ILASA), Consulting Engineers South Africa (CESA), and South African Institute of Architects (SAIA) in making decisions about the criteria to be given priority in providing the relevant support towards the MM of public and private buildings in the South African built environment. It will ultimately enable them to know the vital areas to commit resources towards capacity development, especially in building or developing structures' maintenance processes and operations.

Similarly, as advanced by Dunn (2003), the output of the study will help professionals within the South African built environment change the

1940's	1960's	1980-1990's	2000's	2022's
Fix After Break	Technical Matter	Production Important	Strategic Issue	Behavioural Concern

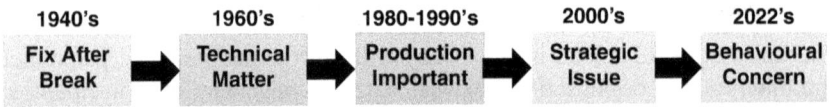

Figure 12.2 Changing focus in maintenance management of buildings.
Source: Author's review (2022).

maintenance focus for their developed and proposed buildings from a strategic position to behavioural concern (see Figure 12.2) based on the influence of the eight constructs as developed by the study.

Furthermore, the integrated, holistic MM framework should be used as a guide to ensure all elements necessary for the South African built environment to achieve the effective MM of buildings after delivery for efficiency and effectiveness of the building. The study offers an opportunity for further research to improve the framework developed in this study and probably refine indicator variables to suit specific environments. Consequently, the recommendations and policy implications for the practice of all these areas in which this study may add value and contribute are presented in the following pages.

Recommendations

The recommendations this study advances are from the integrated MM framework, methodological, theoretical, and practical points of view.

Integrated Maintenance Management Framework

It is recommended that municipalities and educational institutions desirous of achieving effective MM of the municipal buildings in the South African education sector could be guided by the maintenance framework this study has developed. Also, the framework is recommended for built environment-based researchers that seek to assess the MM system of facility management firms, especially in the built environment of developing economies. The framework is recommended for use by non-governmental organisations (NGOs) and government agencies that seek to maintain or renovate the municipal buildings in the South African education sector.

Methodological

It is advocated that a similar study should be conducted with different populations and samples (in other developing economies) to improve the application in developing economies. Likewise, further research should be conducted on the indicator variables to validate the conceptual framework and to establish an improvement in model fit. There is the possibility that more indicator variables could define MM. It should be recognised, thus, that the perfect framework does not exist. Nonetheless, there should be a move to try improve on the current framework rather than develop a new one.

The study recommends that the Delphi survey be encouraged in studies such as the current one, where a test-retest methodology may not be feasible to validate a study. This situation is often common in construction management and engineering studies, where most studies end up as a questionnaire survey and render the generalisation of conclusions, especially on causality, questionable.

Theoretical

The literature showed that there were still different definitions and understandings of how MM is derived. This has led to a limited view and narrow conceptualisation of the MM of buildings in the past. Additionally, there has not been a consensus on how the MM of municipal buildings in the education sector should be measured. However, this study reviewed and synthesised the literature on the determinants of MM of municipal buildings. In conjunction with the experts' knowledge obtained through the Delphi study, an eight-factor MM framework was arrived at for the MM of municipal buildings in the South African education sector. These factors were organisational maintenance policy, maintenance budget factors, human resources management, training factors, monitoring and supervision, maintenance information system, communication among stakeholders, and maintenance culture factors. It is therefore recommended that the developed framework and theory of MM of municipal buildings with particular emphasis on operationalising it, should form the basis for further refinement of the concept, thereby making it beneficial to the municipalities and educational sector in South Africa and other developing economies. It is further recommended that the influence of communication among stakeholders and the maintenance culture factors should be integrated into existing frameworks, as proposed in other studies that have been developed.

Policy Implication and Practical Recommendation

Informed by the identified contributions that this study makes, as revealed by the findings, the following policy implications and practical recommendations have been identified:

- The policy implication suggests that the MM of municipal buildings can be improved through the enhancement of organisational maintenance policy, maintenance budget factors, human resources management, training factors, monitoring and supervision, maintenance information systems, communication among stakeholders, and the maintenance culture factors.
- Similarly, maintenance departments/units, physical planning units, education administrators and management, non-governmental organisations, and other stakeholders in the South African education sector should focus on

developing the capacity of educational institutions' maintenance personnel so that they are responsive to the eight-factor framework, especially the training factors since training factors construct emerged as the most influential construct that would aid educational institutions in achieving effective MM of their municipal buildings.

- Curricula for training maintenance personnel should be responsive to organisational maintenance policy, maintenance budget factors, human resources management, training factors, monitoring and supervision, maintenance information system, communication among stakeholders, and the maintenance culture factors.

Limitation

Although interesting and valuable findings have emerged from this study, it is not without limitations. The following limitations regarding this study should be considered. Firstly, the research was only conducted in South Africa. This is because the education sector in South Africa has been flooded with different educational institutions from both government and private sectors, owing to the liberalisation of the education sector. This has taken place against the mounting pressure on existing and new municipal buildings and a raised awareness of the need to maintain the condition of the municipal buildings in the education sector more effectively. Given enough resources, it would be preferable to conduct a similar research study with the entire population of educational sectors in developing economies. Moreover, given enough resources, conducting a similar research study with a wider population would be preferable.

Also, using additional items or constructs might improve the inherent reliability and validity of the measures used. A review of the research tool would have benefitted the study's findings.

Recommendation for Further Research

The subsequent suggestions for further studies have been identified:

- Further studies should examine factors related to the limitations of this study. Firstly, more rigorous and detailed testing of measurement scales in South Africa and other developing economies would further the knowledge of MM of municipal buildings. It is possible that some scales developed in Western culture and this study may not be suitable for other cultural contexts.
- The framework did not include motivation of maintenance personnel as a separate construct, and including this may have influenced the research results. However, it was used as a latent variable for the maintenance culture factors latent construct.
- These results also need to be replicated with other populations. Important features of the MM of municipal buildings may vary between different regions and cultures, while some might remain similar.

• A validation of the Ogunbayo Integrated Holistic MM Framework presented in this book is recommended.

Summary

An integrated MM framework for municipal buildings was developed using existing MM and other theories grounded in MM studies and the findings of the Delphi study conducted. It was postulated that the overall MM of municipal buildings is directly related to the influence of the exogenous (latent) variables in predicting or determining the overall MM. The finalised empirical framework revealed that the exogenous variables (organisational maintenance policy, maintenance budget factors, human resources management, training factors, monitoring and supervision, maintenance information system, communication among stakeholders, and the main tenance culture factors) had a statistically significant influence on determining the MM of municipal buildings in the South African education sector. It is therefore concluded that the eight-factor framework represents an adequate description of the MM of municipal buildings in the South African education sector.

The results of this study have theoretical, methodological, and policy (practical) values because respondents for the Delphi study were drawn from academia and industry. The respondents for the questionnaire survey were academic lecturers, maintenance managers of public buildings, property managers/developers, facility managers, members of maintenance committees of educational institutions, and registered members of professional bodies in the built environment, such as the Association of South African Surveyors (ASAQS), the South African Council for the Quantity Surveying profession (SACQSP), South African Council for planners (SACPLAN), South African Council for the Property Valuers Profession (SACPVP), The South African Council for the Project and Construction Management Professions (SACPCMP), Association of Construction Project Managers (ACPM), Institute for Landscape Architecture in South Africa (ILASA), Consulting Engineers South Africa (CESA), and South African Institute of Architects (SAIA). Similarly, the respondents had a good working knowledge of the investigated issue.

The result of the study provided information that can inform the municipalities and educational institutions: management, maintenance departments, physical planning units, as well as governmental, corporate institutions, institutional, and policymakers as they plan for and implement maintenance programs designed to enhance the MM of municipal buildings in the South African education sector. Secondly, the study provides indicators that will be a baseline for assessing MM in the built environment within the construction industries of developing economies. Additionally, the conceptual framework of the MM of municipal buildings formulated in this study will provide a reference to researchers who will carry out studies relating to MM in the future.

This study supports previous studies that employed alternative methods to establish the factors that influence the MM of municipal buildings and concluded that the MM of municipal buildings is multi-faceted, as also claimed in this study. The practical implication is that the MM of municipal buildings can be enhanced by improving the organisational maintenance policy, maintenance budget factors, human resources management, training factors, monitoring and supervision, maintenance information systems, communication among stakeholders, and the maintenance culture factors.

References

Campbell, J. D. (1998). *Uptime, strategies for excellence in maintenance management.* Portland, OR: Productivity Press.

Crespo Márquez, A. Moreu de León, P., Gómez Fernández, J. F., Parra Márquez, C., & González, V. (2009). The maintenance management framework: A practical view to maintenance management. *Journal of Quality in Maintenance Engineering, 15*(2), 167–178.

Dunn, S. (2003). The fourth generation of maintenance. In *Proceedings of International Conference of Maintenance Societies (ICOM)*, Perth, 20–23 May 2003.

Hassanain, M. A., Froese, T. M., & Vanier, D. J. (2001). Development of a maintenance management model based on IAI standards. *Artificial Intelligence in Engineering, 15*(2), 177–193.

Lynch, R. L., & Cross, K. F. (1995). *Measure up: How to measure corporate performance.* Malden, MA: Blackwell.

Márquez, A. C. (2007). *The maintenance management framework: Models and methods for complex systems maintenance.* London: Springer-Verlag.

Marquez, A. C., & Gupta, J. N. (2006). Contemporary maintenance management: Process, framework, and supporting pillars. *Omega, 34*(3), 313–326.

Omar, M. F., Ibrahim, F. A., & Omar, W. M. S. W. (2017). Key performance indicators for maintenance management effectiveness of public hospital building. In *MATEC Web of Conferences*, Ho Chi Minh City, Vietnam, 5–6 August 2016, pp. 01056. (Vol.7). EDP Sciences.

Pintelon, L. M., & Gelders, L. F. (1992). Maintenance management decision making. *European Journal of Operational Research, 58*(3), 301–317.

Sharma, S. K. (2013), Maintenance reengineering framework: A case study. *Journal of Quality in Maintenance Engineering, 19*(2), 96–113.

Takata, S., Kirnura, F., van Houten, F. J., Westkamper, E., Shpitalni, M., Ceglarek, D., & Lee, J. (2004). Maintenance: Changing role in life cycle management. *CIRP Annals, 53*(2), 643–655.

Vanneste, S. G., & Van Wassenhove, L. N. (1995). An integrated and structured approach to improve maintenance. *European Journal of Operational Research, 82*(2), 241–257.

Wireman, T. (2005). *Developing performance indicators for managing maintenance.* New York: Industrial Press Inc.

Index

Note: Page numbers in **Bold** refer to tables; and page numbers in *italics* refer to figures

For Product Safety Concerns and Information please contact our EU
representative GPSR@taylorandfrancis.com
Taylor & Francis Verlag GmbH, Kaufingerstraße 24, 80331 München, Germany